Photo Acoustic and Optical Coherence Tomography Imaging, Volume 3

Angiography: an application in vessel imaging

Photo Acoustic and Optical Coherence Tomography Imaging, Volume 3

Angiography: an application in vessel imaging

Edited by

Ayman El-Baz

University of Louisville, Louisville, KY, USA

and

University of Louisville at AlAlamein International University (UofL-AIU), New Alamein City, Egypt

Jasjit S Suri

AtheroPoint LLC, Roseville, CA, USA

IOP Publishing, Bristol, UK

ISBN 978-0-7503-2060-3 (ebook)
ISBN 978-0-7503-2058-0 (print)
ISBN 978-0-7503-2059-7 (myPrint)
ISBN 978-0-7503-2061-0 (mobi)

DOI 10.1088/978-0-7503-2060-3

Version: 20211201

IOP ebooks

British Library Cataloguing-in-Publication Data: A catalogue record for this book is available from the British Library.

Published by IOP Publishing, wholly owned by The Institute of Physics, London

IOP Publishing, Temple Circus, Temple Way, Bristol, BS1 6HG, UK

US Office: IOP Publishing, Inc., 190 North Independence Mall West, Suite 601, Philadelphia, PA 19106, USA

With love and affection to my mother and father, whose loving spirit sustains me still

— Ayman El-Baz

To my late loving parents, immediate family, and children

— Jasjit S Suri

Contents

Preface

The book covers the state-of-the-art techniques of optical coherence tomography angiography (OCTA) imaging for the diagnosis of retinal diseases. Clinical disorders of the retina have been attracting the attention of researchers, aiming at reducing the blindness rate. This includes uveitis, diabetic retinopathy, macular edema, endophthalmitis, proliferative retinopathy, age-related macular degeneration and glaucoma. Currently, most ophthalmologists perform diagnosis by visual observation and interpretation. Treatment is significantly dependent on having early and accurate diagnosis, which can be significantly improved by employing disease-specific computer-aided diagnostic (CAD) systems based on different image modalities such as: optical coherence tomography (OCT), Fundus Imaging and optical coherence tomography angiography (OCTA). This book will focus on OCTA imaging for the diagnosis of any retinal disease. Among the topics discussed in the book are clinical application of OCTA; hyaloid vascular regression in developing mouse eyes; OCTA stripe noise removal and vessel segmentation; OCTA changes in early type 3 neovascularization after anti-vascular endothelial growth factor treatment; OCTA in multiple sclerosis and neuromyelitis optica; diagnosis of polypoidal choroidal vasculopathy; objective assessment of OCTA; and age-related macular degeneration.

In summary, the main aim of this book is to help advance scientific research within the broad field of OCTA imaging for the diagnosis of retinal diseases. The book focuses on major trends and challenges in this area, and it presents work aimed to identify new techniques and their use in biomedical analysis.

Ayman El-Baz
Jasjit S Suri

Acknowledgements

The completion of this book could not have been possible without the participation and assistance of so many people whose names may not all be enumerated. Their contributions are sincerely appreciated and gratefully acknowledged. However, the editors would like to express their deep appreciation and indebtedness particularly to Dr Ali H Mahmoud and Reem Haweel for their endless support.

Ayman El-Baz
Jasjit S Suri

Editor biographies

Ayman El-Baz

Ayman El-Baz is a Distinguished Professor at University of Louisville, Kentucky, United States and University of Louisville at AlAlamein International University (UofL-AIU), New Alamein City, Egypt. Dr El-Baz earned his BSc and MSc degrees in electrical engineering in 1997 and 2001, respectively. He earned his PhD in electrical engineering from the University of Louisville in 2006. Dr El-Baz was named as a Fellow for Coulter, AIMBE and NAI for his contributions to the field of biomedical translational research. Dr El-Baz has almost two decades of hands-on experience in the fields of bio-imaging modeling and non-invasive computer-assisted diagnosis systems. He has authored or coauthored more than 500 technical articles (155 journals, 44 books, 85 book chapters, 255 refereed-conference papers, 196 abstracts, and 36 US patents and Disclosures).

Jasjit S Suri

Jasjit S Suri, PhD, MBA is a Fellow of IEEE, AIMBE, SVM, AIUM, and APVS. He is currently the Chairman of AtheroPoint, Roseville, CA, USA, dedicated to imaging technologies for cardiovascular and stroke. He has nearly ~22,000 citations, has co-authored 50 books, and has an H-index of 72.

List of contributors

Rupesh Agrawal

Tang Tock Seng Hospital, Singapore

Dr Rupesh Agrawal, is practising as Senior Consultant Ophthalmologist at National Healthcare Group Eye Institute, Tan Tock Seng Hospital, Singapore. After completing his medical studies from Nagpur (India), he did his post-graduation and fellowship in Uveitis and Ocular trauma from Sankara Netralaya, Chennai (India). He was subsequently working as Consultant Ophthalmologist (Uveitis, Ocular Trauma and Cataract) at Shri Ganapati Netralaya, Jalna (India). He also worked briefly at L V Prasad Eye Institute before migrating to Singapore in 2009. Subsequently, he has been working in Singapore and involved in significant translational and basic science research projects. He completed his overseas research training fellowship—awarded by National Medical Research Council Scholarship, Ministry of Health, Singapore at University College London on retinal imaging using retinal leukogram and erythrogram projects. He is also working on mechanical properties of the red blood cell and its implications in microvascular disorders.

He has to his credit 200+ peer reviewed publications and 50+ book chapters. He is actively involved in numerous collaborative projects on uveitis. He is recipient of many grants for numerous basic science projects pertinent to ocular inflammation. He is currently secretary of International Society of Ocular trauma and Asia Pacific Ophthalmic Trauma Society.

Minhaj Nur Alam

Department of Biomedical Data Science, Stanford University, Stanford, CA, USA

Minhaj Nur Alam, PhD is currently a postdoctoral fellow at the Stanford University, Department of Biomedical Data Science (Center for Artificial Intelligence in Medical Imaging, AIMI). He is an alumnus of the Biomedical Optics and Functional Imaging lab at the University of Illinois at Chicago. His research focus is in the field of biomedical and optical image processing and medical AI applications. He has extensive experience in incorporating quantitative image processing and machine learning algorithms for computer aided diagnosis in ophthalmology/radiology.

Muhammad Amer Awan

Shifa International Hospital, Islamabad, Pakistan

Muhammad Amer Awan, FRCS, FRCOphth, is a Professor of Ophthalmology and a consultative Vitreoretinal surgeon currently affiliated with Shifa College of Medicine/Shifa International Hospital. He is an avid researcher and has published extensively in peer-reviewed journals. He has a special interest in a 27 gauge micro-incision vitrectomy system to manage various retinal diseases.

Francesco Bandello

Department of Ophthalmology, University Vita-Salute, IRCCS Ospedale San Raffaele, Milan, Italy

Francesco Bandello, MD, FEBO, is Full Professor and Chairman at the Department of Ophthalmology University Vita-Salute, Scientific Institute San Raffaele, Milan, Italy. He is Academic Dean of 'Corso di Laurea Specialistica/Magistrale in Medicina e Chirurgia' at the same University. Professor Bandello is Past President of EURETINA, President of Academia Ophthalmologica Europea and Vice-President of EuroLam. Professor Bandello is Editor-in-Chief of the European Journal of Ophthalmology and former board member of the Club Jules Gonin and Macula Society. He is member of Executive Board of ESASO Foundation (European School for Advanced Studies in Ophthalmology), member of the Academia Ophthalmologica Internationalis and the Accademia Nazionale di Medicina.

Professor Bandello is co-author of 11 books and he serves as a peer reviewer for grant applications for the NEI. He has authored or co-authored 500 PubMed articles and he served as trained Principal Investigator in several clinical trials performed following ICH/GCP and mainly concerning retinal diseases.

Enrico Borrelli

Department of Ophthalmology, University Vita-Salute, IRCCS Ospedale San Raffaele, Milan, Italy

Enrico Borrelli, MD, FEBO, graduated from Vita-Salute San Raffaele University (Milan, Italy) in 2013. He completed his residency in ophthalmology at Gabriele D'Annunzio University (Chieti, Italy) and his clinical research fellowship in medical retina and retinal imaging at Doheny Eye Institute, David Geffen School of Medicine at UCLA (Los Angeles, CA, USA). Currently, he works as retina specialist at San Raffaele Hospital (Milan, Italy). His research interests include age-related macular degeneration and image analysis studies.

Davide Borroni

Eye and Vision Science Department, University of Liverpool, Liverpool, UK

Mr Davide Borroni is an ophthalmologist and researcher in The Venice Eye Bank Foundation, Venice, Italy. He completed his Fellowship in Cornea, Cataract and Refractive Surgery in St Paul's Eye Unit, Royal Liverpool University Hospital, Liverpool, UK. He received the 2018 ESCRS Clinical Research Awards for his studies in metagenomics in ophthalmology. He has experience in corneal transplants and microbiology.

Talisa de Carlo

University of Colorado at Denver, Denver, CO, USA

Dr de Carlo earned her Bachelors of Arts Degree from the University of Pennsylvania, where she graduated with Cum Laude honors. She received her M.D. from Tufts University School of Medicine, graduating with research honors and awarded the William Dameshek Undergraduate Research Award. She completed the Optical Coherence Tomography Research Fellowship at the New England Eye Center and Massachusetts Institute of Technology during a gap year in medical school. Dr de Carlo completed her Internship at the University of Hawaii and her Ophthalmology Residency at the Illinois Eye and Ear Infirmary of the University of Illinois at Chicago. She is currently a Vitreoretinal Surgery Fellow at the University of Colorado at Denver. She has authored over 50 manuscripts and presented her research at 15 international and local ophthalmology meetings. She is also one of the founding members of the International Ocular Circulation Society and an invited reviewer for multiple esteemed ophthalmology journals. Some of her accolades include the University of Illinois at Chicago Hospital Resident of the Year (August 2019), VitreoRetinal Surgery Foundation Research Award (November 2017), and the Retina Research Foundation/Joseph M and Eula C Lawrence Travel Grant of the Association for Research in Vision and Ophthalmology (February 2015).

Carlo Di Biase

Department of Ophthalmology, University Vita-Salute, IRCCS Ospedale San Raffaele, Milan, Italy

Carlo Di Biase, MD, received his Medical degree at University of Genoa, Italy in 2017, and is currently resident, in the second year, of ophthalmology at Vita-Salute San Raffaele University in Milan, Italy.

Dongxu Gao

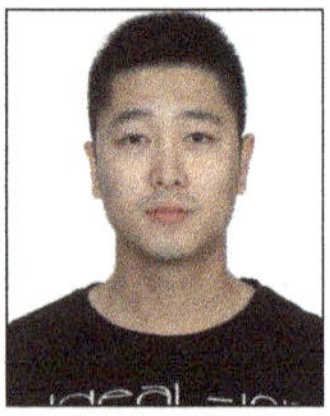

Eye and Vision Science Department, University of Liverpool, Liverpool, UK

Dongxu Gao is a Postdoctoral researcher at the University of Liverpool, UK. He received his PhD degree in computer science from the University of Portsmouth in 2017. He has rich experience in face recognition, object tracking and human motion recognition. His research interests are deep learning, computer vision and their applications.

Onur Gökmen

Yuzuncu Yil University Faculty of Medicine, Ophthalmology Department, Van, Turkey

Dr Onur Gökmen studied medicine in Gazi University Faculty of Medicine, Ankara, Turkey. After completion of his MD degree, he enrolled in the ophthalmology residency program at Baskent University Faculty of medicine ophthalmology department. He fulfiled the government duty at Health Sciences University Van Education and Research Hospital and worked five years. Later, he worked as a doctor lecturer for two years at the university of Health Sciences Samsun Training And Research Hospital. He was an observer at the University of Miami Bascom Palmer Eye Institute and Cleveland Clinic Cole Eye İnstitute. He has acquired the postnominal acronym FEBO in 2016. He is married and has two daughters. He has published a dozen scientific papers in well-known peer-reviewed journals. His specific area of interest is ophthalmic plastic surgery and medical-surgical retina. He is currently working as an academic staff member in Van Yuzuncu Yil University, Faculty of Medicine.

Adekunle Hassan

Eye Foundation Retina Institute and Eye Foundation Hospital, Lagos, Nigeria

Dr Kunle Hassan is an internationally acclaimed ophthalmic and vitreoretinal surgeon with vast experience in the care of vitreoretinal diseases. He has trained several generations of ophthalmic and retinal surgeons and is the founder and chairman of the first ophthalmic group practice, with multiple branches across Nigeria. He has several firsts to his credit, including acquiring the first OCT Angiography technology in the country and the first indigenous state-of-the-art vitreoretinal surgical facility in the region. This center receives vitreoretinal referrals from within the country and the region. Dr Hassan's passion for eye care providers' training is evident as he established the first private ophthalmic residency training program in the country and established training for other cadres of eye care providers.

Smaha Jahangir

The University of Faisalabad, Faisalabad, Punjab, Pakistan

Smaha Jahangir, OD, is a dedicated optometric physician and is specializing in retinal imaging and diseases. She is a post-graduate student at the University of Faisalabad. She is an astute clinician with excellent knowledge and command on ocular imaging. Her research interests are focused on clinical applications of OCT.

Patrick Kelley

Department of Physics, Indiana University-Purdue University, Indianapolis, IN, USA

Patrick Kelley received his BS degree with a dual major in physics and mathematics from Purdue University in 2016; he received his MS degree in educational physics from Purdue University in 2019 performing textual analysis, primarily sentiment analysis, on data accumulated from a social media platform called CourseNetworking (CN) tailored to student interaction. He applied his knowledge of text mining toward machine learning for image processing. He has an active interest in machine learning, deep learning and its future applications. He is currently working on his PhD in experimental physics. The project is on optical cooling of a nanoparticle, implementing machine learning, with the goal to use in ultra-weak gravity force measurements. Of the many skills, he has gathered a strong background in LabView, Python, R, SAS, MatLab, and Mathematica, with some familiarity with HTML, Javascript, CSS, and MarkDown.

Hashim Ali Khan

SEHHAT Foundation Hospital, Gilgit, Pakistan

Hashim Ali Khan, OD, FAAO, is a consultative optometrist, specializing in retinal diseases and retinal imaging. He is an Assistant Professor of Optometry at the University of Faisalabad. With a remarkable background in publishing scientific articles, he serves on the editorial boards of the *Journal of Ocular Infection and Inflammation* and *Journal Inside Optometry*. He is a certified clinical OCT/OCTA grader and provides retinal image reading services to many centers.

Muna Kharel

Nepalese Army Institute of Health Sciences, Kathmandu, Nepal

Dr Muna completed her MBBS from Kathmandu medical college and teaching hospital, Kathmandu Nepal and is currently consultant ophthalmologist at Birat Eye Hospital. She has a keen interest in glaucoma and medical retina and has multiple publications to her name.

Anadi Khatri

Birat Eye Hospital, Biratnagar, Nepal

Dr Anadi Khatri is the Medical Director, co-founder and head of Vitreo-Retina Services at Birat Eye Hospital, Nepal. He is also Associate Professor of Ophthalmology at Birat Medical College. He is also on the editorial boards of *Nepalese Journal of Ophthalmology*, *BMC ophthalmology* and *Birat Journal of Health Sciences*. His interests and field of research are in lasers in ophthalmology, multimodal imaging, Anti-VEGF agents, retinal vasculopathies and innovations in vitreo-retina surgery. He has over 45 publications in reputed journals on the same fields of ophthalmology.

He has multiple awards including American academy of ophthalmology Eyewiiki international ophthalmologist contest winner, ARVO DCERF, Japanese Ophthalmological society young investigator award and SAARC Academy of Ophthalmology Best Research Paper to his name.

Taehoon Kim

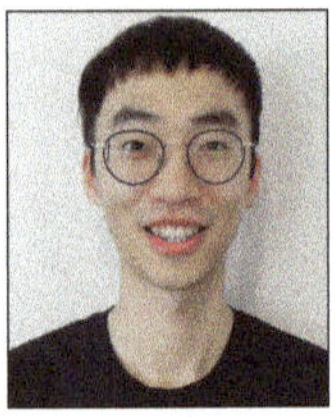

Department of Biomedical Engineering, University of Illinois at Chicago (UIC), Chicago, IL, USA

Taehoon Kim is a PhD candidate at University of Illinois at Chicago (UIC). Taehoon obtained his BS and MS degrees in Biomedical Engineering at Catholic University of Daegu. After graduation, he worked in Texas Tech University as a research scientist for a year. Taehoon then joined the Yao lab at UIC in August 2017 for his doctoral study. His current research focuses on optical imaging of retinal function.

Gregg Kokame

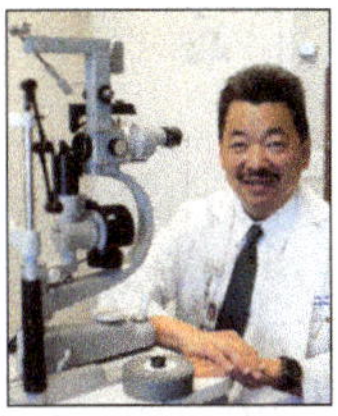

Retina Consultants of Hawaii; University of Hawaii John A Burns School of Medicine, Aiea, HI, USA

Dr Kokame earned his Bachelor of Arts Degree from Pomona College, where he graduated with Phi Beta Kappa and Magna Cum Laude honors. He received his MD from the UCLA School of Medicine, where he graduated with honors in the honor medical society, Alpha Omega Alpha. Dr Kokame completed his ophthalmology residency at the Jules Stein Eye Institute at the UCLA School of Medicine, consistently rated as the top eye hospital in the western USA. He completed his Vitreoretinal Surgery Fellowship at Bascom Palmer Eye Institute at the University of Miami School of Medicine in Miami, Florida, which was recently rated as the top eye surgeon hospital and training program in ophthalmology in the USA. Dr Kokame is Chief of Ophthalmology and Clinical Professor at the University of Hawaii John A Burns School of Medicine, where he teaches international vitreoretinal fellows, residents, interns and medical students about retinal diseases. He has directed The Retina Center at Pali Momi into a nationally and internationally recognized center for retinal research and education. Many new clinical trials bringing new advances over the last decade have initially started in Hawaii as clinical trials at The Retina Center at Pali Momi. He has authored over 100 scientific papers in peer-reviewed journals, and is a consultant and reviewer for all the major journals in ophthalmology. Dr Kokame is a member of the American Ophthalmological Society, Association of University Professors of Ophthalmology, Macula Society, American Academy of Ophthalmology, and American Society of Retina Specialists. He has received the Honor and Senior Honor Award of the American Society of Retinal Specialists, and the Secretariat and Achievement Award and Senior Achievement Award from the American Academy of Ophthalmology. He has given over 70 invited lectures, including the Pettit Lecture at the UCLA Jules Stein Eye Institute Annual Seminar in 2013.

David Le

Department of Biomedical Engineering, University of Illinois at Chicago, Chicago, IL, USA

David Le is a PhD candidate at the University of Illinois at Chicago (UIC). David obtained his BS in Biomedical Engineering from Rutgers University. After graduation, he did research at Rutgers University related to artificial intelligence in biomedical imaging. David then joined the Yao lab at UIC in August 2018 for his doctoral study. His current research focuses on disease classification using retinal imaging.

Jennifer I Lim

Department of Ophthalmology and Visual Sciences, University of Illinois at Chicago, Chicago, IL, USA

Dr Jennifer I Lim holds the Marion H Schenk Esq., Chair in Ophthalmology for Research in the Aging Eye as a Professor of Ophthalmology, Vice-Chair of Ophthalmology for Diversity & Inclusion, and Director of the Retina Service at University of Illinois at Chicago. Her areas of expertise include surgical and medical retinal diseases. She received her MD with Distinction from Northwestern University (Six-Years Honors Program in Medical Education), where she was elected into AOA and received both the Dean's AOA Student Research Award and the Julius Conn Memorial Award. She completed her ophthalmology training at University of Illinois at Chicago (UIC), Illlinois Eye and Ear Infirmary where she received the Resident Research Award and the American College of Surgeons Resident Competition Chicago Chapter Keeshin Prize. She completed medical and surgical retina fellowships at the Wilmer Eye Institute of Johns Hopkins Hospital, where she was a Heed Fellow and a Heed Knapp Fellow Awardee. After completing her training, she joined the faculty of Emory University as an Assistant Professor of Ophthalmology. She was recruited to Doheny Eye Institute of University of Southern California as an Associate Professor with tenure and then promoted to Professor of Ophthalmology. In 2007, she was recruited to University of Illinois at Chicago as the Director of the Retina Service. Her research interests include clinical trials, translational research, retinal imaging for which she has an R01 NIH grant (Co-PI on 'Differential artery-vein analysis in OCT angiography for objective classification of diabetic retinopathy', and retinal vascular disease. She has been principal investigator in over 60 clinical trials, serves on Data and Monitoring Committees and sits on Executive Committees. She collaborates with basic scientists on angiogenesis and imaging research. She has received funding from the National Eye Institute, Macula Society Research Fund, American Cancer Society and numerous industry grants. Current leadership positions include Associate Editor for JAMA Ophthalmology, EyeWiki Deputy Editor-in-Chief, Vice-President of The Retina Society, President of Chicago Ophthalmological Society, IOVS Editorial Board member, AOA UIC Medical School Councilor, and UIC Faculty Senator. She is a past-President of both Women in Ophthalmology and Chinese American Ophthalmology Society and past Chair of Women in Retina.

She has received the AAO Life Achievement Honor Award, AAO Senior Achievement Award, AAO Secretariat Award, American Society of Retinal Specialists Senior Honor Award, Suzanne-Veronneau Troutman Award, ARVO Gold and Silver Fellow Awards, the Paul Henkind Memorial Award, Women in Ophthalmology Scientific Contribution Award, Chinese American Ophthalmology Service awards, USC teaching awards, ASRS Retina Hall of Fame, UIC Departmental Faculty of the Year Award, Mother McAuley Liberal Arts HS Hall of Honor Award, and honored on the inaugural 2021 Newsweek Best Ophthalmologists in America, Chicago Super Docs, Best Doctors, and Top

Doctors. She has authored over 350 articles, 30 book chapters and edited several books including Age-Related Macular Degeneration (now in the Third Edition). She has delivered several named lectures, including the inaugural UIC Distinguished Sweeney Lecture. She has authored or co-authored over 300 articles, 25 book chapters and edited several books.

Savita Madhusudhan

St Paul's Eye Unit, Royal Liverpool University Hospital, Liverpool, UK

Savita Madhusudhan is a consultant ophthalmologist at Royal Liverpool University Hospital, with special interests in medical retina, inherited retinal diseases and retinal imaging, and an honorary senior lecturer at the University of Liverpool. Over the last 10 years, she has been actively involved in clinical research as co-investigator and principal investigator for studies in age-related macular degeneration and ophthalmic genetics. She is the lead for Liverpool Ophthalmic Reading Centre, developing and providing retinal image grading and analysis for various projects both independently and through NetwORC UK. This has included multimodal grading for hydroxychloroquine retinotoxicity, malarial retinopathy, studies in age-related macular degeneration and diabetic retinopathy, and collaborating on development of automated grading and machine learning algorithms through the University of Liverpool. She is an associate editor for the retina section of *BMJ Open Ophthalmology*.

Olwen C Murphy

Johns Hopkins University School of Medicine, Baltimore, MD, USA

Dr Olwen C Murphy is a neurology fellow in Johns Hopkins University School of Medicine. She received her medical degree from University College Dublin, Ireland. After undertaking residency training in general internal medicine at Trinity College Dublin University Hospitals, she completed neurology residency training at Cork University Hospital, Mercy University Hospital (Cork) and Mater Misericordiae University Hospital (Dublin), Ireland. Since 2017 she has been a fellow in neurology at Johns Hopkins University, undertaking specialized training in neuroimmunology and neurological infections (2017–2020), followed by ocular motor and vestibular neurology (2020–2021). Her research interests include non-invasive imaging biomarkers in multiple sclerosis, neurosarcoidosis, and other neuroimmune disorders.

Ogugua Okonkwo

Eye Foundation Retina Institute and Eye Foundation Hospital, Lagos, Nigeria

Dr Ogugua Okonkwo is an ophthalmic and vitreoretinal surgeon. He has several years of experience in the care of complicated vitreoretinal diseases. He has trained several ophthalmologists and vitreoretinal fellows and is passionate about advancing the care of vitreoretinal disease in the African region. He has published several papers on vitreoretinal-related topics and reviews for a long list of journals, several of which are indexed on PubMed. He is the editor in chief of two ophthalmic journals, on the editorial board of several others, and actively mentors juniors in research.

Muhammet Derda Özer

Yuzuncu Yil University Faculty of Medicine, Ophthalmology Department, Van, Turkey

Dr Lecturer Muhammet Derda Özer studied medicine in Istanbul University, Istanbul Faculty of Medicine. After completion of his M.D. degree, he enrolled into the ophthalmology residency program at the same Institution. He fulfiled the government duty at Hakkari State Hospital as a surgeon. He has acquired the postnominal acronym FICO in 2020 and awarded with a three month fellowship by International Council of Ophthalmology. He has published a dozen of scientific papers in well-known peer-reviewed journals. His specific area of interest is ocular immunology and inflammation, medical and surgical retina. He is currently working as an academic staff member in Van Yuzuncu Yil University, Faculty of Medicine.

Araniko Pandey

Civil Service Hospital of Nepal, Kathmandu, Nepal

Dr Araniko Pandey completed his MBBS from Nepal Medical College—Kathmandu University and residency in ophthalmology from Tribhuvan University. He was an attending ophthalmologist at Lumbini Eye Institute and Medical Director at Palpa Lions Lacoul Eye Hospital, a secondary level rural eye care system. He has published more than a dozen scientific papers and has presented his research papers at various national and international scientific meetings including, ARVO and RANZCO.

Dr Pandey is an ECFMG certified ophthalmologist with an experience of International Research Observership at Bascom Palmer Eye Institute, McKnight Vision Research Center in the 'Quantitative Imaging and Eye–Brain Research Laboratory', Miami, USA. He is particularly interested in studies of cost effective, noninvasive, user friendly ophthalmological tools, to identify potential clinical biomarkers.

Eli Pradhan

Tilganga Institute of Ophthalmology, Kathmandu, Nepal

Dr Eli Pradhan is an ophthalmologist by profession currently working in Tilganaga Institute of Ophthalmology. Dr Pradhan is a retina consultant as she is recognized for her skills, knowledge and empathetic response in this field. Her concern in this field lies not only to diagnose the patient but she likes to be involved with the psychological aspects with the treatment. Adding value to the medicine sector, where doctors are often criticized for too much objectivity and less concern with patients, she holds the belief that better communication and proper counseling are very important aspects of any treatment.

She is also the editor in chief of the *Nepalese Journal of Ophthalmology* and president of the Nepal Vitreoretinal Society. Compassionate and empathetic, Dr Pradhan is surely a shining gem in the field of ophthalmology in Nepal. Despite being recognized as a doctor, her interests include writing and editing in various journals, newspapers and magazines including *Healthy Life*. She utilizes her spare time in making portraits and landscapes, sketching, and is involved in humanitarian work through organizations like Rotary.

Gunjan Prasai

Kathmandu Medical College, Kathmandu, Nepal

Dr Gunjan completed her MBBS from KIST Medical college, Tribhuvan University. She is currently a third-year resident in ophthalmology at Kathmandu medical college and teaching hospital, Kathmandu Nepal. She has a keen interest in retina and ocular imaging. She also serves as an editor in *Nepalese Journal of Ophthalmology*.

Francesco Prascina

Department of Ophthalmology, University Vita-Salute, IRCCS Ospedale San Raffaele, Milan, Italy

Francesco Prascina, MD, received his medical degree at the University of Bari (Italy), and his specialization in ophthalmology at the Department of Ophthalmology from Bari (Italy). He is consultant in the Medical Retina and Imaging Unit of the Department of Ophthalmology, University Vita-Salute, Scientific Institute San Raffaele (Milan, Italy). His main research interests include the ophthalmic imaging and management of retinal diseases.

Giuseppe Querques

Department of Ophthalmology, University Vita-Salute, IRCCS Ospedale San Raffaele, Milan, Italy

Giuseppe Querques, MD, PhD, is Associate Professor at University Vita-Salute, Scientific Institute San Raffaele, Milan, Italy. His main topics are medical retina and ophthalmic surgery (retina and cataract). Professor Querques has contributed to more than 300 peer-reviewed articles published mainly in the areas of medical retina (age-related macular degeneration, retinal vascular diseases, hereditary retinal diseases, ophthalmic genetics). His current main area of both clinical and laboratory research is focusing on the diagnosis (imaging) and treatment of age-related macular degeneration, retinal vascular diseases, and hereditary retinal diseases. Doctor Querques is a member of the Italian Society of Ophthalmology (SOI), the French Society of Ophthalmology (SFO), the European Association of Retinal Specialists (EURETINA), the American Academy of Ophthalmology (AAO), the Association for Research in Vision and Ophthalmology (ARVO), the Retina Society, the Club Jules Gonin, and the Macula Society. Doctor Querques serves as Editor in Chief of *Ophthalmology @ Point Of Care*, and as Editorial Board Member of *European Journal of Ophthalmology*, *Acta Diabetologica*, *Ophthalmologica*, *Ophthalmology and Therapy*, *Retinal Cases* and *Brief Reports*, and *American Journal Case Report*.

Lea Querques

Department of Ophthalmology, University Vita-Salute, IRCCS Ospedale San Raffaele, Milan, Italy

Lea Querques, MD, received her medical degree and specialization in ophthalmology at the Department of Ophthalmology from the University Vita-Salute San Raffaele, Milan (Italy). She is consultant in the Medical Retina and Imaging Unit of the Department of Ophthalmology, University Vita-Salute, Scientific Institute San Raffaele (Milan, Italy). Dr Querques has published more than 100 articles in Pub-Med journals mainly related to retinal diseases. Her main research interests include the ophthalmic imaging and management of retinal diseases.

Julie Rodman

Nova Southeastern University, Fort Lauderdale, FL, USA

Dr Julie Rodman is the Chief of the Broward Eye Care Institute in Fort Lauderdale, FL. Her research interests include Optical Coherence Tomography/Angiography and Vitreoretinal Disease. Dr Rodman has authored over thirty publications with an emphasis on retinal disease. She recently published *Optical Coherence*

Tomography Atlas: A Case Study Approach, the first reference book on this topic written by an optometrist. Dr Rodman is a member of the American Optometric Association, American Academy of Optometry, Florida Optometric Association, Optometric Retina Society and Association for Research in Vision and Ophthalmology. Dr Rodman is a member of the Optovue Advisory Board where she serves as a lecturer and consultant. She holds her Oral Pharmaceuticals Certification and is Laser/Surgical Certified as well.

Dr Rodman has been the recipient of numerous teaching awards, including the Golden Apple Award for Excellence in Clinical Precepting, and Preceptor of the Year. She was recognized as a Primary Care Optometry News 'Top 300' Optometrists recognizing excellence in innovation and furthering of the profession.

FYI: In her free time, Dr Rodman loves spending time with her husband and two daughters … just doing anything with them!!!!

Riccardo Sacconi

Department of Ophthalmology, University Vita-Salute, IRCCS Ospedale San Raffaele, Milan, Italy

Riccardo Sacconi, MD, FEBO, received his medical degree from the University Vita-Salute San Raffaele, Milan (Italy) in 2012. He completed his residency in Ophthalmology at University of Verona (Italy), and a clinical research fellowship in medical retina and retinal imaging at the department of Ophthalmology of San Raffaele Hospital, Milan (Italy). Currently, he works as retina specialist at San Raffaele Hospital and he is completing his PhD program at the Vita-Salute San Raffaele University. His main research interests include the ophthalmic imaging and management of retinal diseases, contributing to more than 120 peer-reviewed articles. He is editorial assistant of *European Journal of Ophthalmology* and *Ophthalmology @ Point Of Care*.

Shiv Saidha

Johns Hopkins University School of Medicine, Baltimore, MD, USA

Dr Shiv Saidha specializes in the diagnosis, management, and treatment of multiple sclerosis (MS), as well as other neuro-immunological disorders of the central nervous system, including those afflicting the brain, spinal cord, and optic nerves.

Dr Saidha received his medical degree, as well as his postgraduate Doctor of Medicine, from the National University of Ireland, Galway (NUIG), Ireland. He completed residency training in general internal medicine at Galway University Hospitals, Ireland. Subsequent to this, he completed residency training in neurology at Galway, Cork, and Beaumont (located in Dublin) University Hospitals, Ireland. He then completed a further three years of specialized fellowship training in neuroimmunology and neuroinfectious diseases at Johns Hopkins University School of Medicine, Baltimore, Maryland. He is currently an Associate Professor of Neurology at Johns Hopkins University School of Medicine.

Dr Saidha's research interests to date have predominantly focused on the non-invasive interrogation of retinal structures using optical coherence tomography (OCT) and OCT angiography (OCTA) in multiple sclerosis (MS), in order to identify and investigate novel outcome strategies for assessing and monitoring neuroprotection and neurorestoration in MS, as well as to further our understanding of the pathobiology of MS, including how retinal pathology in MS may be related to more global central nervous system disturbances. He has been at the fore of discovering and demonstrating that inner and outer retinal layer pathology occur within the unmyelinated retina in MS (both quantitatively and qualitatively). He is committed to continuing to delineate and ascertain the mechanisms involved in the development of these findings, the clinical relevance of these changes (both locally in the retina, as well as more globally), and determine if this information may be used to guide the development of new, and potentially unique/novel effective MS therapies. His work to date has been primarily structurally focused (both within the retina using OCT and OCTA, as well as more globally within the central nervous system through the assessment of brain substructure volumes, diffusion tensor imaging and magnetization transfer ratio metrics within specific pathways in MS). He also has expertise in the functional assessment of the anterior visual pathway in MS through the application of multifocal electroretinography, pupillometry, and retinal function imaging techniques, amongst others, as well as of global and regional brain metabolism. He has published first author papers in *Lancet*, *Lancet Neurology*, *Annals of Neurology*, *Brain*, *Neurology*, *JAMA neurology*, and *Multiple Sclerosis Journal* amongst others.

Muhammad Aamir Shahzad

Aziz Fatimah Medical and Dental College, Faisalabad, Punjab, Pakistan

Muhammad Aamir Shahzad, MBBS, FRCS, is an Assistant Professor of Ophthalmology and a Consultant Ophthalmic surgeon at Aziz Fatimah Medical and Dental College/Aziz Fatimah Hospital. He is a dedicated retinal physician and researcher. His main areas of expertise are epidemiology and management of diabetic retinopathy and retinal imaging.

Kinsuk Singh

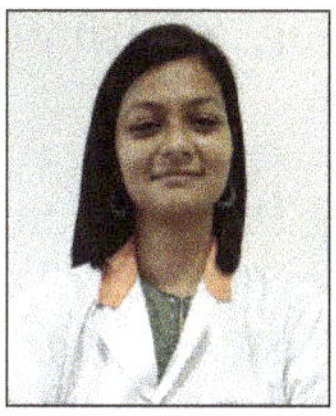

Vasan Eye Centre, New Delhi, India

Dr Kinsuk completed her MBBS from Institute of Medicine, Tribhuvan University, Kathmandu, Nepal. She went on to complete her MD in Ophthalmology from B.P. Koirala Lions Centre for Ophthalmic Studies, Institute of Medicine, Tribhuvan University, Kathmandu, Nepal in the year 2016. She completed her training in anterior segment surgery from Biratnagar Eye Hospital, Biratnagar, Nepal in the year 2018. She is presently a Member of Royal College of Surgeons of Edinburgh (MRCSEd). She is currently working at Vasan Eye Care, India.

Taeyoon Son

Department of Biomedical Engineering, University of Illinois at Chicago, Chicago, IL, USA

Taeyoon Son, PhD is a research assistant professor in the Biomedical Optics and Functional Imaging lab at the University of Illinois at Chicago. His primary interest is in the development of human and animal OCT/OCTA systems. He is currently developing the technology for oximetry based artery-vein classification in humans. Dr Son is also leading the project for a portable wide-field fundus camera for examination of retinopathy of prematurity (ROP). With a strong background in programming languages such as MATLAB, C, and LabVIEW he has developed graphical user interfaces for clinical implementation of different algorithms. He has a PhD in Biomedical Engineering from Yonsei University, Korea.

Meysam Tavakoli

Radiation Oncology Department, University of Texas Southwestern Medical Center, Dallas, TX, USA

Meysam Tavakoli received the BS degree in physics, Tehran University, MS in Medical Physics from Mashhad Medical School, and PhD in physics from Purdue University. Since 2006 he has worked on medical image processing, especially retinal images. He applied his knowledge toward machine and deep learning for image processing. His research interests include medical physics, medical image processing, computer vision, pattern recognition, and statistical models.

Xiyin Wu

College of Big data and Information Engineering, Guizhou University, Guiyang, China

Xiyin Wu is a lecturer at College of Big data and Information Engineering, Guizhou University, China. She obtained her PhD degree in pattern recognition and intelligence systems from Nanjing University of Science and Technology in 2020. She obtained her Bachelor's degree in information security (2012) and the Master's degree in computer technology (2015) from Jiangnan University. Her research interests include image processing, computer vision, pattern recognition and medical image analysis.

Xincheng Yao

Department of Biomedical Engineering and Department of Ophthalmology and Visual Sciences, University of Illinois at Chicago (UIC), Chicago, IL, USA

Xincheng Yao, PhD is a Richard and Loan Hill Professor of Biomedical Engineering and Ophthalmology and Visual Sciences, University of Illinois at Chicago (UIC). Dr Yao received his PhD in optics from the Institute of Physics, Chinese Academy of Sciences in 2001. This was followed by his postdoctoral research in the Biophysical Group at Los Alamos National Laboratory (LANL) from 2001 to 2004. He held a LANL Technical Staff appointment from 2004 to 2006 and served at CFD Research Corporation as a Senior Research Scientist from 2006 to 2007. He was a faculty of University of Alabama at Birmingham (UAB) from 2007 to 2014 and joined UIC faculty in 2014. Dr Yao's research interests include biomedical optics instrumentation, ultra-wide field fundus photography, functional optical coherence tomography (OCT), OCT angiography, super-resolution ophthalmoscopy, and machine learning based image classification. He was elected as a fellow of the International Society for Optics and Photonics (SPIE) and the American Institute for Medical and Biological Engineering (AIMBE) in 2019.

Yalin Zheng

Eye and Vision Science Department, University of Liverpool, Liverpool, UK

Yalin Zheng was awarded a PhD in Electronics and Computer Science from the University of Southampton, Southampton, UK, in 2003. He is currently Reader in Ophthalmic Imaging at the Department of Eye and Vision Science, University of Liverpool, Liverpool, UK. His research interests include image processing, computer vision, artificial intelligence and medical image analysis. He has published over 150 conference and journal papers.

Ilaria Zucchiatti

Department of Ophthalmology, University Vita-Salute, IRCCS Ospedale San Raffaele, Milan, Italy

Ilaria Zucchiatti, MD, received her medical degree at the University of Udine (Italy), and her specialization in ophthalmology at the Department of Ophthalmology, University Vita-Salute San Raffaele, Milan (Italy). She is consultant in the Medical Retina and Imaging Unit of the Department of Ophthalmology, University Vita-Salute, Scientific Institute San Raffaele (Milan, Italy). Her main research interests include the ophthalmic imaging and management of retinal diseases.

Chapter 1

Clinical application of optical coherence tomography angiography in retinal diseases

Hashim Ali Khan, Muhammad Aamir Shahzad, Muhammad Amer Awan and Smaha Jahangir

Optical coherence tomography angiography (OCTA) is a functional extension of optical coherence tomography (OCT) that has made *in vivo* vascular imaging possible using motion contrast. Being a recently introduced modality it is rapidly evolving and so is our understanding of OCTA. In this chapter we review the main principles behind acquiring ocular blood flow information using OCTA, commercialization of OCTA, and compare it with existing ocular vascular imaging techniques. We discuss clinical interpretation of retinal OCTA, related artifacts and limitations of the system. Further, OCTA features of important retinal and choroidal diseases are also reviewed.

1.1 Introduction

In everyday clinical practice, we encounter a significant number of patients where we need to investigate the retinochoroidal vasculature and hemodyanamics of different lesions, diagnose the condition and guide treatment plans. The retina is a unique ocular tissue where the vasculature can be directly visualized non-invasively *in vivo*. Most conditions and lesions can be examined clinically in office settings and no further testing is required. But, oftentimes we may need to study the vascular properties and hemodynamics of the retinal tissue to differentiate between similar looking conditions or to tailor our treatment plan. In addition, clinical examination has its own limitations and many diseases may exist and progress before they become clinically evident.

Fundus fluorescein angiography (FA) and indocyanine green angiography (ICGA) are dye based ocular vascular imaging techniques that have been used for a long time. Retinal vasculature is better visible on FA while ICGA displays the

doi:10.1088/978-0-7503-2060-3ch1

choroidal vasculature more clearly. Both techniques have been the mainstay of ophthalmic vascular imaging for a long time. However, both are invasive and associated with significant adverse events.

Optical coherence tomography angiography (OCTA) is a newer and emerging, non-invasive vascular imaging technique, based on motion contrast extracted from high speed optical coherence tomography (OCT) images and produces high resolution, depth resolved images of retinochoroidal vasculature.

1.2 Comparison of OCTA with FA and ICGA

FA and ICGA have been in practice for decades. Both are dye based tests and require intravenous administration of indocyanine green and fluorescein sodium. Generally, both agents are considered safe; but they are associated with a significant number of adverse events and rarely severe reactions and even death. The risk of adverse effects increases with each repetition, limiting frequent administration of FA & ICGA. Conversely, OCTA uses motion contrast to extract vascular information of the tissue under study and is completely non-invasive. It does not cause any adverse events and can be repeated as many times as required.

During ICGA and FA, we need to study the different phases of dye transit through the ocular tissue to get useful information. Therefore, these tests last longer and need significant time for each patient to be studied. OCTA, on the other hand, is time efficient and gathers vascular images in less than a minute. Both FA and ICGA are dynamic tests and each phase of dye transit corresponds to a particular vascular bed at different planes of choroid and retina. Conversely, OCTA is a static test where each vascular network can be displayed at any desired depth and thickness of retinochoroidal tissue from volumetric data. In addition, structural information from corresponding structural OCT scans is also available for any region of interest on OCT angiograms.

FA and ICGA provide a wide field of view (up to 200 degrees) while OCTA has a limited field of view typically limited to the posterior pole. However, montaging multiple OCTA images can produce up to 100-degree field of view (table 1.1). But with increased field of view the fine details and resolution of image is compromised. However, multiple OCT angiograms from multiple smaller areas of the retina can be acquired and combined into one image for extended field of view and with high detail.

Artifacts are more common in OCTA and are minimal on ICGA or FA. These may significantly compromise the image quality and details of retinal vasculature. The common OCTA artifacts include projection artifacts, shadowing and motion artifacts.

1.3 Principles and methods of OCTA

OCTA is a functional extension of OCT. The conventional OCT devices acquire structural information of tissue based on partial coherence interferometry. Each OCT B-Scan consists of several A scans and multiple B Scans are used to reconstruct *en face* and three-dimensional structure of the retinal tissue. Each OCT B scan has

Table 1.1. Comparison between FA, ICGA and OCTA.

Parameter	FA	ICGA	OCTA
Invasive	Yes	Yes	No
Label	Sodium fluorescein	Indocyanine green	Label free
Time Needed	10–20 min	Up to 35 min	Up to 5 min
Ideal for	Retinal vasculature	Choroidal vasculature	Both
Adverse events	Relatively common	Less common	None
Field of view	Up to 200 degrees	Up to 200 degrees	Up to 100 degrees
Artifacts	Minimal	Minimal	Common
Structural information	Minimal	Minimal	Great
Objective features	No	No	Multiple

FA = fluorescein angiography: ICGA = indocyanine green angiography; OCTA = optical coherence tomography angiography

phase and intensity/ amplitude information of the sample. Based on different approaches, OCTA uses motion contrast to extract the flow characteristics and vascular patterns of the tissue of interest. These strategies detect variations in phase, amplitude or both, caused by moving blood cells against a static background (retina), from backscattered OCT signals. The methods can be divided into phase-based methods, amplitude based and complex methods [1–3].

1.3.1 Phase based strategies

The moving blood cell in vasculature causes a phase shift in backscattered OCT signals. The shift in phase can be used to compute flow measurements. Initial approaches of phase based OCTA extracted flow information using the phase shifts between successive A-scans. The initial phase-based approach was employed by Leitgeb *et al* [4] in their Doppler OCT (DOCT) using phase information from two consecutive A-scan lines [2]. Without dedicated scanning protocols and hardware modifications in DOCT, only axial flow can be measured. Also, the DOCT systems are subject to smaller motions caused by eye movements. Different attempts have been made to compensate it. These included histograms based bulk motion correction between consecutive A-scans and cross correlation of A-lines between B scan frames. With the continuous advancement in hardware and speed of OCT systems, the phase information has shifted from comparison of consecutive A scans to sequential B scans, making detection of slower flow possible. Phase variance technique by Fingler *et al* [5–7] using phase analysis between sequential, repeated B scans at the same location demonstrated vascular imaging comparable to FA. Phase gradient analysis (PGA) and split-spectrum phase gradient angiography (SSPGA) are recent phase based methods that map the vasculature without the need for correction of bulk axial movement. These methods provide a better signal-to-noise ratio (SNR) and contrast [1–3, 8].

1.3.2 Intensity/amplitude based strategies

These approaches use intensity and speckle information from OCT signals. contrary to phase-based methods, amplitude-based methods are less susceptible to phase noise [1–3]. During an OCT scan, a random speckle noise is generated as result of OCT backscattering. However, the speckle patterns generated by moving backscatterers (moving erythrocytes) is different from static scatterers (the static retinal tissue) [1–3]. Initial approaches included algorithms based on speckle variance. Further improvements were introduced in speckle variance technique by adjusting the number of repetitions of B scans at a given location [9–11]. Alternative approaches based on intensity include differential logarithmic intensity variance, intensity difference between successive B Scans and comparison of vessel reflectivity features from non-vascular areas [12–14]. Correlation mapping method generates two-dimensional and three-dimensional flow maps by computing the correlation between successive B scans [15, 16]. Split-spectrum amplitude-decorrelation (SSADA), described by Jia *et al* [17] is an algorithm based on the time-varying speckle. Contrary to correlation mapping strategies, it computes a decorrelation signal between sequential B scans. The OCT spectrum is split into narrow bands and each is processed separately generating multiple flow signals from the same B-scan. The algorithm improves SNR. Due to reduced axial resolution it is less susceptible to axial bulk motion, making it very useful for everyday clinical ocular blood flow imaging [2, 17]. OCTA ratio analysis (OCTARA) is another intensity based proprietary algorithm of Topcon (Topcon Corporation, Tokyo, Japan) paired with Topcon's Swept Source (SS) OCT. It is not based on amplitude decorrelation and the spectrum is not split, thus preserving axial resolution and optimized for visualization of both retina and choroid [18].

1.3.3 Complex signal-based OCTA

Phase based strategies are less sensitive to transverse flow while intensity-based approaches have poor performance in detecting slow flow. Complex methods combine information both from phase and amplitude thus providing better results [1–3]. Optical microangiography (OMAG), eigen-decomposition of OCT signals (based on the orthogonality) and the most recent split-spectrum amplitude and phase gradient analysis (SSAPGA) utilize both phase and amplitude information to generate vascular flow maps in high resolution. SSAPGA performance is superior to PGA, SSPGA, and SSADA in terms of contrast and SNR [8].

1.4 Commercial devices

Currently the following OCTA systems are commercially available for ophthalmic use. Optovue AngioVue™ OCTA System, Topcon Triton SS OCT Angio™, ZEISS AngioPlex™; Heidelberg Spectralis® OCTA, and Nidek AngioScan.

1.4.1 RTVue XR Avanti

AngioVue is OCTA software deployed on RTVue XR Avanti, a spectral domain (SD) OCT device. The software is based on SSADA algorithm and uses motion correction technology (MCT) for correcting motion artifacts and 3D projection artifact (PAR) removal algorithm to correct projection artifacts. The instrument has an A scan rate of 70000 per second and utilizes a light source centered on 840 nm with a bandwidth of 50 nm. The *en face* OCT angiograms may cover up to an area of 12 × 9 mm. The angiovue provide many qualitative metrics such as area and perimeter of foveal avascular zone (FAZ), the flow area and vessel density mapping [19].

1.4.2 Topcon Triton SS OCTAngio™

Topcon's (Topcon Inc., Tokyo, Japan) OCTA is built into their Topcon Triton SS OCT based on their proprietary intensity-based algorithm called OCTARA. It has a speed of 100000 A scans per second and uses a light source of 1050 nm which provides greater tissue penetration and visualization of choroid without losing details of vitreous. Scan size is up to 6.0 × 6.0 mm [18].

1.4.3 ZEISS AngioPlex™

Zeiss angioplex (Carl Zeiss Meditec AG, Germany) is an OCT angiography system based on OMAG algorithm deployed on the CIRRUS™ HD-OCT Model 5000 SD OCT and PLEX Elite 9000 SSOCT. Cirrus HD OCT 5000 has a light source centered on 840 nm and has a speed of 68000 A scans per second while PLEX Elite 9000 has a tunable laser centered on 1040 and 1060 nm with a speed of 100000 A scans per second. Angioplex provides a field of view of up to 8 × 8 mm on Cirrus 5000 and 12 × 12 mm on Plex Elite 9000. The field of view can be further extended by montaging multiple angiograms [20, 21].

1.4.4 Heidelberg Spectralis® OCTA

This is based on the amplitude decorrelation OCTA algorithm provided on the Heidelberg's (Heidelberg Engineering GmbH, Germany) Spectralis SD OCT platform. The device uses a light source of 870 nm and bandwidth of 50 nm with speed of up to 85000 A scans/second. The field of view for *en face* angiogram is up to 8.8 × 4.4 mm [22].

1.4.5 Nidek AngioScan

Angioscan is available on the Nidek RS 3000 Advance (Nidek, Aichi, Japan) SD OCT, which utilizes a light source of 880 nm. The device speed is up to 53000 A scans per second and it utilizes a modified OMAG algorithm. The field of view is up to 9 × 9 mm and can be further extended to 12 × 9 by montaging multiple scans. It can display vascular details of retina and choroid as well as providing vascular density maps and FAZ area/shape analysis.

1.5 OCTA of normal retina

1.5.1 Retinal and choroidal vasculature

The ocular blood supply is provided by the ophthalmic artery which branches into the central retinal artery (CRA), long and short posterior ciliary arteries and anterior ciliary arteries. The choroid is supplied by ciliary arteries which form arterioles of outer choroid and continuous single layer of wide lumen fenestrated capillaries of choriocapillaris (CC). The choroid has three vascular networks; the outer Haller's layer of larger vessels, inner Sattler's layer of smaller vessels and CC which contains fine network of capillaries [23].

The major vascular networks in human retina are superficial vascular plexuses (SVP) of major retinal arteries, veins and their corresponding branches and capillaries and is primarily located in the ganglion cell layer (GCL). The two deeper networks, intermediate capillary plexus (ICP) and deep capillary plexus (DCP) receive their supply through vertical anastomoses from the SVP. The ICP is a three-dimensional capillary network at the level of the inner plexiform (IPL) and superficial portion of the inner nuclear layer (INL) while the DCP is located along a single plane in a deeper portion of the INL and outer plexiform layer (OPL) [1–3, 24–27].

1.5.2 Default OCTA segmentations

By default, most OCTA devices display SVP, DCP, avascular outer retina (AOR) and CC as *en face* angiograms along with corresponding cross-sectional B Scans. ICP is also displayed by some platforms by default. Nonetheless, segmentations may be manually adjusted or selected from corresponding B scans to display vascular networks at desired depth planes [1–3].

The superficial vascular network appears as a homogenous mesh of white lines (representing vessels) against a dark background (avascular area) distributed across the entire retina, in a regular centripetal pattern, converged on the FAZ.

The Intermediate capillary plexus has a spider web-like channel and has a well-defined and smaller FAZ.

The deep capillary plexus appears as a web of fine, densely packed capillaries, centripetally converged around the perifovea.

AOR segmentation extends from the outer nuclear layer (ONL) to Bruch's membrane (BM) and is presented as a dark slab on OCTA.

The CC appears as a homogenous, texture of fine capillaries evenly distributed across the scan and located between the BM and the choroid. Figure 1.1 shows the OCTA segmentations and their respective depth locations on structural OCT.

1.5.3 Artifacts

Unlike FA and ICGA, OCTA is typically prone to a wide range of artifacts. These may be technical artifacts, operator/patient induced artifacts inherent to OCTA algorithms. OCTA angiograms are acquired using OCT platforms and may suffer from all artifacts inherent to structural OCT such as movement artifacts, blink artifacts, segmentation artifacts, etc. Artifacts specific to OCTA are projection artifacts and shadow artifacts.

Holmen *et al* [28] graded 406 OCTA images from 234 eyes and it was found that at least one artifact was present in 97% of the scans. Severe artifacts, associated with reliability of quantitative measures were present in up to 53%. Shadowing, defocus and movement were most prevalent in their study.

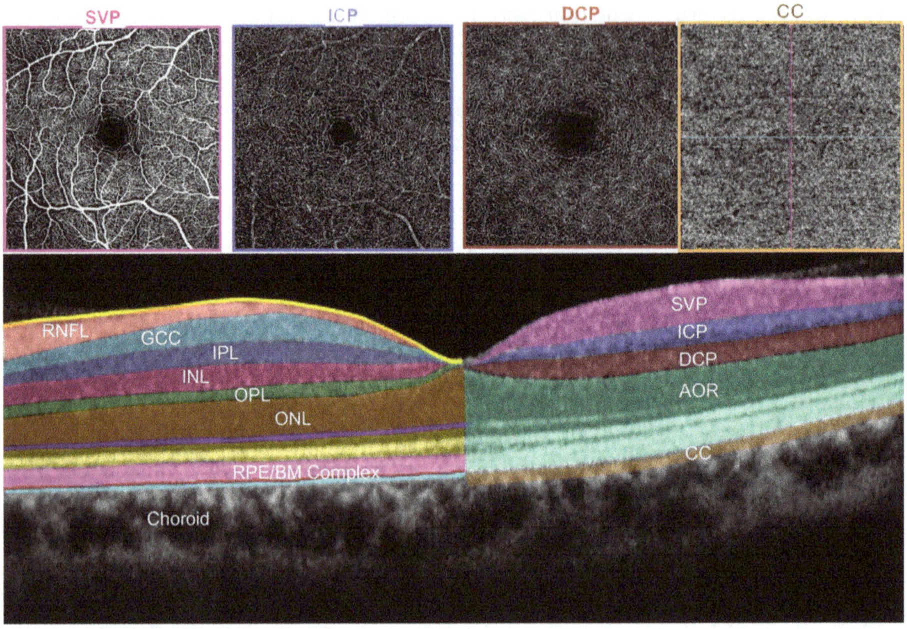

Figure 1.1. Bottom panel shows major layers on structural OCT on the left side and OCTA segmentations on the right. Top panel shows the color-coded *en face* angiograms corresponding to SVP, ICP, DCP and CC *en face* angiograms. SVP, ICP and DCP copied and edited from [26] under creative common license.

It is necessary to understand the artifacts, their impact on qualitative and quantitative results on OCT angiograms, before interpreting OCTA results.

1.5.3.1 Motion artifacts

As OCTA generates flow maps using motion contrast strategies, movements of the eye or within the eye in axial or transverse directions give rise to significant motion artifacts. The movements in axial direction are usually induced by respiration and cardiac rhythm. The false flow induced by axial motion is difficult to discern from actual low flow. However, some algorithms like SSADA are less susceptible to axial bulk motion.

The transverse ocular movements, as a result of saccades and ocular drifts, are still a problem. The movements appear as white lines on *en face* angiograms. Many manufactures use eye-tracking and algorithms to correct for transverse motion. But often these corrections, gives rise to another set of artifacts such as vessel doubling, stretching and crisscrossing.

Ocular blinks momentarily block the OCT signals and can be identified as horizontal black lines on *en face* angiograms. Figure 1.2 shows blink and transverse motion artifacts.

1.5.3.2 Shadow / masking artifacts

OCT signal attenuation from generalized media opacities (corneal disorders, cataracts and vitreous opacities) and poor beam alignment may result in low contrast and false low flow or flow voids on angiograms. Most devices display an

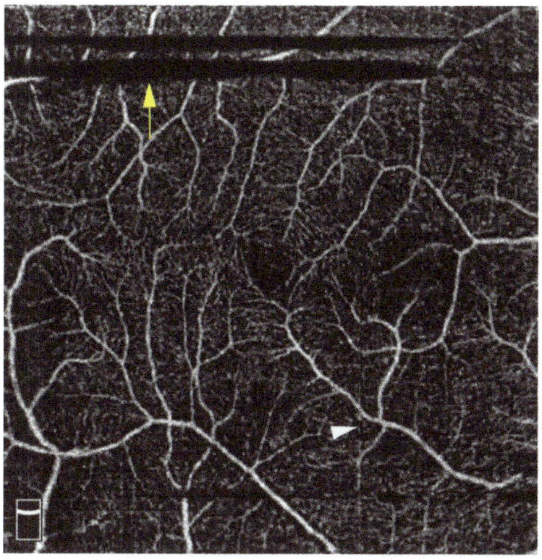

Figure 1.2. Black horizontal lines (arrow) representing blink artifacts and vertical faint white lines (arrow head) corresponding to transverse motion artifacts.

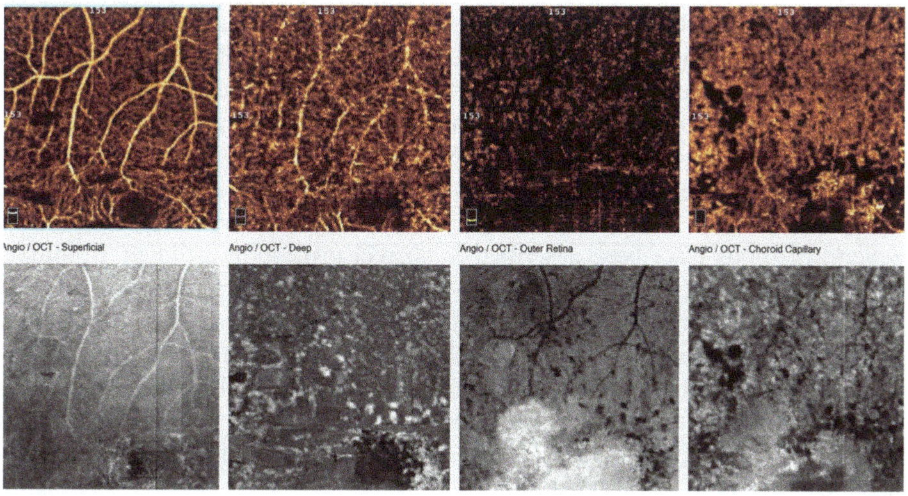

Figure 1.3. Shadowing from vitreous visible on *en face* OCT and OCTA (all slabs) and from intraretinal exudates to most posterior angiograms.

SNR which may be taken into account while interpreting the scans. A high SNR is always desired.

Vitreous floaters, tufts of vitreous hemorrhage, inflammatory cells, etc may interfere with the OCT signals resulting in localized signal attenuation and corresponding pseudo low flow on OCT angiograms. Side by side evaluation of OCT shadowgram, *en face* structural and cross-sectional images are helpful in differentiating true low flow from false low flow [2]. Figure 1.3 shows shadow artifacts.

1.5.3.3 Segmentation artifacts

Several approaches are used to identify retinal layer boundaries in OCT to generate *en face* angiograms of desired layers. These include layer reflectivity, texture, use of deep learning, etc. Layer continuity is frequently employed implied in segmentation strategies. Usually, these algorithms demonstrate excellent performance towards identifying the layer boundaries in normal retinae. However, the algorithms frequently fail segmenting the boundaries accurately in diseased eyes. Eyes with macular edema, choroidal neovascularization, high myopia, preretinal membrane and retinal traction pose a special challenge to segmentation algorithms [29].

On *en face* angiograms, the given slab may have information from other layers. Concomitant cross-sectional image analysis and manual correction of segmentation can help in such cases; but, it is a tiring and time-consuming job to correct segmentations across multiple B Scans.

1.5.3.4 Projection artifacts

These artifacts are by far the most common and inherent to all OCTA scans. When an OCTA beam passes through blood vessels, it may be reflected back, scattered, absorbed or refracted. The OCTA beam passing through these blood vessels and reaching deeper layers suffers time varying fluctuations. The fluctuating beam, subsequently, highlights the posterior structure. Some of these structures including retinal pigment epithelium (RPE), plexiform layers and choriocapillaris act as reflective surfaces. The reflection of this, time varying fluctuating signal across B-scans is, falsely, computed as flow signal coming from the reflecting surfaces and displayed as duplication of superficial flow maps on deeper slabs. Small reflective structures, such as drusen, exudates and pigment can also give rise to localized projection artifacts.

Several methods have been described to correct projection artifacts. Commonly employed is subtraction of superficial flow maps from deeper slabs; but it gives rise to another troublesome artifact. The pattern of bright vessels is replaced by dark lines. At present, no method can effectively correct these artifacts.

Clinically, uncorrected projection artifacts appear as duplication of superficial flow maps on deeper slabs and duplication of superficial vascular patterns as dark lines on posterior structures suggest a corrected projection artifact. Figure 1.4 shows projection artifacts.

1.6 OCTA in retinal vascular diseases

OCTA, being a non-invasive vascular imaging method, is increasingly finding its role in diagnosis in management of retinal vascular diseases. It can also be used as a standalone vascular imaging technique in most cases and as adjunct to FA in some situations. Management of most of the vascular diseases of the retina needs structural OCT information for establishing the diagnosis, severity grading and decisions to treat. OCTA being available on the same platform can be acquired quickly and studied alongside structural OCT scans.

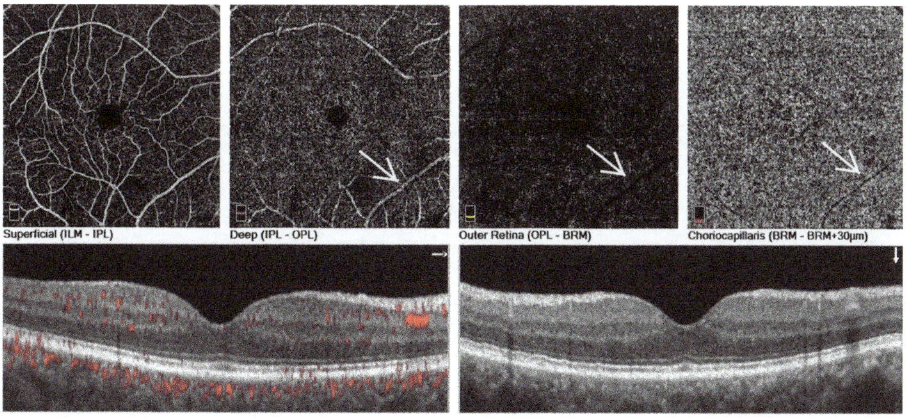

Figure 1.4. Projection of SVP vessels (indicated by white arrow) on DCP, outer retina and CC.

1.6.1 Diabetic retinopathy

OCTA has comparable results to FA in demonstrating the classical clinical findings of diabetic retinopathy (DR) [2, 30]. There exists a great agreement between FA and OCTA in measurement of the FAZ diameter, a potential new disease biomarker, as well as in detecting microaneurysms, vascular beading, capillary non-perfusion (CNP) and vascular proliferation [30, 31]. Nonetheless, OCTA is non-sensitive to vascular leakage, hence it may display the actual extent of retinal neovascularization which is obscured by leakage on FA [2].

OCTA provides multiple various qualitative metrics for assessing and monitoring DR including vessel density, intercapillary distance, measurement of CNP area at different vascular networks of the retina. Rosen *et al* [32] suggested higher perfused capillary density (PCD) as a sign of preclinical diabetic retinopathy and that it warrants the use of OCTA in early detection of DR. Incessant capillary loss in DR causes a low PCD in eyes with clinical diabetic retinopathy. As FA displays vascular information as a composite map of all vascular beds, we cannot visualize each individual vascular tree. OCTA allows us to view individual as well composite flow maps and provides better contrast due to absence of background choroidal flush. Figure 1.5 shows OCTA findings in diabetic retinopathy.

1.6.2 Retinal vein occlusion (RVO)

The severity of visual symptoms and prognosis of both branch and central RVO depends on the severity of macular edema and extent of vascular insufficiency. OCT is the most common test used for guiding treatment in eyes with venous occlusions. OCTA can delineate the area of vascular insufficiency, CNP, FAZ changes and other potential complications of RVO can be monitored [33–36]. Nonetheless, due to marked macular edema in venous occlusions, correct segmentation of retinal slabs often becomes a challenge. However, the composite full thickness OCTA maps can be used to obtain useful vascular flow information [37].

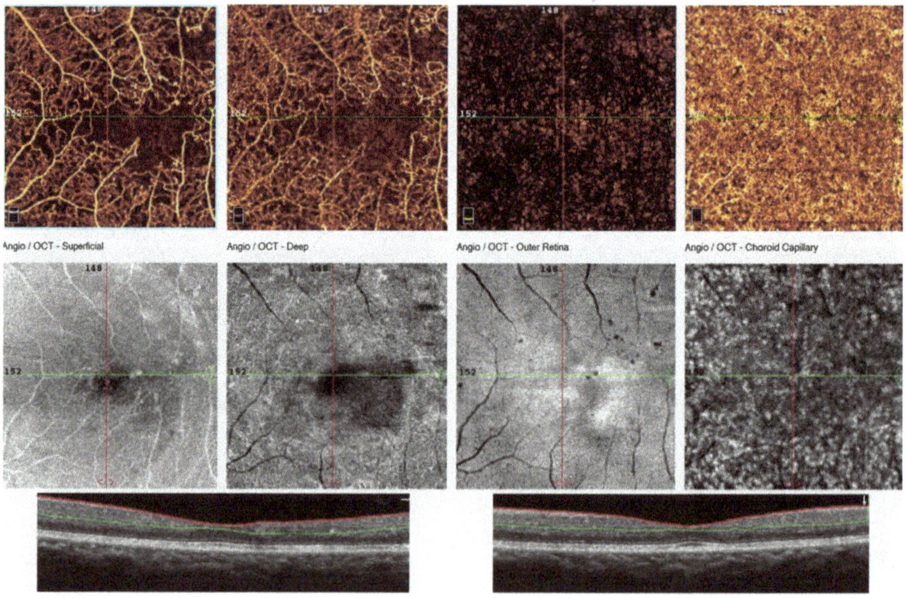

Figure 1.5. Multiple microaneurysms, irregular and widened foveal avascular zone and capillary dropout in SVP and DCP. Projection of SVP vessels and localized shadowing on CC can be seen.

Studies of venous occlusions, using OCTA, have demonstrated dilated venous segment within areas of retinal edema. Furthermore, it was found that vascular insult was more profound in DCP compared to SVP [38–41]. Figure 1.6 shows dilated veins, flow deficits and macular edema in the case of central retinal vein occlusion.

OCTA has proven to be comparable to FA in documenting vascular changes in RVO, especially when using 3 × 3 mm and HD angio scans. However, with increasing field of view the performance is reduced. Peripheral vascular non-perfusion is a common feature in eyes with RVO and is an important prognostic factor. The smaller field of view on OCTA limits its utility in documenting peripheral ischemia [2, 3].

1.6.3 Retinal artery occlusion

Retinal artery occlusion (RAO) manifests as occlusion of one of the branches, hemiretinal arteries or central retinal artery secondary to thrombosis, arterial embolism or increased intraocular pressure. The severity of occlusion ranges from partial reduction to complete loss of arterial flow.

A limited number of case reports and series about OCTA findings in RAO have been published. Overall, good agreement exists between FA and OCTA in studying the vascular status in RAO [2].

de Castro-Abeger *et al* [42] demonstrated reduced flow in the area distal to the embolus in branch artery occlusion. In their study, flow reduction was more profound in corresponding SVP compared to DCP. Other studies have reported

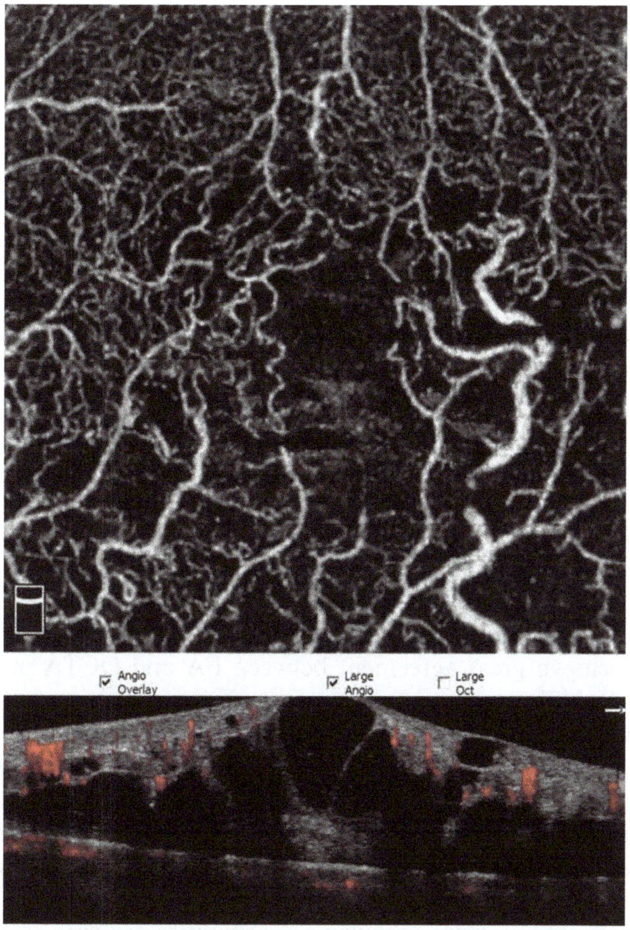

Figure 1.6. Dilated veins with large flow voids and macular edema in the case of retinal vein occlusion.

the severe perfusion defects in both superficial and deep vascular networks in acute cases and severity of low perfusion substantially improving over time in DCP in chronic cases [43, 44].

Optic disc perfusion may be relatively preserved in acute cases progressing to diffuse attenuation of radial peripapillary capillaries in chronic phase [2, 45, 46].

1.6.4 Central serous chorioretinopathy

Central serous chorioretinopathy (CSCR) is mostly a self-limiting retinal disease characterized by serous separation of the neurosensory retina. A relatively small proportion of CSCR becomes chronic in course and may further complicate by developing choroidal neovascularization (CNV). Often, chronic cases are difficult to distinguish from age-related macular degeneration and other causes of CNV.

OCTA in acute stages shows a normal flow profile in both SVP and DCP. However, CC textural alteration and flow impairments, either due to impaired flow or shadowing from overlying subretinal fluid, have been described in acute stages [47, 48].

Further, OCTA may be a great tool in monitoring chronic CSCR for development of CNV and differentiating it from similar looking diseases. OCTA showed a high sensitivity in detecting CNV in eyes with CSCR [2].

1.6.5 Macular telangiectasia

Macular telangiectasia (MacTel) is an acquired vascular disorder of retinal vasculature in the elderly. The clinical manifestations include loss of foveal translucency in earliest stages. With the advancement of disease there may be retinal cavitation, volume loss, vascular alterations, right angled venules, intraretinal crystals, RPE pigmentations and vascular remodeling. It may further progress to and is complicated by CNV formation [49].

The earlier clinical findings may be very subtle to detect clinically and may be missed. FA is the current gold standard in diagnosing MacTel. It highlights the telangiectatic vasculature in early phases of angiogram followed by diffuse hyperfluorescence in later phases [49].

Structural OCT is very useful in evaluating the retinal cavitation, photoreceptor disruptions, changes in retinal reflectivity and potential development of CNV [49].

Studies have shown great agreement between FA and OCTA in exhibiting the features of MacTel [2].

OCTA shows dilatation of temporal juxtafoveal capillaries of DCP in early stages of the disease. The photoreceptor disruption on OCT corresponds to the area of vascular alterations in DCP [49–56].

Multiple aneurysmal dilatations of DCP vessels extending to outer retina with a concordant area of ellipsoid zone loss on structural OCT are characteristics of the intermediate stage [49–56].

In proliferative stage, severe vascular abnormalities are seen in both SVP and DCP. Vascular anastomosis is displayed as abnormal vessels in outer retinal slabs. Submacular CNV can be seen in this stage [49–56]. Figure 1.7 shows OCTA displaying changes in early and advanced macular telangiectasia.

Rarefaction of retinal capillaries, widened intercapillary distances and reduction in capillary density, both in SVP and DCP have been demonstrated in multiple studies [49–56]. In a study by Toto *et al* [49] parafoveal capillary density in MacTel was reduced in both DCP and SVP. However, the difference between MacTel and normal eyes was significant only in SVP.

1.6.6 Other retinal diseases

OCTA may be useful in diagnosing and managing a variety of other retinal diseases including retinal arterial macroaneurysms, radiation retinopathy, Coats' disease, foveal hypoplasia, Purtscher's retinopathy, retinitis pigmentosa, Stargardt disease, best vitelliform dystrophy, etc. Not only can it aid in diagnosis but it also delivers additional insights into our understanding of these diseases by providing details of each individual vascular network in retina and choroid [2].

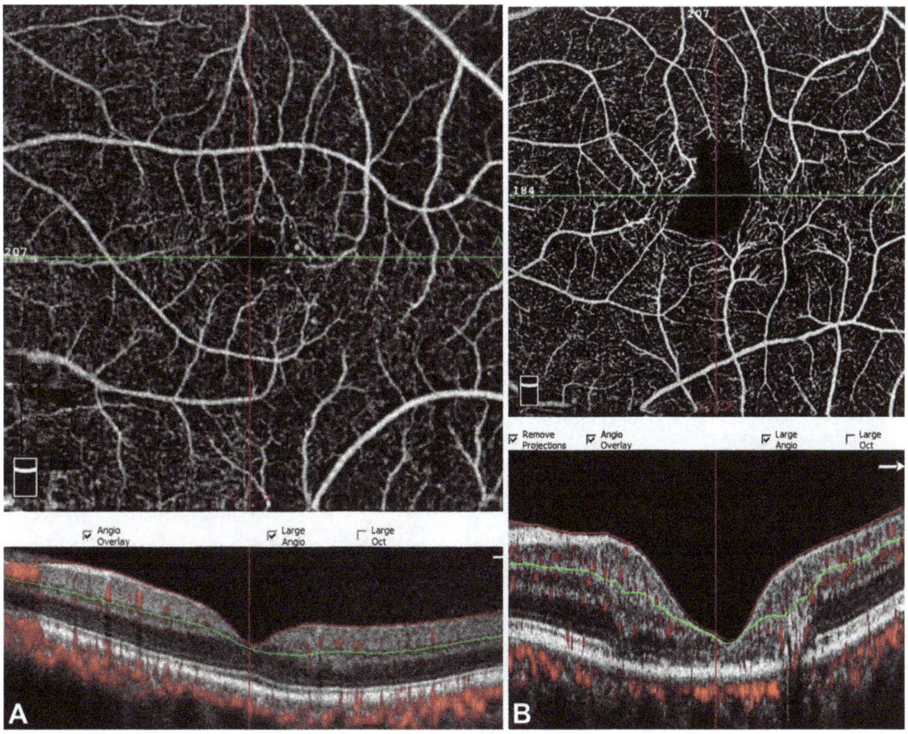

Figure 1.7. Composite *en face* OCTA and Flow B scan (a) of a case with early macular telangiectasia demonstrating microaneurysms in temporal juxtafovea. OCTA of and advanced case of MacTel (b) with vascular remodeling and irregular FAZ on OCTA and volume loss on B-Scan.

1.7 Choroidal diseases

1.7.1 Dry age-related macular degeneration and variants

Age-related macular degeneration (AMD) is the most common retinal degeneration in the elderly. Drusen under RPE with disruptions/alteration of overlying photoreceptors are the earliest sign. The disease further progresses to pigmentary abnormalities, more marked photoreceptor loss and geographic atrophy (GA) of chorioretinal tissue. CC inadequacy has been implicated as a mechanism behind development of drusen [57, 58].

OCTA has demonstrated flow impairment in CC under the drusen. However, it is unclear if it is true flow impairment or the shadowing/signal attenuation caused by drusen leads to a false low flow. SS OCTA have demonstrated lesser signal loss compared to SD OCTA. Yet, it is believed that CC flow impairment in part may be contributed by signal attenuation and imparted by true flow impairment [59–62]. CC flow alterations and capillary rarefaction under and beyond the margins of GA have also been described using conventional OCTA and the variable interscan time analysis (VISTA) algorithm, which is particularly helpful in differentiating true low flow from false low flow. Regardless of any system and algorithms used CC under drusen and GA demonstrated flow impairment. Reduced vascular density in SVP and DCP of eyes with early and moderate AMD have also been reported [62–64]. Figure 1.8 shows features of geographic atrophy on OCTA and structural OCT.

Although GA in advanced dry AMD can be well visualized clinically and on structural OCT, OCTA can be used to delineate the extent of GA and study the

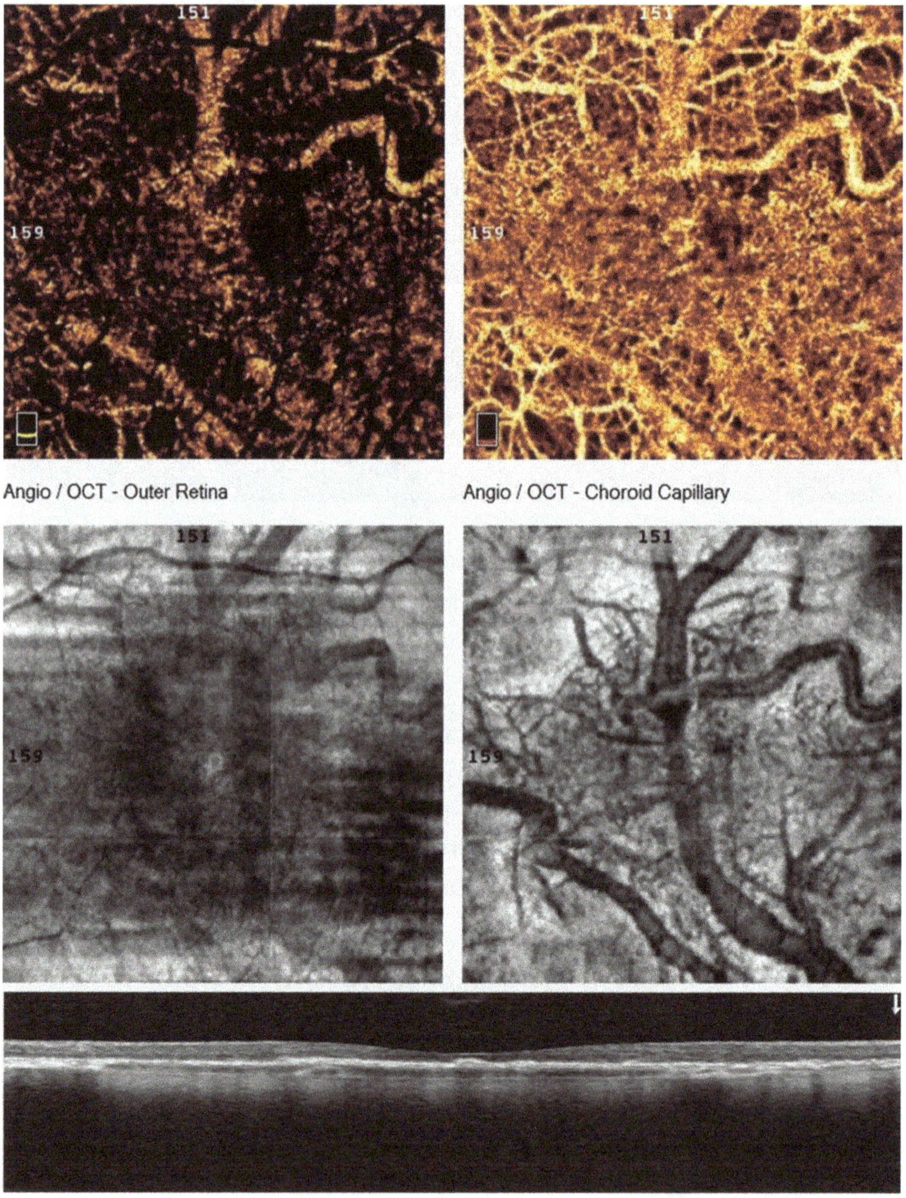

Figure 1.8. severe retinal thinning on OCT and complete absence of choriocapillaris on OCTA in a case of GA. Larger choroidal vessels are visible on CC slabs due to increased OCT signal penetration.

alteration in CC and deeper choroidal vessels (visible through CC defects). At this stage research has demonstrated reduced vascular density in both SVP and DCP [65].

In addition, OCTA offers further benefits by resolving individual vascular tissue in the retina which is not possible with FA and ICGA. CC, in AMD and other conditions, can be better resolved and visualized using OCTA than with FA or ICGA.

1.7.2 Wet AMD and variants

CNV is the most important feature of wet AMD. Typing of CNV, differentiation between CNV variants and early intervention may yield better results and visual outcomes. Other features of wet AMD include pigment epithelial detachment (PED), subretinal hemorrhaging, RPE tears, retinal edema and subsequent chorioretinal scarring and irreversible visual loss.

FA and ICGA are often used to diagnose and differentiate between types of CNV and its variants, while OCT is routinely employed to guide treatment decisions.

Origin and anatomic relationships of CNV are important in differentiating between the three types of CNV. A less distinct FA lesion located and limited to sub RPE space is Type I (or occult) CNV, whereas type II (classic) CNV has proliferated onto and grows in subretinal space. Type III CNV, or retinal angiomatous proliferation (RAP) originates intraretinally and manifests with intraretinal hemorrhaging, exudation and edema. Anastomosis between intraretinal and subretinal components may be with type III lesions [66].

1.7.2.1 Type I CNV

It is often difficult to detect on OCT and has an indistinct appearance on FA. ICGA may be required in some cases to diagnose type I CNV. On OCTA, it has a distinctive appearance as a highly vascular complex of a pruned vascular tree or vascular loop or a tangled network on CC slabs. A hyperreflective halo surrounding the neovascular tree is often seen which may be due to shadowing from RPE or more likely a CC hypoperfusion defect [1–3].

OCTA may have a sensitivity similar to or better than FA, ICGA and OCT in detecting type I CNV. However, CNV size may appear smaller on OCTA compared to ICGA, likely because it detects the only flow column in CNV, while in ICGA the vascular walls as well as lumen are both visualized [2, 67, 68].

1.7.2.2 Type II CNV

A glomerular or medusa patterned hyperflow lesion is seen in outer avascular retina slabs of OCT angiograms. A thicker feeder trunk from the neovascular complex communicates with the deeper choroid is usually seen. Similar to type I CNV, a prelesional hypoflow halo is frequently visible [1–3]. Figure 1.9 shows a very subtle small type II CNV visible on CC and AOR slabs. The lesion was barely perceptible on structural OCT on careful inspection.

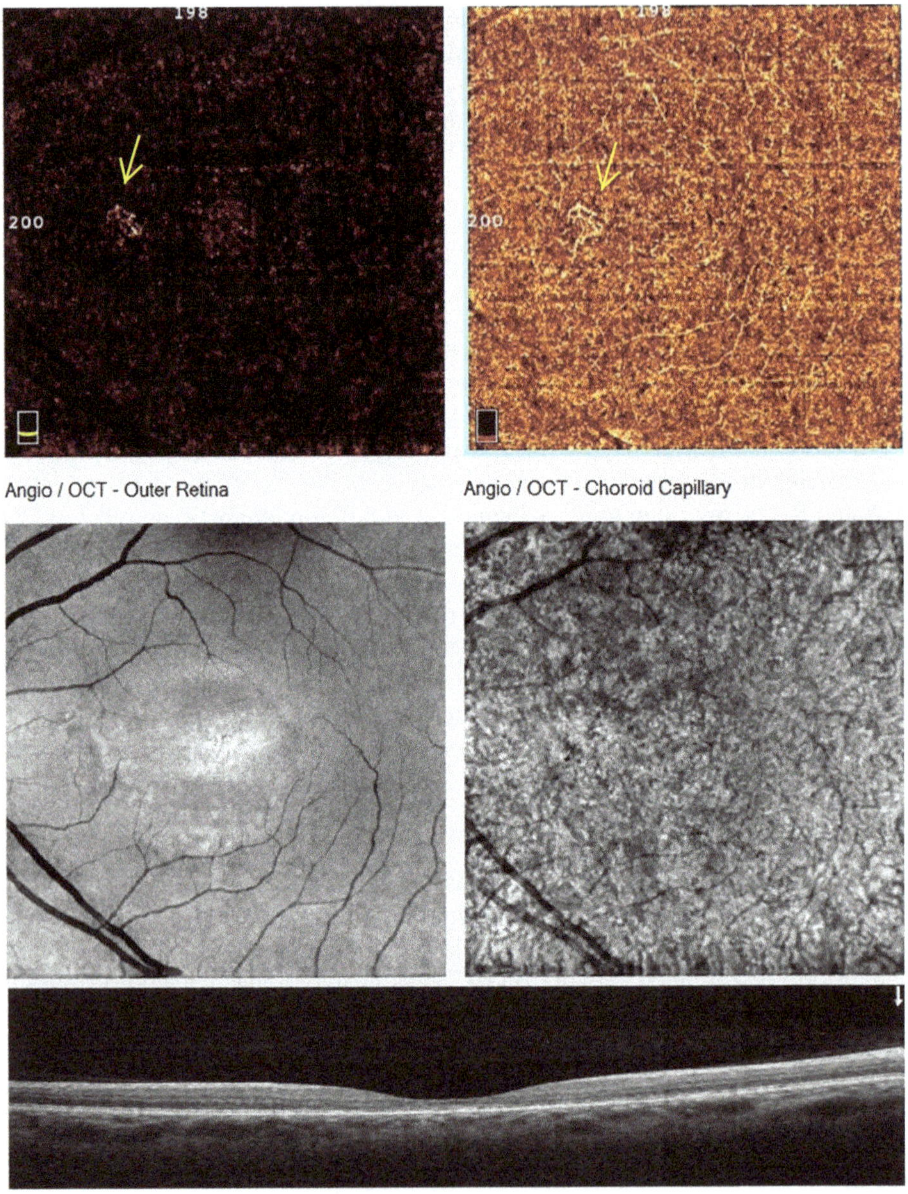

Figure 1.9. A very small choroidal neovascular complex is visible in the outer retina and CC (top row). The lesion can be barely noticed on careful inspection of OCT B scan (bottom row).

OCTA may be a great tool in baseline assessment of CNV as well as guiding us throughout the treatment course of CNV, although some clinicians are more comfortable with FA/ICGA for guiding treatment response and decision. Nonetheless, unlike FA/ICGA, OCTA only highlights CNV with patent flow (active part) along with structural information from reflectivity profiles, so we may use it confidently. However, in certain cases FA/ ICGA are required to better

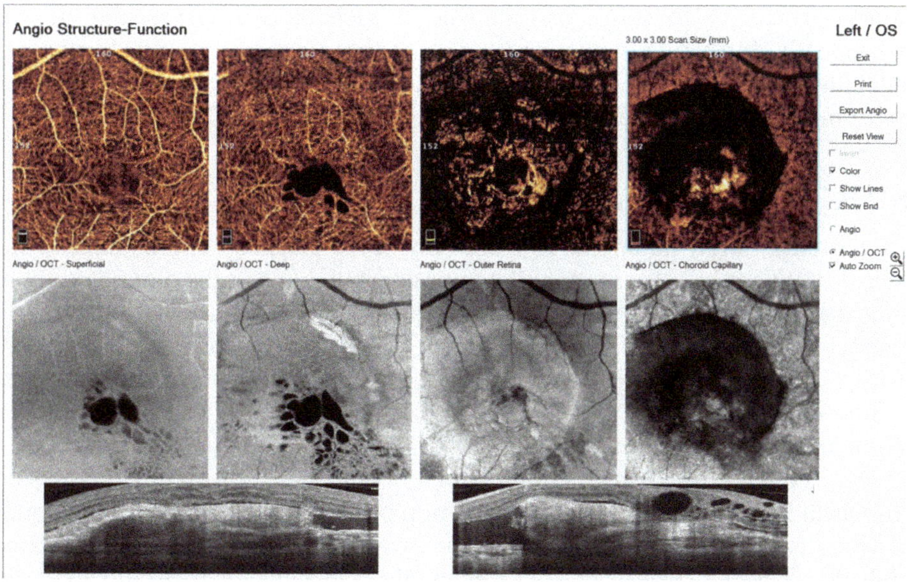

Figure 1.10. A large subretinal hyperreflective lesion on OCT (bottom row) with RPE rip and cystic changes in retina. The neovascular complex can be seen in outer retinal and CC slabs. Multiple artifacts including segmentation artifacts and projection artifacts are noticeable.

understand the nature and activity of lesions. Figure 1.10 shows advanced type II CNV with retinal edema, RPE rip and cystic changes. Segmentation errors are apparent.

1.7.2.3 Type III CNV

Retinal angiomatous proliferation is an AMD variant which may be difficult to distinguish from classic CNV in some cases.

Type III lesions appear as a hyperflow signal in outer retinal slabs as a tuft of small, curvilinear vessels anastomosing DCP through a feeder vessel. A small, corkscrew like lesion in CC slab is commonly seen.

1.7.2.4 Polypoidal choroidal vasculopathy

Polypoidal choroidal vasculopathy (PCV) manifests as bilateral, multiple, recurrent serosanguineous and serous RPE detachments secondary to branching vascular network (BVN) and polypoidal lesions in choroid [66, 69].

Polyps appear as round hyperflow lesions in a nodular, ring or clustered conformation in CC segmentations. The central lesion is usually surrounded by a dark halo, while BVN assumes configuration of medusa or seafan of tangled bright vascular signal in CC.

Compared to ICGA, BVN may be better visualized on OCTA; however, ICGA is superior in detecting and delineating polyps [1–3].

At present, OCTA may not replace the current gold standard vascular imaging techniques for CNV diagnosis and management. Nevertheless, it provides us with a unique opportunity to study the changes in each individual vascular network independently, furthering our understanding of these vascular lesions. Further, the availability of structural OCT information, in addition to flow characteristics equips us with more information compared to FA or ICGA alone.

1.8 OCTA in ocular inflammation

OCTA is increasingly finding its role not only in the diagnosis and management of posterior uveitides but also in expanding our understanding of the disease processes.

The common use of FA, ICGA and structural OCT in ocular inflammation includes documentation of macular edema, vascular leakage, vasculitis, retinal ischemia, choroidal neovascularization and differentiation between similar looking conditions.

Retinal vasculitis, a common sign in a variety of ocular and systemic diseases, is characterized by diffuse/segmental inflammation and leakage of retinal arteries and veins which further progresses to capillary non perfusion, retinal neovascularization and vitreous hemorrhage. OCTA, although, cannot detect leakage from inflamed vessels that is well displayed by FA; it highlights the actual changes in vasculature because the view is not obscured by dye leakage, unlike FA. It further, helps in delineating the areas of non-perfusion and retinal vascular proliferation. OCTA has been utilized to study the features of many retinal vasculitis spectra including Idiopathic Retinal Vasculitis, Aneurysms and Neuroretinitis (IRVAN) syndrome. In IRVAN eyes it can detect retinal aneurysms with a sensitivity similar to FA. However, most cases of this disease spectrum are associated with peripheral CNP which can be missed by OCTA due to smaller field of view. Montaging multiple images from different fields can be used to generate OCT angiograms with wider field of view.

In many of the posterior uveatides, hypofluorescence of inflammatory foci in early phases of FA are typical. The origin of the hypofluorescence and hypocyananee on ICGA has been long debated to originate from the CC hypoperfusion or choroidal masking/ shadowing from the inflammatory foci or cloudy swelling of retinal pigment epithelium. OCTA also has confirmed the dark spots on CC slabs corresponding to true low flow or signal attenuation because of inflammatory foci. The CC perfusion defect is believed to be true flow void because, on co-registered B-Scans and *en face* images, the CC perfusion defect sizes were larger than the focus of inflammation. Further, alterations in CC texture have been described in a number of eyes after complete resolution of inflammation in primary inflammatory choriocapillaropathies and Vogt–Koyanagi–Harada's disease [70–73].

Paracentral acute middle maculopathy (PAMM), poses a clinical challenge for the diagnosis. It is characterized by parafoveal hyperreflective band at the level of OPL to IPL, on structural OCT, in the absence of any FA correlates. OCTA, however, has revealed perfusion abnormalities in both SVP and DCP. The hypoperfusion, as may be expected, is more marked in DCP. The visual deficit in PAMM corresponds well with the severity and location of non-perfusion on OCTA. The persistent flow defects on OCTA explain the progressive thinning of retina in PAMM [2, 70].

1.9 Conclusion

FA and ICGA have remained the gold standard of vascular imaging for decades. However, associated adverse events were a major factor limiting their frequent use. OCTA is an emerging vascular imaging modality based on motion contrast capable of depth resolved imaging of ocular vasculature using motion contrast, non-invasively.

It has shown excellent results comparable to FA and ICGA in most of the situations. Its unique capability of resolving individual vascular networks provides more information about the origin of a disease process as well as tracing the vascular beds from where a particular disease process actually begins.

Despite its non-invasive nature, high resolution, excellent reproducibility, OCTA algorithms are prone to a significant number of different artifacts limiting their usefulness in everyday clinical practice. A good understanding of these artifacts and their impact on the displayed angiograms is necessary to derive useful information.

Even though, OCTA has broadened our understanding about ocular disease, its clinical usefulness and applicability is still evolving; we recommend it as a supplement to existing vascular imaging instead of using it as an exclusive tool. With future improvements in scan acquisition speeds, improved hardware and newer OCTA algorithms we expect to obtain more useful information with fewer artifacts in shorter time.

References

[1] Kashani A H, Chen C L and Gahm J K *et al* 2017 Optical coherence tomography angiography: A comprehensive review of current methods and clinical applications *Prog. Retin. Eye Res* **60** 66–100

[2] Khan H A, Mehmood A and Khan Q A *et al* 2017 A major review of optical coherence tomography angiography *Expert Rev. Ophthalmol* **12** 373–85

[3] Spaide R F, Fujimoto J G and Waheed N K *et al* 2018 Optical coherence tomography angiography *Prog. Retin. Eye Res* **64** 1–55

[4] Leitgeb R A, Werkmeister R M and Blatter C *et al* 2014 Doppler optical coherence tomography *Prog. Retin. Eye Res* **41** 26–43

[5] Fingler J, Readhead C and Schwartz D M *et al* 2008 Phase-contrast OCT imaging of transverse flows in the mouse retina and choroid *Investig. Ophthalmol. Vis. Sci.* **49** 5055–59

[6] Fingler J, Schwartz D and Yang C *et al* 2007 Mobility and transverse flow visualization using phase variance contrast with spectral domain optical coherence tomography *Opt. Express* **15** 12636–53

[7] Fingler J, Zawadzki R J and Werner J S *et al* 2009 Volumetric microvascular imaging of human retina using optical coherence tomography with a novel motion contrast technique *Opt. Express* **17** 22190–200

[8] Liu G, Jia Y and Pechauer A D *et al* 2016 Split-spectrum phase-gradient optical coherence tomography angiography *Biomed. Opt. Express* **7** 2943–54

[9] Barton J and Stromski S 2005 Flow measurement without phase information in optical coherence tomography images *Opt. Express* **13** 5234–9

[10] Mariampillai A, Standish B A and Moriyama E H *et al* 2008 Speckle variance detection of microvasculature using swept-source optical coherence tomography *Opt. Lett.* **33** 1530–2

[11] Mariampillai A, Leung M K and Jarvi M *et al* 2010 Optimized speckle variance OCT imaging of microvasculature *Opt. Lett.* **35** 1257–9
[12] Motaghiannezam S M, Koos D and Fraser S E 2012 Differential phase-contrast, swept-source optical coherence tomography at 1060 nm for *in vivo* human retinal and choroidal vasculature visualization *J. Biomed. Opt.* **17** 026011
[13] Blatter C, Klein T and Grajciar B *et al* 2012 Ultrahigh-speed non-invasive widefield angiography *J. Biomed. Opt.* **17** 070505
[14] Schmoll T, Singh A S and Blatter C *et al* 2011 Imaging of the parafoveal capillary network and its integrity analysis using fractal dimension *Biomed. Opt. Express* **2** 1159–68
[15] Enfield J, Jonathan E and Leahy M 2011 *In vivo* imaging of the microcirculation of the volar forearm using correlation mapping optical coherence tomography (cmOCT) *Biomed. Opt. Express* **2** 1184–93
[16] Jonathan E, Enfield J and Leahy M J 2011 Correlation mapping method for generating microcirculation morphology from optical coherence tomography (OCT) intensity images *J. Biophotonics* **4** 583–7
[17] Jia Y, Tan O and Tokayer J *et al* 2012 Split-spectrum amplitude-decorrelation angiography with optical coherence tomography *Opt. Express* **20** 4710–25
[18] Stanga P E, Tsamis E and Papayannis A *et al* 2016 Swept-source optical coherence tomography angio (Topcon Corp, Japan): technology review *Dev. Ophthalmol* **56** 13–7
[19] Huang D, Jia Y and Gao S S *et al* 2016 Optical coherence tomography angiography using the Optovue device *Dev. Ophthalmol* **56** 6–12
[20] ZEISS CIRRUS OCT with AngioPlex – Making the revolutionary, routine. – ZEISS Medical Technology: ZEISS International [2 May 2020]. Available from https://zeiss.com/meditec/int/specialties/retina/diagnostics/optical-coherence-tomography/oct-optical-coherence-tomography/cirrus-oct-with-angioplex.html
[21] Rosenfeld P J, Durbin M K and Roisman L *et al* 2016 ZEISS Angioplex™ spectral domain optical coherence tomography angiography: technical aspects *Dev. Ophthalmol* **56** 18–29
[22] Rocholz R, Teussink M M and Dolz-Marco R *et al* 2018 SPECTRALIS Optical Coherence Tomography Angiography (OCTA): Principles and Clinical Applications https://heidelbergengineering.com/media/e-learning/Totara/Dateien/pdf-tutorials/210111-001_SPECTRALIS%20OCTA%20-%20Principles%20and%20Clinical%20Applications_EN.pdf
[23] Nickla D L and Wallman J 2010 The multifunctional choroid *Prog. Retin. Eye Res.* **29** 144–68
[24] Tan P E Z, Yu P K and Balaratnasingam C *et al* 2012 Quantitative confocal imaging of the retinal microvasculature in the human retina *Investig. Ophthalmol. Vis. Sci.* **53** 5728–36
[25] Tan P E Z, Yu P K and Cringle S J *et al* 2014 Quantitative assessment of the human retinal microvasculature with or without vascular comorbidity *Investig. Ophthalmol. Vis. Sci.* **55** 8439–52
[26] Hirano T, Chanwimol K and Weichsel J *et al* 2018 Distinct retinal capillary plexuses in normal eyes as observed in optical coherence tomography angiography axial profile analysis *Sci. Rep.* 2018/06/20 **8** 9380
[27] Campbell J P, Zhang M and Hwang T S *et al* 2017 Detailed vascular anatomy of the human retina by projection-resolved optical coherence tomography angiography *Sci. Rep.* **7** 42201
[28] Holmen I C, Konda S M and Pak J W *et al* 2020 Prevalence and severity of artifacts in optical coherence tomographic angiograms *JAMA Ophthalmol* **138** 119–26

[29] Spaide R F and Curcio C A 2017 Evaluation of segmentation of the superficial and deep vascular layers of the retina by optical coherence tomography angiography instruments in normal eyes *JAMA Ophthalmol* **135** 259–62
[30] Khadamy J, Abri Aghdam K and Falavarjani K G 2018 An update on optical coherence tomography angiography in diabetic retinopathy *J. Ophthalmic. Vis. Res* **13** 487–97
[31] Enders C, Baeuerle F and Lang G E *et al* 2020 Comparison between findings in optical coherence tomography angiography and in fluorescein angiography in patients with diabetic retinopathy *Ophthalmologica.* **243** 21–6
[32] Rosen R B, Andrade Romo J S and Krawitz B D *et al* 2019 Earliest evidence of preclinical diabetic retinopathy revealed using optical coherence tomography angiography perfused capillary density *Am. J. Ophthalmol.* **203** 103–15
[33] Kuehlewein L, An L and Durbin M K *et al* 2015 Imaging areas of retinal nonperfusion in ischemic branch retinal vein occlusion with swept-source OCT microangiography *Ophthalmic. Surg. Lasers Imaging Retina* **46** 249–52
[34] Nobre Cardoso J, Keane P A and Sim D A *et al* 2015 Systematic evaluation of optical coherence tomography angiography in retinal vein occlusion *Am. J. Ophthalmol.* **163** 93-107. e6
[35] Rispoli M, Savastano M C and Lumbroso B 2015 Capillary network anomalies in branch retinal vein occlusion on optical coherence tomography angiography *Retina* **35** 2332–8
[36] Sakimoto S, Gomi F and Sakaguchi H *et al* 2015 Analysis of retinal nonperfusion using depth-integrated optical coherence tomography images in eyes with branch retinal vein occlusion *Invest. Ophthalmol. Vis. Sci.* **56** 640–6
[37] Novais E A and Waheed N K 2016 Optical coherence tomography angiography of retinal vein occlusion *OCT Angiography Retin. Macular Diseases* vol 56 (Basel: Karger Publishers) 132–38
[38] Adhi M, Filho M A B and Louzada R N *et al* 2016 Retinal capillary network and Foveal avascular zone in eyes with vein occlusion and fellow eyes analyzed with optical coherence tomography angiographyretinal capillary network and FAZ in RVO with OCTA *Investig. Ophthalmol. Vis. Sci.* **57** OCT486–94
[39] Casselholmde Salles M, Kvanta A and Amrén U *et al* 2016 Optical coherence tomography angiography in central retinal vein occlusion: correlation between the Foveal avascular zone and visual acuityOCT angiography in CRVO *Investig. Ophthalmol. Vis. Sci.* **57** OCT242–6
[40] Coscas F, Glacet-Bernard A and Miere A *et al* 2016 Optical coherence tomography angiography in retinal vein occlusion: evaluation of superficial and deep capillary plexa *Am. J. Ophthalmol.* **161** 160–71
[41] Tsuboi K, Ishida Y and Kamei M 2017 Gap in capillary perfusion on optical coherence tomography angiography associated with persistent macular edema in branch retinal vein occlusion *Investig. Ophthalmol. Vis. Sci.* **58** 2038–43
[42] de Castro-Abeger A H, de Carlo T E and Duker J S *et al* 2015 Optical coherence tomography angiography compared to fluorescein angiography in branch retinal artery occlusion *Ophthalmic Surg. Lasers Imaging Retina* **46** 1052–4
[43] Yu S, Pang C E and Gong Y *et al* 2015 The spectrum of superficial and deep capillary ischemia in retinal artery occlusion *Am. J. Ophthalmol.* **159** 53–63
[44] Bhanushali D R, Yadav N K and Dabir S *et al* 2016 Spectral domain optical coherence tomography angiography features in a patient of central retinal arterial occlusion before and after paracentesis *Retina* **36** e36–8

[45] Bonini Filho M A, Adhi M and de Carlo T E *et al* 2015 Optical coherence tomography angiography in retinal artery occlusion *Retina* **35** 2339–46
[46] Lee A Y, Zhang Q and Baughman D M *et al* 2016 Evaluation of bilateral central retinal artery occlusions with optical coherence tomography-based microangiography: a case report *J. Case Rep* **10** 307
[47] Aggarwal K, Agarwal A and Deokar A *et al* 2017 Distinguishing features of acute Vogt–Koyanagi–Harada disease and acute central serous chorioretinopathy on optical coherence tomography angiography and en face optical coherence tomography imaging *J. Ophthalmic Inflamm. Infect* **7** 3
[48] Costanzo E, Cohen S Y and Miere A *et al* 2015 Optical coherence tomography angiography in central serous chorioretinopathy *J. Ophthalmol* **2015** 134783
[49] Toto L, Di Antonio L and Mastropasqua R *et al* 2016 Multimodal imaging of macular telangiectasia Type 2: focus on vascular changes using optical coherence tomography angiography imaging macular telangiectasia type 2 by OCTA *Invest. Ophthalmol. Vis. Sci.* **57** OCT268–76
[50] Chidambara L, Gadde S G and Yadav N K *et al* 2016 Characteristics and quantification of vascular changes in macular telangiectasia type 2 on optical coherence tomography angiography *Br. J. Ophthalmol.* **100** 1482–8
[51] Engelbert M and Yannuzzi L A 2012 Idiopathic macular telangiectasia type 2: the progressive vasculopathy *Eur. J. Ophthalmol* **23** 1–6
[52] Gaudric A, Krivosic V and Tadayoni R 2015 Outer retina capillary invasion and ellipsoid zone loss in macular telangiectasia type 2 imaged by optical coherence tomography angiography *Retina* **35** 2300–6
[53] Roisman L and Rosenfeld P 2016 Optical coherence tomography angiography of macular telangiectasia type 2 *OCT Angiography Retin. Macular Diseases* vol 56 (Basel: Karger Publishers) 146–58
[54] Spaide R F, Klancnik J M Jr and Cooney M J 2015 Retinal vascular layers in macular telangiectasia type 2 imaged by optical coherence tomographic angiography *JAMA Ophthalmol* **133** 66–73
[55] Spaide R F, Klancnik J M Jr and Cooney M J *et al* 2015 Volume-rendering optical coherence tomography angiography of macular telangiectasia type 2 *Ophthalmology* **122** 2261–9
[56] Thorell M R, Zhang Q and Huang Y *et al* 2014 Swept-source OCT angiography of macular telangiectasia type 2 *Ophthalmic Surg. Lasers Imaging Retina* **45** 369–80
[57] Mullins R F, Johnson M N and Faidley E A *et al* 2011 Choriocapillaris vascular dropout related to density of drusen in human eyes with early age-related macular degeneration *Invest. Ophthalmol. Vis. Sci.* **52** 1606–12
[58] Lengyel I, Tufail A and Hosaini H A *et al* 2004 Association of drusen deposition with choroidal intercapillary pillars in the aging human eye *Invest. Ophthalmol. Vis. Sci.* **45** 2886–92
[59] Chu Z, Gregori G and Rosenfeld P J *et al* 2019 Quantification of choriocapillaris with optical coherence tomography angiography: a comparison study *Am. J. Ophthalmol.* **208** 111–23
[60] Chatziralli I, Theodossiadis G and Panagiotidis D *et al* 2018 Choriocapillaris vascular density changes in patients with drusen: cross-sectional study based on optical coherence tomography angiography findings *Ophthalmol. Ther* **7** 101–7

[61] Lane M, Moult E M and Novais E A *et al* 2016 Visualizing the choriocapillaris under drusen: comparing 1050-nm swept-source versus 840-nm spectral-domain optical coherence tomography angiography *Invest. Ophthalmol. Vis. Sci.* **57** OCT585–90
[62] Moreira-Neto C A, Moult E M and Fujimoto J G *et al* 2018 Choriocapillaris loss in advanced age-related macular degeneration *J. Ophthalmol.* **2018** 8125267
[63] Lauermann J L, Eter N and Alten F 2018 Optical coherence tomography angiography offers new insights into choriocapillaris perfusion *Ophthalmologica.* **239** 74–84
[64] Moult E M, Waheed N K and Novais E A *et al* 2016 Swept-source optical coherence tomography angiography reveals choriocapillaris alterations in eyes with nascent geographic atrophy and drusen-associated geographic atrophy *Retina* **36Suppl 1** S2–S11
[65] Toto L, Borrelli E and Di Antonio L *et al* 2016 Retinal vascular plexuses' changes in dry age-related macular degeneration, evaluated by means of optical coherence tomography angiography *Retina* **36** 1566–72
[66] Yannuzzi L A 2010 *The Retinal Atlas* 1st edn (Edinburgh: Elsevier Saunders)
[67] Inoue M, Jung J J and Balaratnasingam C *et al* 2016 A comparison between optical coherence tomography angiography and fluorescein angiography for the imaging of type 1 neovascularization *Invest. Ophthalmol. Vis. Sci.* **57** OCT314–23
[68] Costanzo E, Miere A and Querques G *et al* 2016 Type 1 choroidal neovascularization lesion size: Indocyanine green angiography versus optical coherence tomography *Invest. Ophthalmol. Vis. Sci.* **57** OCT307–13
[69] Ryan S J 2013 *Retina* 5th edn (London: Elsevier)
[70] Dingerkus V L S, Munk M R and Brinkmann M P *et al* 2019 Optical coherence tomography angiography (OCTA) as a new diagnostic tool in uveitis *J. Ophthalmic Inflamm. Infect* **9** 10
[71] Khan H A, Iqbal F and Shahzad M A *et al* 2019 Textural properties of choriocapillaris on OCTA in healed inflammatory choriocapillaropathies *Ophthalmic Surg. Lasers Imaging Retina* **50** 566–72
[72] Pichi F, Invernizzi A and Tucker W R *et al* 2020 Optical coherence tomography diagnostic signs in posterior uveitis *Prog. Retin. Eye Res.* **75** 100797
[73] Pichi F, Sarraf D and Arepalli S *et al* 2017 The application of optical coherence tomography angiography in uveitis and inflammatory eye diseases *Prog. Retin. Eye Res.* **59** 178–201

IOP Publishing

Photo Acoustic and Optical Coherence Tomography Imaging, Volume 3

Angiography: an application in vessel imaging

Ayman El-Baz and Jasjit S Suri

Chapter 2

Longitudinal optical coherence tomography and angiography of hyaloid vascular regression in developing mouse eyes

Tae-Hoon Kim and Xincheng Yao

The hyaloid vascular system transiently nourishes ocular components in the embryonic eye. Failure of natural regression of the hyaloid vessels results in congenital ocular disorders. However, the exact mechanism of hyaloid vascular regression and its correlation with eye development are not well understood. Previous studies typically relied on end-point examinations with biochemically fixed tissues, lacking a full description of natural dynamics of the hyaloid vessels along with eye development. Non-invasive *in vivo* monitoring is thus required for better understanding of the dynamic relationships between the hyaloid vessels and the surrounding ocular tissues. In this chapter, we demonstrate the feasibility of *in vivo* longitudinal optical coherence tomography (OCT) and OCT angiography (OCTA) monitoring of the hyaloid vessels in the developing mouse eyes. OCT allows structural monitoring of the vessels, and OCTA enables functional assessment of the vessels. In addition, we will report comparative OCT and OCTA monitoring of hyaloid vascular regression in wild-type and retinal degeneration 10 (rd10) mice. Longitudinal OCTA monitoring revealed accelerated regression of the hyaloid vessels correlated with retinal degeneration in rd10, confirming the close relationship between the hyaloid vascular system and retinal neurovascular development. Further OCT/OCTA study would provide in-depth understanding of the complex mechanism of hyaloid vascular regression in normal and diseased eyes, which is not only important for advanced study of the nature of the visual system but can also provide insights into the development of better treatment protocols of congenital ocular disorders.

doi:10.1088/978-0-7503-2060-3ch2

2.1 Introduction

The hyaloid vessels are the transient sources of nutrients for immature intraocular tissues in the embryonic eye [1]. Four vascular branches form an entire circulatory system called the hyaloid vascular system: (1) the hyaloid artery (HA) that arises from the dorsal ophthalmic artery and enters the eyecup through the embryonic fissures and divides into (2) the vasa hyaloidea propria (VHP), capillary branches extending to the posterior surface of the lens. The VHP further divides to form (3) the tunica vasculosa lentis (TVL), a dense capillary network enclosing the posterior and lateral surfaces of the lens; (4) the pupillary membrane (PM), which is an anastomosis of the TVL and the anterior ciliary arteries, covers the anterior surface of the lens [2, 3]. The HA is a typical arteriole consisting of smooth muscle cells and endothelial cells connected by tight junctions. The VHP, TVL, and PM are capillaries consisting of a single layer of non-fenestrated endothelial cells with a continuous basement membrane and an incomplete layer of pericytes [4]. The hyaloid vessels are all arteries; thus, blood runs from the HA to the annular vessel and drains into the choroid vein [5].

The hyaloid vessels are programmed to naturally regress in the human fetus by 34 weeks of gestation, coinciding with development of the retinal vasculature [6]. Failure of apoptosis of the hyaloid vessels results in a congenital ocular anomaly called persistent fetal vasculature (PFV) that leads to retinal detachment, cataract, and intraocular hemorrhage [7–9]. Usually, 95% of PFV cases are unilateral, and few bilateral cases have been reported in conjunction with congenital syndromes such as Norrie disease and Walker–Warburg syndrome [10]. A study on childhood blindness and visual loss showed that PFV accounts for about 5% of blindness in childhood in the United States [11]. In addition, 95% of premature infants and 3% of full-term infants have persistent hyaloid vessels which can cause multiple ocular disorders [12, 13]. There are more than 450 000 preterm deliveries each year in the United States [14], accounting for more than 10% of the 3.9 million newborns annually [15]. The exact mechanism of the PFV is still unknown as most PFV cases in humans are sporadic [16, 17]. Only few reports were published on inherited forms of PFV, and there are no candidate genes identified in humans [10]. Advanced knowledge of the regression mechanism is thus necessary for a better understanding of developmental ocular disorders as well as for a better assessment of therapeutic interventions to prevent vision loss of premature infants due to the PFV.

As experiments on the living fetus are almost impossible, there have been so far no studies on the mechanism of the postnatal spontaneous regression of the hyaloid vessels in humans [18]. Instead, experimental studies on the hyaloid vessels are typically performed with neonatal mice [19–21]. The highly vascularized mouse eye offers several advantages. In contrast to humans, the mouse pups have persistent hyaloid vessels at birth, which are known to regress mostly at around postnatal day 16 (P16) [22, 23]. In addition, ocular vascular systems can be genetically manipulated. Several gene mutations have been shown to exhibit the clinical features of the PFV in mouse models [10]. Moreover, the prenatal development of the mouse eye is very similar to that of the human eye, and the gestational period of the mouse is rather short, which makes the mouse an efficient experimental model for the study of early development of the mammalian eye in line with the hyaloid vascular regression [24, 25]. However, experimental study using mouse eyes has relied on conventional fixed-tissue analysis. Since histological examination can only provide snapshots of vessel remodeling, a dynamic description of vessel remodeling correlated with eye development can hardly be appreciated. Another caveat of histology is that postmortem eyeballs can be significantly altered in shape within a few minutes after extraction [26]. These issues limit the opportunity to investigate coherent relationships between the hyaloid vessels and the surrounding ocular tissues such as the retina, the lens, and the vitreous. The ability to monitor biological processes in natural context is important to advance in-depth knowledge of the underlying mechanism of hyaloid regression.

Thus, *in vivo* imaging techniques have been extensively explored. Scanning laser confocal microscopy has visualized intraocular vasculatures including the hyaloid vessels [23]. High frequency ultrasound microscopy revealed a progressive decrease in blood flow velocity in the hyaloid vessels with age [27]. Micro-computed tomography with microvascular corrosion casting technique demonstrated three-dimensional (3D) imaging of the hyaloid vessels in neonatal mice [28]. Optical coherence tomography (OCT) successfully visualized parts of the hyaloid vessels in mouse embryos [29]. With advancement in ophthalmic imaging techniques, fundus photography, fluorescein angiography, Doppler ultrasound, and OCT have been used to examine the persistent HA in clinics as well [30–34]. In a recent case report, OCT angiography (OCTA) was for the first time used to check the presence of blood flow of the persistent HA [35]. As a new modality of OCT, OCTA provides a label-free method to visualize the 3D details of ocular vasculatures at high resolution. OCTA has been widely used for constructing a detailed view of the retinal and choroidal vasculature both in human and animal eyes [36–42]. Recently, we have demonstrated, for the first time, the feasibility of *in vivo* OCT/OCTA monitoring of hyaloid vascular regression in developing mouse eyes. OCT enabled structural imaging of the hyaloid vessels, while OCTA allowed functional assessment of the hyaloid vessels with active blood flow [43–45]. In addition, non-invasive *in vivo* OCT/OCTA imaging can not only reduce litter size but also effectively follow up phenotypic changes, allowing studies of the progression of ocular diseases and the prenatal effects of pharmacological agents.

In this chapter, we will describe OCT/OCTA imaging of the hyaloid vessels in the mouse eye and demonstrate immature hyaloid vascular regression due to retinal degeneration during ocular development.

2.2 Methods

2.2.1 OCT/OCTA imaging system

A custom-designed OCT system was developed for *in vivo* hyaloid vessel imaging (figure 2.1). A superluminescent diode (SLD) with a central wavelength λ of 850 nm and a bandwidth $\Delta\lambda$ of 100 nm (D840, Superlum, Cork, Ireland) was used as the OCT light source, providing 3 μm axial resolution. The lateral resolution was

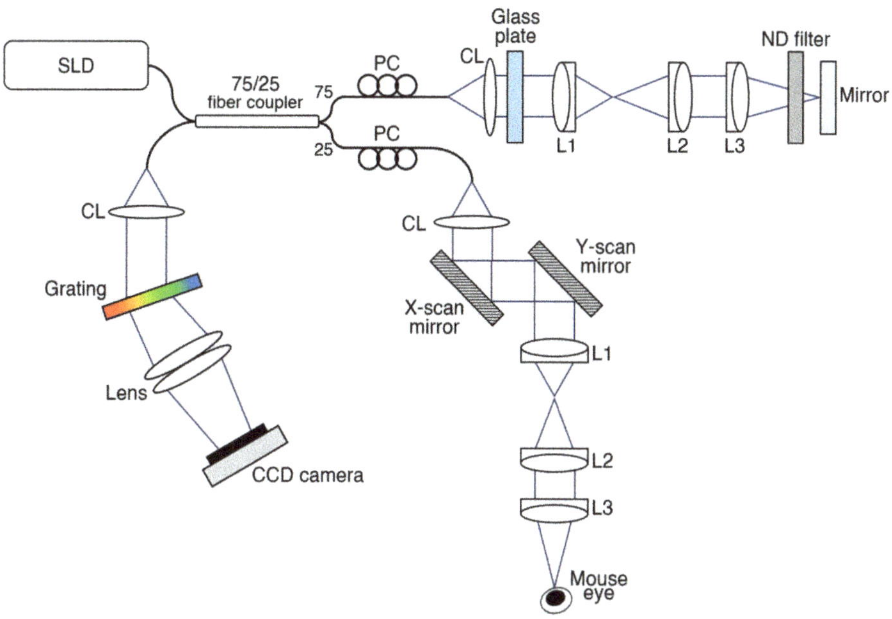

Figure 2.1. Optical diagram of the OCT setup. CL: collimation lens; L1, L2, L3: lens; PC: polarization controller; SLD: superluminescent diode (λ = 850 nm). (Reprinted from [44] with permission from SAGE.)

estimated at 12 μm. The light from the SLD was divided into the sample and reference arms by a fiber coupler with a splitting ratio of 75:25. The reference arm had a glass plate for balancing dispersion between the sample and reference paths, and the sample arm had two scanning mirrors to allow two-dimensional (2D) scanning. A custom-designed spectrometer was used for OCT recording. A linear CCD camera (AViiVA EM4; e2v Technologies, Chelmsford, UK) used in the spectrometer provided a A-scan rate up to 70 kHz. The incident light power on the mouse cornea was set to 0.95 mW. Technical details are also reported in our previous publication [46].

2.2.2 OCT/OCTA image acquisition

3D OCT/OCTA volumes were acquired over 1.2 mm (width) × 1.2 mm (length) × 1.4 mm (height), containing parts of the posterior surface of the lens, the vitreous, the hyaloid vessels, and the retina (figure 2.2). The optic nerve head (ONH), where the HA extends from, was located at the center of the imaging area. Four consecutive OCT B-scans were obtained at each imaging position to compute speckle variances (SV) for OCTA construction [47]. Thus, each OCTA volume consisted of a total of 4 × 500 × 500 A-scans.

2.2.3 Quantitative image analysis

Hyaloid vessel density, floater density, retinal thickness, vitreous chamber (VC) length, and vitreoretinal (VR) length were comprehensively analyzed to see the

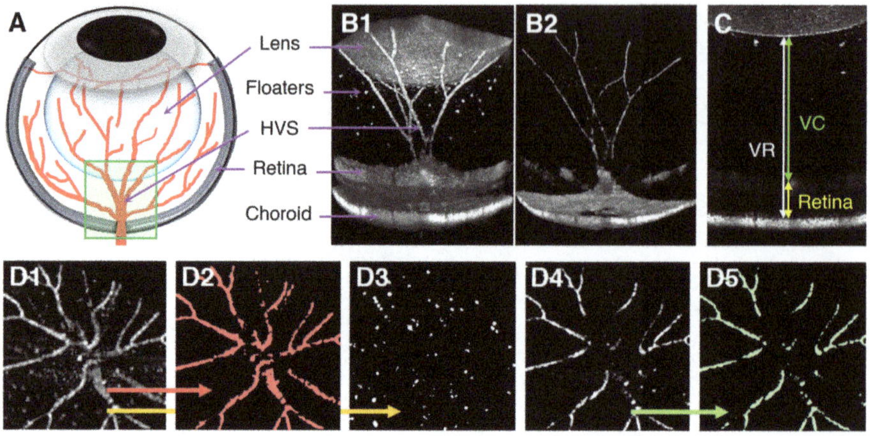

Figure 2.2. (A) Schematic diagram of the hyaloid vascular system (HVS) in the mouse eye. The green box indicates the OCT/OCTA imaging area. Representative OCT (B1) and OCTA (B2). In this imaging field, the hyaloid vessels can be clearly captured together with neighboring ocular tissues including the lens, the retina, and the vitreous with floaters. (C) Axial length measurements of the posterior segment. Vitreoretinal (VR) length (white line), vitreous chamber (VC) length (green line) and retinal thickness (yellow line) were separately measured. (D) Image processing diagram. Constructed *en face* OCT (D1) and OCTA (D4) images for morphological and functional vessel density analysis, respectively. Applied image processing ultimately binarizes hyaloid vasculatures (D2), floaters (D3), and functional hyaloid vessels (D5) for density measurements. (Reprinted from [43], copyright (2019) with permission from Springer (CC BY 4.0).)

relationship between the hyaloid vessels and the surrounding ocular tissues during eye development. Retinal thickness (from the retinal nerve fiber layer (RNFL) to the retinal pigment epithelium (RPE)), VC axial length (from the posterior pole of the lens to the RNFL), and VR axial length (from the posterior pole of the lens to the RPE) were separately measured, as shown in figure 2.2C. The hyaloid vessel density was separately analyzed into two categories: (1) 'structural' vessel density which is processed with the OCT dataset and (2) 'functional' vessel density which is processed with the OCTA dataset. Floaters are mainly macrophages and vessel fragments in the vitreous. Figure 2.2D illustrates the image processing for the vessel density and floater density analysis.

For the structural vessel density, four repeated OCT B-scans at each scanning position were first averaged, and an OCT B-scan volume was sliced parallel to the B-scan cross-section to create a C-scan (*en face*) stack. Contrast level was adjusted to individual C-scan images. Next, maximum intensity projection was implemented in the C-scans (figure 2.2D1). During the projection process, the retinal vasculature was not included. The Otsu's automatic thresholding method was then applied for image binarization [48], followed by floater removal (figure 2.2D2). Binarized floater images were also produced following a similar procedure to the vessel image processing, but before the binarization, the vasculature was first removed, followed by the Otsu's automatic thresholding (figure 2.2D3).

For the functional vessel density, OCTA volumetric images constructed by calculating SV from four repeated OCT B-scans were sliced parallel to the B-scan cross-section to create a C-scan stack. Then the maximum intensity projection was implemented in the OCTA C-scans (figure 2.2D4). Next, binarization processing was implemented by using the Otsu's automatic thresholding method (figure 2.2D5). Fiji software (http://fiji.sc/Fiji) was used for all the image processing. The vessel density was defined as the percentage of pixel area occupied by the vasculature in the binarized image.

All data are expressed as the mean ± standard deviation (SD). Unpaired t-tests between the animal groups were performed using either Student's t-test (equal variances) or Welch's t-test (unequal variances), and a P-value <0.05 was considered statistically significant.

2.2.4 Animals

Wild-type mice (WT: C57BL/6J, The Jackson Laboratory, Bar Harbor, ME) and retinal degeneration 10 mice (Rd10: homozygous for the $Pde6b^{rd10}$ on C57BL/6J background, The Jackson Laboratory, Bar Harbor, ME) were used in this study. Rd10 carries a spontaneous point mutation in PDE6B gene (cGMP phosphodiesterase 6B, rod photoreceptor), and progressive rod photoreceptor death begins at around P16 at which hyaloid vascular regression is still in progress [49]. Thus, rd10 model provides a unique time window to study correlations of hyaloid vascular regression with ocular tissue deformities during eye development.

OCT/OCTA measurements were longitudinally conducted. The imaging was conducted in a room under ambient light condition. Anesthesia was induced intraperitoneally by injecting a mixture of 100 mg kg^{-1} ketamine and 5 mg kg^{-1} xylazine. A drop of ophthalmic mydriatic (1% atropine sulfate ophthalmic solution, Akorn, Lake Forest, IL) was applied to the imaging eye. The mouse was then placed in an animal holder integrated with a bite bar and ear bars to minimize motion artefacts. A cover glass (12–545–80; Microscope cover glass, Thermo Fisher Scientific, Waltham, MA) along with a drop of eye gel (Severe; GenTeal, Novartis, Basel, Switzerland) was placed on the imaging eye. After 10 min for acclimatization and full pupil dilation, the mouse head was fixed for OCT/OCTA recording. During the experiment, a heating pad was wrapped around the animal holder to maintain body temperature.

All animal care and experiments were performed in accordance with the Association for Research in Vision and Ophthalmology (ARVO) statement for the use of animals in ophthalmic and vision research. All experiments were performed following the protocols approved by the Animal Care Committee at the University of Illinois at Chicago.

2.3 Results

2.3.1 OCTA monitoring of hyaloid vascular regression

In this section, we will demonstrate the feasibility of non-invasive *in vivo* OCT/OCTA monitoring of the hyaloid vessels in developing mouse eye.

OCT/OCTA images in figure 2.3 show three branches of the hyaloid vascular system, that is, HA, VHP, and TVL. Multiple capillaries of VHP and vessel fragments

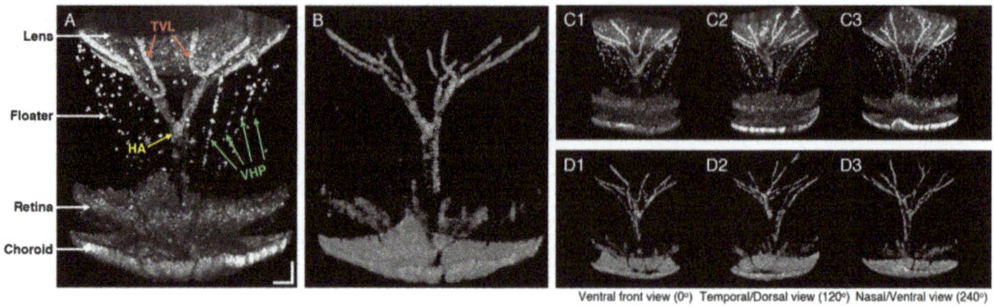

Figure 2.3. Representative OCT (A) and OCTA (B) images of a wild-type mouse eye at P14. Ventral (C1 and D1), temporal/dorsal (C2 and D2), and nasal/ventral (C3 and D3) views of the OCT (C) and OCTA (D) images. Scale bars: 100 μm. (Reprinted from [44] with permission from SAGE.)

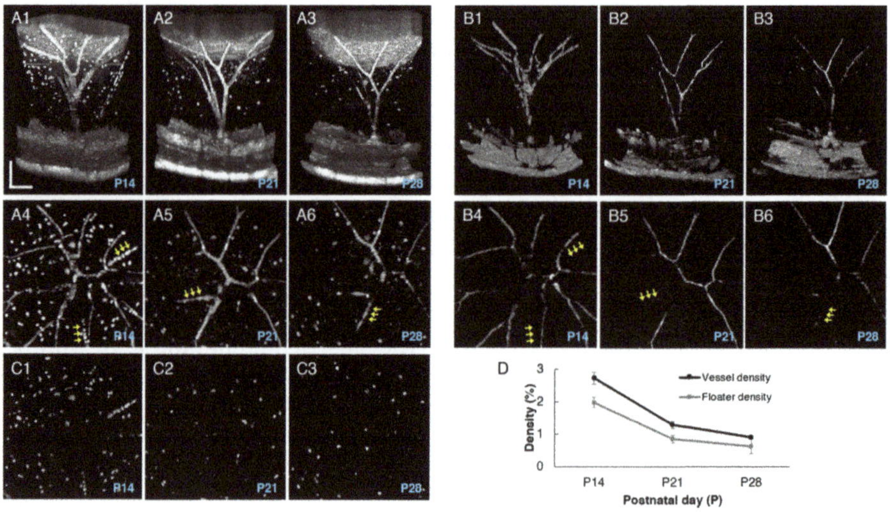

Figure 2.4. Comparative OCT (A) and OCTA (B) images from a wild-type mouse eye measured at P14, P21, and P28. A1, A2, and A3 illustrate side views of OCT images and A4, A5, and A6 illustrate *en face* OCT projection images at the three time points. B1, B2, and B3 illustrate side views of OCTA images and B4, B5, and B6 illustrate *en face* OCTA projection images at the three time points. Yellow arrows indicate capillaries in functional loss. C1, C2, and C3 show floater images, which isolated from A4, A5, and A6, respectively. (D) Hyaloid vessel density and floater density measurements. Scale bars: 200 μm. (Reprinted from [44] with permission from SAGE.)

were clearly observed in the OCT image (figure 2.3A), but not in the OCTA image (figure 2.3B), indicating a lack of blood flow in the VHP. Views from different angles further confirmed this observation (figures 2.3(C) and (D)). Since the VHP supplies oxygen to the inner retina until the formation of retinal vasculature [50], it is supposed to regress by around P14 [22, 27] at the time retinal vasculatures are being formed [6].

Figure 2.4 shows representative OCT/OCTA images longitudinally recorded at P14, P21, and P28. Comparative C-scan images of OCT (figures 2.4A4–A6) and OCTA (figures 2.4B4–B6) disclose functionally impaired hyaloid vessels (yellow arrows). A sequence of regression process also observed that atrophy of the VHP begins first, followed by regression of capillaries of the TVL, and finally the HA. In addition, in OCT images, many floaters were observed, and their numbers decreased with age (figure 2.4C). The floaters are mostly retrolental cell mass and macrophages [21] and cleaned by phagocytosis [51]. Thus, the vitreous became clear with age.

For quantitative analysis, functional vessel density and floater density were computed based on OCTA and OCT images, respectively, longitudinally acquired at P14, P21, and P28 of WT mice (figure 2.4D). The result shows that the functional vessel density gradually decreased with age (2.72 ± 0.2% [P14] to 0.89 ± 0.1% [P28]), and the floater density also steadily decreased with age (1.97 ± 0.2% [P14] to 0.62 ± 0.2% [P28]). It is observed that a later phase of regression (P21 to P28) proceeds slower than a fast regression period (P14 to P21).

2.3.2 Accelerated regression of hyaloid vessels in retinal degeneration mice

In this section, we will demonstrate comparative OCT/OCTA monitoring of hyaloid vascular regression in WT and rd10 mice. With the onset of photoreceptor degeneration in rd10, a significant retinal capillary dysfunction was observed [38]. As disturbed retinal development can accompany the persistent hyaloid vessels [7], it was expected that rd10 mice would reveal a different regression pattern compared to the healthy control group. Three-dimensional OCT/OCTA images of the hyaloid vessels were longitudinally acquired in WT and rd10 mice, and the axial length of VC and VR as well as the retinal thickness were measured to examine a potential correlation with hyaloid vascular regression (figure 2.2C).

First, the body weights of WT and rd10 mice were compared as the growing body size is highly correlated with the growth rate of the eye in juvenile animals [52, 53], and no significant difference was found between two groups (figure 2.5A). There was also no significant difference in the retinal thickness between two groups at P14. However, rd10 mice showed a significant reduction of retinal thickness from P17 (figure 2.5B) due to progressive photoreceptor degeneration [38, 54]. It was also observed that, in conjunction with retinal degeneration, the axial length of the vitreous abnormally and rapidly changed in rd10. At P14, the VC length was similar between WT and rd10 mice; however, the VC length of P17 rd10 mice significantly increased along with the onset of photoreceptor cell death (Rd10: 797.3 ± 38 μm and WT: 726.3 ± 24.7 μm, $P = 0.0014$). This result indicates that the inner retinal layers settled down on the outer retina due to the photoreceptor layer collapsing. After P17, rd10 mice narrowed the gap of the VC as the VR length rapidly decreased with age (figures 2.5C and 5D). The VR length of rd10 mice became much shorter than that of WT mice at P28 ($P = 0.0036$). These abrupt changes of the vitreoretinal geometry demonstrate an active emmetropization in rd10 mice, adjusting the axial length of the eyeball to the focal plane of the ocular optics [55].

Figure 2.6 shows representative OCT/OCTA images of the hyaloid vessels from a WT mouse and a rd10 mouse. OCTA clearly visualizes the contrast enhanced vasculatures (figures 2.6B and E). Binarized C-scans of structural OCT vessel image and functional OCTA vessel image were prepared for quantitative analysis. Composite images (figures 2.6C and F) represent the binarized structural vessels

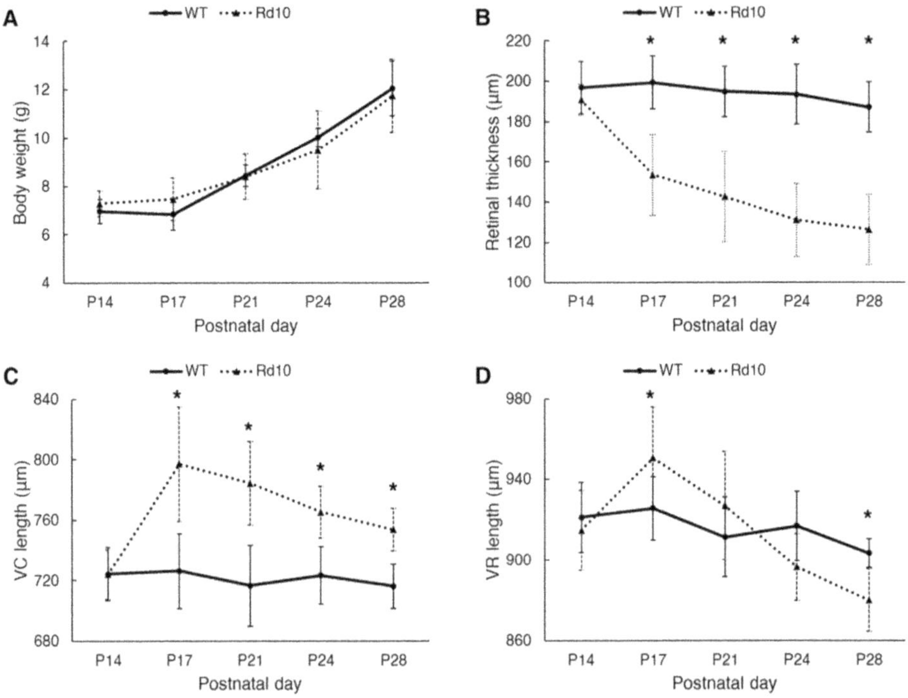

Figure 2.5. (A) Body weight measurement. (B) Retinal thickness measurement. Photoreceptor degeneration resulted in significant thinning of the retinal thickness in rd10 mice. (C) VC length measurement. Onset of photoreceptor degeneration induced rapid extension of the VC length (724.0 ± 16.3 μm [P14] to 797.3 ± 38.0 μm [P17]). Subsequently, the VC length of rd10 mice steadily decreased with aging: 784.3 ± 27.5 μm (P21) and 753.7 ± 14.1 μm (P28). WT showed little change in the VC length throughout the experimental period: 724.3 ± 17.7 μm (P14) and 716.2 ± 14.7 μm (P28). (D) VR length measurement. Rd10 mice initially had a wide VR length in response to the onset of photoreceptor degeneration, followed by a sharp narrowing phase. The VR length in WT mice showed a slight declining trend (921.2 ± 17.3 μm [P14] to 903.4 ± 7.2 μm [P28]). $N = 7$ for WT and $N = 7$ for rd10. P values for an unpaired t-test between WT and rd10 mice are indicated by asterisks: $^*P < 0.05$. (Reprinted from [43], copyright (2019) with permission from Springer (CC BY 4.0).)

(red color), floaters (white color), and functional vessels (green color). It is readily observed that vascular branches and floaters decreased with age in two mouse groups. Although the structural vessel density (figure 2.6G) and floater density (figure 2.6I) were similar between the two groups until P24, rd10 mice showed a rapid vascular dropout at P28 (1.96 ± 0.9%) (figure 2.6G). Notably, almost complete functional loss was found in rd10 mice at P24 before obliteration of the hyaloid vasculature (figure 2.6H). While, WT mice revealed a gradual declining trajectory in vessel density with age, and some vessels were still functional even at P28 (4.11 ± 0.5%). In rd10 mice, the floater density abruptly increased at P28 due to the accelerated hyaloid vascular regression (figure 2.6I). These results demonstrate that retinal degeneration during ocular development affects the hyaloid vascular regression process.

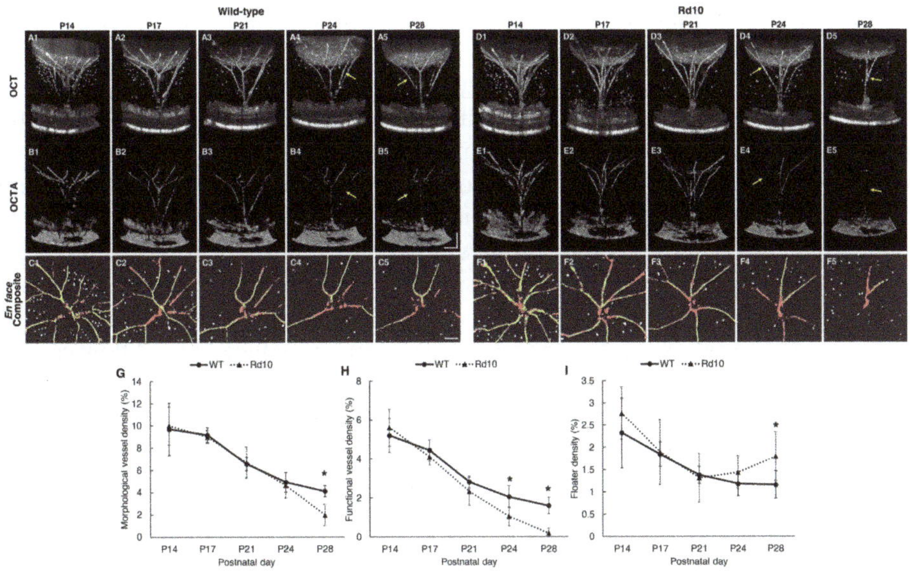

Figure 2.6. Representative OCT (A, D) and OCTA (B, E) images from a WT mouse and a rd10 mouse, longitudinally recorded at P14, P17, P21, P24, and P28. In binarized *en face* composite images (C) and (F), red, white, and green colors represent the hyaloid vasculature, floaters, and functional vessels, respectively. The retinal and choroidal vasculature was excluded from *en face* image construction. Yellow arrows indicate capillaries in functional loss, that is, a lack of blood flow. Scale bars: 200 μm. (G) Hyaloid vasculature density measurements from OCT datasets. Accelerated regression of hyaloid vessel occurred in rd10 mice compared to WT mice (4.11 ± 0.5% [WT at P28] and 1.98 ± 0.9% [Rd10 at P28], $P = 0.0002$). (H) Functional hyaloid vessel density measurements from the OCTA datasets. Functional OCTA revealed even earlier physiological dysfunction in rd10 mice before obliteration of the hyaloid vasculature (2.06 ± 0.6% [WT at P24] and 1.05 ± 0.5% [Rd10 at P24], $P = 0.0061$). (I) Floater density measurements. Floater density steadily decreased in WT with age. $N = 7$ for WT and $N = 7$ for rd10. P values for an unpaired t-test between WT and rd10 mice are indicated by asterisks: $*P < 0.05$. (Reprinted from [43], copyright (2019) with permission from Springer (CC BY 4.0).)

2.4 Discussion

In this chapter, we introduced OCT/OCTA imaging of the hyaloid vessels in the developing mouse eyes. OCT imaging allowed morphological monitoring of the hyaloid vessels and surrounding tissues. OCTA imaging enabled functional assessment of the hyaloid vessels, that is, insufficient or lacking blood flow, before morphological changes occur. Accelerated hyaloid vascular regression was also observed in rd10 mice, following the onset of rod cell degeneration. This premature regression is a rare case as most mutant mouse models have shown persistent hyaloid vessels in the case of malformation of retinal vasculatures [56]. On the other hand, all WT mice had functional hyaloid vessels over P28, which is consistent with a previous fluorescein angiographic study where the hyaloid vessels were found in all WT mice at 8–25 weeks old [57]. Since the primary function of the hyaloid vessels is to nurture the developing ocular components, the sustaining functional vessels in WT mice may be still involved in later developmental phases of the eye [58–60].

Accordingly, the premature regression of the hyaloid vessels in rd10 mice could impinge on the normal eye development. In fact, rd10 mice have shown shorter axial length of the eyeball and lighter eye weight compared to WT mice [61, 62].

The exact mechanism of hyaloid vascular regression under the normal condition is not fully understood yet. Multiple contributing factors are suggested to be directly involved in the regression process. Meeson *et al* proposed two stages in regression. The first stage is the apoptosis of a single endothelial cell, which restricts the capillary lumen and subsequently imposes either a temporary or permanent block to blood flow. In the second stage, the cessation of blood flow consequently causes synchronous apoptosis of downstream endothelial cells in the affected segment [63]. Reduction of shear stress on the endothelial cells due to reduced blood flow can also stimulate expression of vasoconstrictors [64], and vasoconstriction of the hyaloid vessels appears in the early phase of regression process [65]. Mitchell *et al* demonstrated widespread apoptosis of the endothelial cells in the TVL, and the vascular regression was associated with loss of capillary integrity and phagocytosis of the apoptotic endothelium by macrophages [66]. Macrophages contribute significantly in hyaloid vascular regression involving the induction of apoptosis, the occlusion, and obliteration of atrophic vessels [3, 19–21]. Kishimoto *et al* further confirmed a narrowed vessel width during regression of the VHP and found severe vascular congestions due to aggregated neutrophilic leukocytes together with erythrocytes in the lumen of hyaloid vessels [21]. The interaction between macrophages and blood vessels was found to involve Wnt7b signaling derived from macrophages via FZD4 receptor and LRP5 co-receptor on endothelial cells. As a result, mice with FZD4 or LRP5 null mutations or a hypomorphic Wnt7b allele were accompanied by the disruption of hyaloid vascular regression [20, 67, 68]. In our study, cessation of blood flow was clearly demonstrated in the OCTA images (figures 2.4 and 2.6). In addition, fragmented threadlike hyaloid vessels were frequently observed in OCT images, which may indicate the presence of aggregated blood cells in the lumen (figure 2.3). This result supports that the reduced blood flow by vascular occlusion with blood cells is the major triggering source of the hyaloid vascular regression under the natural context.

The hyaloid vascular regression is also considerably affected by retinal oxygen levels that regulate the expression of vascular endothelial growth factor (VEGF), a signal protein related to the formation of the hyaloid vessels as well as its maintenance [69]. It was reported that low oxygenation in the retina due to malformation of retinal vasculatures induced upregulation of VEGF levels, subsequently resulting in persistent hyaloid vessels [70]. The bilateral PFV occurs in humans with Norrie disease [71], and mice lacking Norrin showed upregulation of VEGF levels in neurons, and reducing VEGF levels corrected the vascular abnormalities in mice [68, 72]. In contrast, high oxygenation can lead to downregulation of VEGF levels that can cause premature regression of the hyaloid vessels [73]. It was confirmed that inactivation of VEGF accelerated hyaloid vascular regression [69]. A study demonstrated that melanopsin, a light-sensitive molecule found in a small subset of retinal ganglion cells, responds to light exposure and restrains VEGF-A expression, which consequently allows hyaloid vascular regression to occur [74]. In our study, rd10 mice manifested accelerated hyaloid vascular regression (figure 2.6), and it is plausible to speculate that the retinal hyperoxia in rd10 mice may influence the regression process. In fact, significant attenuation of retinal blood flow in the deep capillary plexus occurred simultaneously with rod photoreceptor apoptosis [38], which enhances oxygen diffusion towards the vitreous due to a lack of autoregulating capacity in the choroid circulation [75]. Another intriguing factor to consider is neurotransmitter dopamine released by dopaminergic amacrine cells (DACs) in the retina, which can directly inhibit VEGF receptor-2 signaling in hyaloid vascular endothelial cells and consequently accelerate hyaloid vascular regression [56]. The inner retinal neurons in rd10 mice undergo extensive remodeling after loss of rod photoreceptors, which makes DACs exhibiting aberrant activity leading to impaired dopamine metabolism [76]. It was found that the dopamine levels in rd10 mouse retinas were significantly higher than those in WT mouse retinas [62]. Thus, the abnormal dopamine levels due to the inner retinal remodeling can be another triggering source of the accelerated hyaloid regression in rd10 mice. Overall, the rd10 study suggests an unknown signaling pathway between the retinal and hyaloid vessels [5]. It has been noted that the disturbed formation of the deep capillary plexus in mutant mouse retinas directly affects hyaloid vessel regression [7].

2.5 Conclusion

In summary, we have demonstrated the feasibility of longitudinal *in vivo* OCT/OCTA monitoring of the hyaloid vascular regression in developing mouse eyes. Comparative OCT and OCTA revealed accelerated regression of the hyaloid vessels in rd10 mice due to rod photoreceptor degeneration during ocular development, which clearly indicates a functional link between the retinal neurovascular system and the hyaloid vascular system. We expect that further longitudinal OCT/OCTA investigation of the retinal and hyaloid vessels in normal and mutant mouse models will be valuable to unravel the unknown mechanisms of the PFV. Label-free OCT/OCTA will also provide an imaging platform for longitudinal monitoring of therapeutic treatments of the animal models, fostering drug developments for effective treatment of PFV.

References

[1] Usui Y, Westenskow P D, Murinello S, Dorrell M I, Scheppke L, Bucher F, Sakimoto S, Paris L P, Aguilar E and Friedlander M 2015 Angiogenesis and eye disease *Annu. Rev. Vis. Sci.* **1** 155–84

[2] Yee K M P, Feener E P, Gao B, Aiello L P, Madigan M C, Provis J, Ross-Cisneros F N, Sadun A A and Sebag J I D 2014 Vitreous cytokines and regression of the fetal hyaloid vasculature ed J Sebag *Vitreous: in Health and Disease* (New York: Springer) 41–55

[3] Zhang H, Tse J, Hu X, Witte M, Bernas M, Kang J, Tilahun F, Hong Y K, Qiu M and Chen L 2010 Novel discovery of LYVE-1 expression in the hyaloid vascular system *Investig. Ophthalmol. Vis. Sci.* **51** 6157–61

[4] Sebag J 2015 *Vitreous: In Health and Disease* (Berlin: Springer)

[5] Saint-Geniez M and D'Amore P A 2004 Development and pathology of the hyaloid, choroidal and retinal vasculature *Int. J. Dev. Biol.* **48** 1045–58
[6] Stahl A *et al* 2010 The mouse retina as an angiogenesis model *Investig. Ophthalmol. Vis. Sci.* **51** 2813–26
[7] Fruttiger M 2007 Development of the retinal vasculature *Angiogenesis.* **10** 77–88
[8] Silbert M and Gurwood A S 2000 Persistent hyperplastic primary vitreous *Clin. Eye Vis. Care* **12** 131–7
[9] Goldberg M F 1997 Persistent fetal vasculature (PFV): an integrated interpretation of signs and symptoms associated with persistent hyperplastic primary vitreous (PHPV). LIV Edward Jackson Memorial Lecture *Am. J. Ophthalmol.* **124** 587–626
[10] Chen C, Xiao H and Ding X 2019 Persistent fetal vasculature *The Asia-Pacific J. Ophthalmol.* **8** 86–95
[11] Mets M B 1999 Childhood blindness and visual loss: an assessment at two institutions including a 'new' cause *Trans. Amer. Ophthal. Soc* **97** 653–96
[12] Fineman M S and Ho A C 2012 Color Atlas and Synopsis of Clinical Ophthalmology: Retina 2nd edn (Baltimore, MD: Wolters Kluwer/Lippincott Williams & Wilkins)
[13] Hartnett M E, Trese M and Capone A 2015 *Pediatric Retina: Medical and Surgical Approaches* (Philadelphia, PA: Wolters Kluwer)
[14] Liegl R, Hellstrom A and Smith L E 2016 Retinopathy of prematurity: the need for prevention *Eye Brain* **8** 91–102
[15] Facts About Retinopathy of Prematurity (ROP) https://neinihgov/health/rop/rop
[16] Vasavada A R, Vasavada S A, Bobrova N, Praveen M R, Shah S K, Vasavada V A, Pardo A J, Raj S M and Trivedi R H 2012 Outcomes of pediatric cataract surgery in anterior persistent fetal vasculature *J. Cataract Refract. Surg.* **38** 849–57
[17] Khaliq S, Hameed A, Ismail M, Anwar K, Leroy B, Payne A M, Bhattacharya S S and Mehdi S Q 2001 Locus for autosomal recessive nonsyndromic persistent hyperplastic primary vitreous *Investig. Ophthalmol. Vis. Sci.* **42** 2225–8
[18] Li J, Zhang J and Lu P 2019 Regression of fetal vasculature and visual improvement in nonsurgical persistent hyperplastic primary vitreous: a case report *BMC Ophthalmol.* **19** 161
[19] Lang R A and Bishop J M 1993 Macrophages are required for cell death and tissue remodeling in the developing mouse eye *Cell.* **74** 453–62
[20] Lobov I B *et al* 2005 WNT7b mediates macrophage-induced programmed cell death in patterning of the vasculature *Nature* **437** 417–21
[21] Kishimoto A, Kimura S, Nio-Kobayashi J, Takahashi-Iwanaga H, Park A M and Iwanaga T 2018 Histochemical characteristics of regressing vessels in the hyaloid vascular system of neonatal mice: novel implication for vascular atrophy *Exp. Eye Res.* **172** 1–9
[22] Ito M and Yoshioka M 1999 Regression of the hyaloid vessels and pupillary membrane of the mouse *Anat. Embryol.* **200** 403–11
[23] Ritter M R, Aguilar E, Banin E, Scheppke L, Uusitalo-Jarvinen H and Friedlander M 2005 Three-dimensional *in vivo* imaging of the mouse intraocular vasculature during development and disease *Investig. Ophthalmol. Vis. Sci.* **46** 3021–6
[24] Pei Y F and Rhodin J A 1970 The prenatal development of the mouse eye *Anatomical Rec.* **168** 105–25
[25] Pardue M T, Stone R A and Iuvone P M 2013 Investigating mechanisms of myopia in mice *Exp. Eye Res.* **114** 96–105

[26] Jiang X, Kurihara T, Kunimi H, Miyauchi M, Ikeda S I, Mori K, Tsubota K, Torii H and Tsubota K 2018 A highly efficient murine model of experimental myopia *Sci. Rep.* **8** 2026
[27] Brown A S, Leamen L, Cucevic V and Foster F S 2005 Quantitation of hemodynamic function during developmental vascular regression in the mouse eye *Investig. Ophthalmol. Vis. Sci.* **46** 2231–7
[28] Atwood R C, Lee P D, Konerding M A, Rockett P and Mitchell C A 2010 Quantitation of microcomputed tomography-imaged ocular microvasculature *Microcirculation* **17** 59–68
[29] Larina I V, Syed S H, Sudheendran N, Overbeek P A, Dickinson M E and Larin K V 2012 Optical coherence tomography for live phenotypic analysis of embryonic ocular structures in mouse models *J. Biomed. Opt.* **17** 081410–1
[30] Sundararajan M, Jansen M E and Gupta M 2018 Unilateral persistence of the hyaloid artery causing vitreopapillary and vitreomacular traction *JAMA Ophthalmol.* **136** e180221
[31] Delaney W V Jr Prepapillary hemorrhage and persistent hyaloid artery *Am. J. Ophthalmol.* 1980**90** 419–21
[32] Chen T L and Yarng S S 1993 Vitreous hemorrhage from a persistent hyaloid artery *Retina* **13** 148–51
[33] Gonçalves A, Cruysberg J R, Draaijer R W, Sellar P W, Aandekerk A L and Deutman A F 1996 Vitreous haemorrhage and other ocular complications of a persistent hyaloid artery *Doc. Ophthalmol. Adv. Ophthalmol.* **92** 55–9
[34] Azrak C, Campos-Mollo E, Lledó-Riquelme M, Ibañez F A and Toldos J J 2011 Vitreous haemorrhage associated with persistent hyaloid artery *Archivos de la Sociedad Espanola de Oftalmologia* **86** 331–4
[35] Jeon H, Kim J and Kwon S 2019 OCT angiography of persistent hyaloid artery: a case report *BMC Ophthalmol.* **19** 141
[36] Park J R, Choi W, Hong H K, Kim Y, Jun Park S, Hwang Y, Kim P, Joon Woo S, Hyung Park K and Oh W Y 2016 Imaging laser-induced choroidal neovascularization in the rodent retina using optical coherence tomography angiography *Investig. Ophthalmol. Vis. Sci.* **57** Oct331–40
[37] Liu W, Li H, Shah R S, Shu X, Linsenmeier R A, Fawzi A A and Zhang H F 2015 Simultaneous optical coherence tomography angiography and fluorescein angiography in rodents with normal retina and laser-induced choroidal neovascularization *Opt. Lett.* **40** 5782–5
[38] Kim T H, Son T, Lu Y, Alam M and Yao X 2018 Comparative optical coherence tomography angiography of wild-type and rd10 mouse retinas *Transl. Vis. Sci. Technol* **7** 42
[39] Borrelli E, Sarraf D, Freund K B and Sadda S 2018 OCT angiography and evaluation of the choroid and choroidal vascular disorders *Prog. Retin. Eye Res.* **67** 30–55
[40] Yao X, Alam M N, Le D and Toslak D 2020 Quantitative optical coherence tomography angiography: a review *Exp. Biol. Med.* **245** 301–12
[41] Jia Y *et al* 2015 Quantitative optical coherence tomography angiography of vascular abnormalities in the living human eye *Proc. Natl. Acad. Sci. USA* **112** E2395–402
[42] Hwang T S, Jia Y, Gao S S, Bailey S T, Lauer A K, Flaxel C J, Wilson D J and Huang D 2015 Optical coherence tomography angiography features of diabetic retinopathy *Retina* **35** 2371–6
[43] Kim T H, Son T, Le D and Yao X 2019 Longitudinal OCT and OCTA monitoring reveals accelerated regression of hyaloid vessels in retinal degeneration 10 (rd10) mice *Sci. Rep.* **9** 16685

[44] Kim T H, Son T and Yao X 2019 Feature article: functional OCT angiography reveals early physiological dysfunction of hyaloid vasculature in developing mouse eye *Exp. Biol. Med.* **244** 819–23

[45] Kim T-H, Son T and Yao X 2019 Longitudinal optical coherence tomography angiography of hyaloid vessels in the developing mouse eye *Proc. SPIE 10858, Ophthalmic Technologies XXIX 108581N*

[46] Son T, Wang B, Thapa D, Lu Y, Chen Y, Cao D and Yao X 2016 Optical coherence tomography angiography of stimulus evoked hemodynamic responses in individual retinal layers *Biomed. Opt. Express* **7** 3151–62

[47] Mahmud M S, Cadotte D W, Vuong B, Sun C, Luk T W, Mariampillai A and Yang V X 2013 Review of speckle and phase variance optical coherence tomography to visualize microvascular networks *J. Biomed. Opt.* **18** 50901

[48] Otsu N 1979 A threshold selection method from gray-level histograms *IEEE Trans. Syst. Man Cybern.* **9** 62–6

[49] Chang B *et al* 2007 Two mouse retinal degenerations caused by missense mutations in the beta-subunit of rod cGMP phosphodiesterase gene *Vis. Res* **47** 624–33

[50] Lutty G A and McLeod D S 2018 Development of the hyaloid, choroidal and retinal vasculatures in the fetal human eye *Prog. Retin. Eye Res* **62** 58–76

[51] Vrolyk V, Haruna J and Benoit-Biancamano M O 2018 Neonatal and Juvenile Ocular development in Sprague-Dawley Rats: a histomorphological and immunohistochemical study *Vet. Pathol.* **55** 310–30

[52] Prashar A, Hocking P M, Erichsen J T, Fan Q, Saw S M and Guggenheim J A 2009 Common determinants of body size and eye size in chickens from an advanced intercross line *Exp. Eye Res.* **89** 42–8

[53] Wang D, Ding X, Liu B, Zhang J and He M 2011 Longitudinal changes of axial length and height are associated and concomitant in children *Investig. Ophthalmol. Vis. Sci.* **52** 7949–53

[54] Lu Y, Kim T-H and Yao X 2020 Comparative study of wild-type and rd10 mice reveals transient intrinsic optical signal response before phosphodiesterase activation in retinal photoreceptors *Exp. Biol. Med.* **245** 360–7

[55] Wallman J and Winawer J 2004 Homeostasis of eye growth and the question of myopia *Neuron* **43** 447–68

[56] Nguyen M T *et al* 2019 An opsin 5-dopamine pathway mediates light-dependent vascular development in the eye *Nat. Cell Biol.* **21** 420–9

[57] McLenachan S, Magno A L, Ramos D, Catita J, McMenamin P G, Chen F K, Rakoczy E P and Ruberte J 2015 Angiography reveals novel features of the retinal vasculature in healthy and diabetic mice *Exp. Eye Res.* **138** 6–21

[58] Zhou X, Shen M, Xie J, Wang J, Jiang L, Pan M, Qu J and Lu F 2008 The development of the refractive status and ocular growth in C57BL/6 mice *Investig. Ophthalmol. Vis. Sci.* **49** 5208–14

[59] Tkatchenko T V, Shen Y and Tkatchenko A V 2010 Analysis of postnatal eye development in the mouse with high-resolution small animal magnetic resonance imaging *Investig. Ophthalmol. Vis. Sci.* **51** 21–7

[60] Schmucker C and Schaeffel F 2004 A paraxial schematic eye model for the growing C57BL/6 mouse *Vis. Res* **44** 1857–67

[61] Wisard J, Chrenek M A, Wright C, Dalal N, Pardue M T, Boatright J H and Nickerson J M 2010 Non-contact measurement of linear external dimensions of the mouse eye *J. Neurosci. Methods* **187** 156–66

[62] Park H, Tan C C, Faulkner A, Jabbar S B, Schmid G, Abey J, Iuvone P M and Pardue M T 2013 Retinal degeneration increases susceptibility to myopia in mice *Mol. Vis.* **19** 2068–79
[63] Meeson A, Palmer M, Calfon M and Lang R 1996 A relationship between apoptosis and flow during programmed capillary regression is revealed by vital analysis *Development* **122** 3929–38
[64] Lobov I B *et al* 2011 The Dll4/Notch pathway controls postangiogenic blood vessel remodeling and regression by modulating vasoconstriction and blood flow *Blood.* **117** 6728–37
[65] Browning J, Reichelt M E, Gole G A and Massa H 2001 Proximal arterial vasoconstriction precedes regression of the hyaloid vasculature *Curr. Eye Res.* **22** 405–11
[66] Mitchell C A, Risau W and Drexler H C 1998 Regression of vessels in the tunica vasculosa lentis is initiated by coordinated endothelial apoptosis: a role for vascular endothelial growth factor as a survival factor for endothelium *Dev. Dyn.* **213** 322–33
[67] Kato M *et al* 2002 Cbfa1-independent decrease in osteoblast proliferation, osteopenia, and persistent embryonic eye vascularization in mice deficient in Lrp5, a Wnt coreceptor *J. Cell Biol.* **157** 303–14
[68] Xu Q *et al* 2004 Vascular development in the retina and inner ear: control by Norrin and Frizzled-4, a high-affinity ligand-receptor pair *Cell.* **116** 883–95
[69] Yoshikawa Y, Yamada T, Tai-Nagara I, Okabe K, Kitagawa Y, Ema M and Kubota Y 2016 Developmental regression of hyaloid vasculature is triggered by neurons *J. Exp. Med* **213** 1175–83
[70] Claxton S and Fruttiger M 2003 Role of arteries in oxygen induced vaso-obliteration *Exp. Eye Res* **77** 305–11
[71] Ye X, Wang Y and Nathans J 2010 The Norrin/Frizzled4 signaling pathway in retinal vascular development and disease *Trends Mol. Med.* **16** 417–25
[72] Rattner A, Wang Y, Zhou Y, Williams J and Nathans J 2014 The role of the hypoxia response in shaping retinal vascular development in the absence of Norrin/Frizzled4 signaling *Investig. Ophthalmol. Vis. Sci.* **55** 8614–25
[73] Penn J S, Li S and Naash M I 2000 Ambient hypoxia reverses retinal vascular attenuation in a transgenic mouse model of autosomal dominant retinitis pigmentosa *Investig. Ophthalmol. Vis. Sci.* **41** 4007–13
[74] Rao S, Chun C, Fan J, Kofron J M, Yang M B, Hegde R S, Ferrara N, Copenhagen D R and Lang R A 2013 A direct and melanopsin-dependent fetal light response regulates mouse eye development *Nature* **494** 243–6
[75] Yu D Y and Cringle S J 2005 Retinal degeneration and local oxygen metabolism *Exp. Eye Res.* **80** 745–51
[76] Ivanova E, Yee C W and Sagdullaev B T 2016 Disruption in dopaminergic innervation during photoreceptor degeneration *J. Comp. Neurol.* **524** 1208–21

IOP Publishing

Photo Acoustic and Optical Coherence Tomography Imaging, Volume 3

Angiography: an application in vessel imaging

Ayman El-Baz and Jasjit S Suri

Chapter 3

Stripe noise removal and vessel segmentation of OCTA images

Xiyin Wu, Dongxu Gao, Davide Borroni, Savita Madhusudhan and Yalin Zheng

Optical coherence tomography angiography (OCTA) is a kind of retinal imaging technology that provides important information on fine retinal vessels in a non-invasive and non-contact way. The effective analysis of retinal blood vessels is valuable for the investigation and diagnosis of vascular and vascular-related diseases, for which accurate vessel segmentation is a vital first step. OCTA images are always affected by some stripe noises artifacts, which will impede correct segmentation and should be removed. To address this issue, we present a two-stage strategy in this chapter, in which stripe noise removal is addressed by image decomposition and vessel segmentation is solved by an active contours approach. Besides, a joint model is also given in this chapter to improve the speed of the algorithm while retaining the quality of the segmentation and destriping. The experimental results are presented on both simulated and real retinal imaging data, demonstrating the effective performance of the joint model for segmenting vessels from the OCTA images corrupted by stripe noise. Furthermore, the stripe direction influence is tested to prove the flexibility of the joint model.

3.1 Introduction

Retinal blood vessel health is integral to high-quality human vision and provides useful clinical information. Changes in the retinal vasculature have a close relationship with many ophthalmological and cardiovascular diseases, such as diabetic retinopathy (DR) and age-related macular degeneration (AMD) [1]. To acquire imagery of fine retinal vessels, the commonly used techniques are fluorescein angiography (FA) and indocyanine green angiography (ICGA) [2]. Both FA and ICGA require intravenous dye injections, which can have adverse side effects and still only provide information relating to superficial blood vessels. Recently,

doi:10.1088/978-0-7503-2060-3ch3

an innovative technology called optical coherence tomography angiography (OCTA) has emerged as an effective way of visualizing the retinal vessels up to the capillary level [3], which previously can only be achieved by invasive and potentially life-threatening angiographic techniques [4]. OCTA is a functional extension of OCT imaging, allowing us to visualize the microvasculature of the eye, without the need for invasive and potentially harmful dyes [3]. Unlike FA and ICGA, OCTA is a non-invasive, fast, depth-resolved approach. OCTA can fully visualize choroidal neovascularization in AMD and small retinal neovascularization in DR, which are difficult to identify with FA. Therefore, the usage of OCTA in the diagnosis of vascular diseases is expected to increase significantly soon.

The acquisition process of OCTA images is illustrated in figure 3.1. It can be viewed as a functional extension of well-established optical coherence tomography (OCT) imaging to visualize the microvasculature of the eye [5]. OCTA reconstructs blood vessels by detecting moving particles such as red blood cells in the tissue by deriving phase [6] or amplitude [7] variances from the repeated OCT scans at the same location. A 3D OCTA map can be produced by taking sequential cross-sectional OCTA scans of the scanned tissue [8]. For the ease of visualizing the retinal vessels, projections of the acquired 3D OCTA map into 2D *en face* images are often used. As shown in figures 3.1(d) and (e), retinal vessels at a certain depth of the retina can be generated by selecting certain layers of the retina from the projection. It has the advantage of acquiring three-dimensional datasets, allowing us to selectively view each layer of the retinal capillary plexuses and choroidal vasculature. For example, the vessels of the whole retina (WR), superficial vascular plexus (SVP),

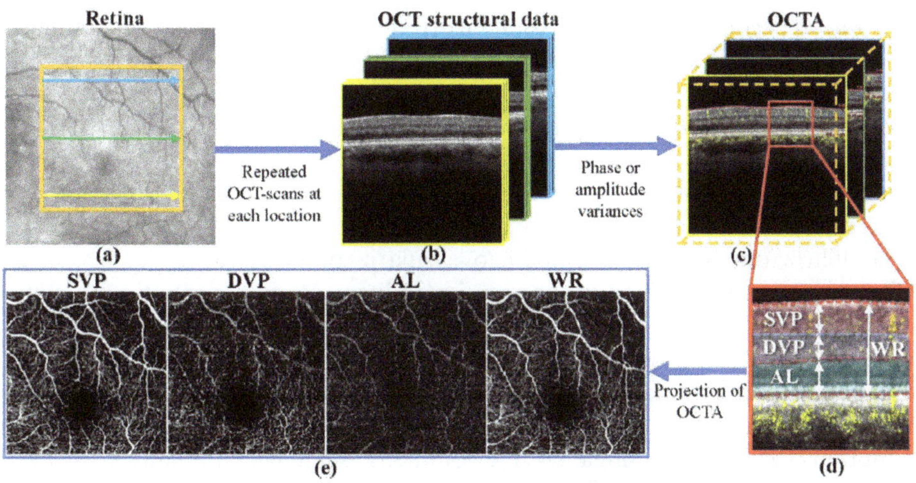

Figure 3.1. The acquisition process of an OCTA image. (a) Infrared image illustrating the scanning position of OCT. (b) OCT structural volume with repeated scans at each scanning location. (c) OCTA volume data (yellow dots) superimposed on the OCT image. OCTA is acquired by phase or amplitude variances from repeated OCT scans at each location. (d) A zoomed-in region of (c). (e) Projection of OCTA map by selecting certain layers: SVP, DVP, AL and WR.

deep vascular plexus (DVP) and avascular layer (AL) are most often used in practice:

- Whole retina (WR): from the internal limiting membrane to the Bruch's membrane.
- Superficial vascular plexus (SVP): from the internal limiting membrane to the inner plexiform layer.
- Deep vascular plexus (DVP): from the inner plexiform layer to the outer plexiform layer.
- Avascular layer (AL): from the outer plexiform layer to the Bruch's membrane.

Retinal vessel segmentation is the first and most important step to detect the retinal vasculature. The segmentation of retinal vessels is a valuable precursor for further processing and analysis, such as retinal image registration, feature extraction and localization of retinal structures like the fovea and the optic disk [9–12]. As a relatively new modality, few studies exist analyzing the retinal vessels in OCTA and the OCTA vessel segmentation is still at an early stage in its development. Eladawi *et al* [13] presented a joint Markov–Gibbs random field method to segment blood vessels from OCTA scans. The authors further estimated three local features from the segmented vessels to distinguish the status of DR patients [14]. Compared with OCTA, research in retinal vessel segmentation with color fundus images has a longer history. Many methods have been proposed over the last two decades, such as active contour models, wavelets methods, Gaussian mixture models, Adaboost and support vector machine, to name a few. One of the most commonly used segmentation approaches is active contours, such as the Chan–Vese model [15]. The Chan–Vese model applied global statistics to the extraction of objects and was useful for objects with homogeneous intensity. To handle the non-uniform situation of the Chan–Vese model, Sum *et al* [16] developed a modified version by combining the local image contrast into a level set based active contour. Bashir *et al* [17] grew a Ribbon of Twins active contour model to locate vessel edges under different conditions. Zhao *et al* [18] segmented retinal vessels by developing an infinite perimeter active counter model with hybrid region information. However, color fundus images cannot provide depth information, limiting the analysis of choroidal neovascularization and small retinal neovascularization.

There is a main limitation in the mentioned literature in contributing to clinical diagnostics. Previous works on OCTA images have neglected the unavoidable noise problem. Additional image noise can arise from the OCTA image acquisition, eye motion and image pre-processing strategies. One of the most common types of noise is white horizontal or vertical stripe noise (see figure 3.1), which results from patient eye motion [19]. During a single OCTA volumetric scan, which usually lasts several seconds (depending on the OCT devices, e.g. 3–5 s [20]), the involuntary eye movement of patients gives rise to motion artifacts, including tremors, micro-saccades and drift [20]. These eye motions cause intensity distribution shift within B-scans at each position, and produce visible horizontal stripes in the OCTA images [21]. These stripes artifacts affect the visualization and interpretation of OCTA images and even the clinical diagnosis, and need to be tackled before segmenting the retinal vessels to achieve effective, good-quality segmentation. Therefore, commercial OCTA devices adopt some motion correction methods to reduce this problem, such as Angioplex and Spectralis OCT (Heidelberg Engineering, Heidelberg, Germany) [22]. A scanning laser ophthalmoscope (SLO) has been used to resume the scan by adjusting the scan to the right location [23, 24]. Orthogonal registration combines the information in horizontal and vertical OCT volume scans to produce one OCTA map [25]. Although motion correction methods adopted in these devices can improve the scan quality, residual non-homogeneous lines and distorted lines remain. Therefore, new image processing solutions are required to address this inherent problem of OCTA.

To overcome this limitation, the stripe noise removal ('destriping') technique is introduced to alleviate this inherent nature of OCTA images during the acquisition process. Although there is scarce literature about removing stripe noise in OCTA images, images of other modalities suffer from similar noise and lots of related works have been explored. These techniques generally based on transform domain or spatial domain. The stripe noise is considered as specific frequencies in transform domain based methods. These methods are very simple to implement and are used widely in lots of imaging modalities [26–28]. However, most of their theory is based on the assumptions that the stripes are periodic and therefore their performance is limited in specific cases. The spatial based methods usually employ deep networks or construct an optimization model. The deep based spatial methods learn the characters of stripe noise in a supervised way and differentiate the stripe from the image structure from a global view [29]. The limitation of these methods is that they take a long time to train models, and the training often requires numerous images as well as their corresponding ground truth images, which are hard to obtain in real applications. The optimization based spatial methods consider stripe removal as an ill-posed inverse problem, in which the stripe free image and stripes are characterized based on prior information. These methods are making momentous progress in different applications, such as remote sensing images and microscopy imaging [30–32]. The destriping models based on optimization theory inspire us to take a similar approach to OCTA imaging.

In this chapter, we will first present a two-stage strategy for stripe noise removal and retinal vessel segmentation in OCTA images. This strategy removes stripe noise and Gaussian noise from the image before segmenting the denoised image. Then a joint model will be given to achieve the two tasks simultaneously. For stripe noise removal, the optimization based spatial theory inspires us to take a similar approach in OCTA imaging. For vessel segmentation, a good choice is active contour models, which can not only provide smooth and closed contours but also achieve subpixel accuracy on vessel boundaries.

3.2 Two-stage strategy (TSS)

A two-stage strategy is introduced for segmenting OCTA images affected by stripe noise. In this strategy, we first remove the stripe noise by an optimization based model and subsequently segment the retinal vessels with an active contour model.

3.2.1 The first stage: stripe noise removal of OCTA images

3.2.1.1 Image observation formula

Considering the stripe acquired along with the image content, we proceed with an image decomposition strategy, i.e., the given noisy image is decomposed to the desired clean image, the stripe noise and small noise. The model of the observed image can be formulated as:

$$I = C + S + N \tag{3.1}$$

where I is the given noisy image; C denotes the desired clean image; S is the additive stripe component; and N represents the linear assumption error and small noise. Stripe noise removal aims to estimate both C and S simultaneously from I.

3.2.1.2 Priors of stripe noise and clean OCTA image

As shown in figure 3.1, the stripe noise in OCTA images has a salient structural characteristic, showing in either the horizontal or vertical direction. To analyze the characteristic of stripe noise, we constructed an image with horizontal stripes (see figure 3.2(b)) and vertical stripes (seen figure 3.2(d)). It is found that the rank of horizontal stripe matrix is equal to 1 and that the rank of the vertical stripe matrix is also equal to 1. The rank value of the stripe matrix indicates that its spanned subspace can be expressed as a low-rank constraint, which is expressed as rank(S). Moreover, we studied the rank change before and after the OCTA image was degraded by stripe noise. A 256 × 256 OCTA image (see figure 3.2(a)) is corrupted by adding the horizontal stripes in figure 3.2(b) (or the vertical stripes in figure 3.2(d)) and the resulting degraded image is shown in figure 3.2(c) (or figure 3.2(e)). Their corresponding natural logarithms of singular values are depicted in figure 3.2(h) (or figure 3.2(j)). While the singular values of both the clean OCTA image and degraded image approximate to zero at rank 256 slowly, the singular value of the stripe image approximates to zero at rank 1 rapidly. The trends of clean

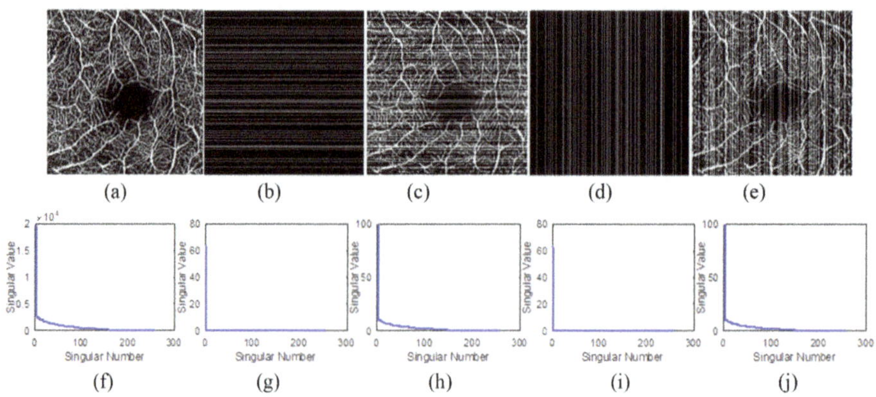

Figure 3.2. Low rank prior of stripes. (a) Simulated OCTA image. (b) Horizontal stripes. (c) Image (a) degraded by horizontal stripes (b). (d) Vertical stripes. (e) Image (a) degraded by Vertical stripes (d). (f)–(j) Singular values of (a)–(e).

and degraded OCTA images in figures 3.2(f) and (h) (or figure 3.2(j)) are similar and their ranks are both equal to 256. We can conclude that the rank of OCTA image is higher than the rank of stripes, and the low-rank constraint of stripe noise will not affect the image content.

Total variation (TV) is a widely used regularizer for reducing noise while preserving edges in images [33]. For OCTA images, anisotropic TV regularization can obtain sharper boundaries in the clean image, which is important for subsequent processing, such as vessel segmentation. As the stripe noise exists in the horizontal or vertical direction, the anisotropic TV regularization can adapt these two situations by loosening constraints on the stripe noise direction. The formulation of anisotropic TV regularization is given as $\|C\|_{\mathrm{TV}} = \|\nabla_x C\|_1 + \|\nabla_y C\|_1$, where $\|\cdot\|_1$ is L_1 norm indicating the sum of the absolute value of the matrix elements. ∇_x, ∇_y are the row and column derivative operators respectively.

3.2.1.3 Stripe noise removal model

As stated above, the low-rank constraint is used for the stripe component and the anisotropic total variation (TV) regularization is used to achieve sharper boundaries in the clean image, which will be important for segmenting vessels accurately. The stripe noise removal model [32] is given as follows:

$$\min_{C,S} \frac{1}{2} \|C + S - I\|_F^2 + \tau \|C\|_{\mathrm{TV}} + \lambda \mathrm{rank}(S) \tag{3.2}$$

where τ and λ are two positive regularization parameters balancing the three terms. Due to the non-convexity of the rank constraint, the nuclear norm is introduced to replace the low-rank constraint as its convex substituted function [34]. The nuclear norm based image decomposition model is given by

$$\min_{C,S} \frac{1}{2} \|C + S - I\|_F^2 + \tau \|C\|_{\mathrm{TV}} + \lambda \|S\|_* \tag{3.3}$$

The model (3.3) can optimize two variables C and S simultaneously by using an alternatively minimizing strategy as follows.

(a) **Stripe component** S: Fixing other arguments, S is evaluated by minimizing the following function:

$$\hat{S} = \operatorname*{argmin}_{S} \frac{1}{2} \|C + S - I\|_F^2 + \lambda \|S\|_* \tag{3.4}$$

which can be solved by the following soft-thresholding operation [35]:

$$\begin{cases} S^{k+1} = U\left(\mathrm{shrink}L_*\left(\sum,\lambda\right)\right)V^{\mathrm{T}} \\ \mathrm{shrink}L_*\left(\sum,\lambda\right) = \mathrm{diag}\left\{\max\left(\sum_{ii} -\lambda,\, 0\right)\right\}_i \end{cases} \tag{3.5}$$

where $Y - I = U\sum V^{\mathrm{T}}$ is the singular value decomposition of $Y - I^k$, and $\sum_{ii}$ is the diagonal element of the singular value matrix $\sum$.

(b) **Clean image** C: Keeping other arguments fixed and minimizing with respect to C, we have

$$\hat{C} = \underset{C}{\operatorname{argmin}} \frac{1}{2} \|C + S - I\|_F^2 + \tau \|C\|_{\mathrm{TV}} \tag{3.6}$$

where $\|C\|_{\mathrm{TV}}$ can be decomposed along with the directions x and y:

$$\|C\|_{\mathrm{TV}} = \tau_x \|\nabla_x C\|_1 + \tau_y \|\nabla_y C\|_1 \tag{3.7}$$

where the weight τ along directions x and y is different due to the directional characteristics of the stripe component. The alternative direction multiplier method (ADMM) is introduced to solve the problem (3.6) efficiently. We convert this problem into two sub-problems. Let $D_x = \nabla_x C$, $D_y = \nabla_y C$, $D = [D_x, D_y]^{\mathrm{T}}$, $\nabla = [\nabla_x, \nabla_y]^{\mathrm{T}}$, $\tau = [\tau_x, \tau_y]^{\mathrm{T}}$, (3.6) equals the following problem:

$$\{\hat{C}, \hat{D}\} = \underset{C,D}{\operatorname{argmin}} \frac{1}{2} \|C + S - I\|_F^2 + \tau \|D\|_1 + \frac{\gamma}{2} \left\| D - \nabla C - \frac{J}{\gamma} \right\|_F^2 \tag{3.8}$$

The C-related sub-problem is followed by

$$\hat{C} = \underset{C}{\operatorname{argmin}} \frac{1}{2} \|C + S - I\|_F^2 + \frac{\gamma}{2} \left\| D - \nabla C - \frac{J}{\gamma} \right\|_F^2 \tag{3.9}$$

We can obtain a closed form of (3.9) via the fast 2-D Fourier transform (FFT):

$$C^{k+1} = \mathcal{F}^{-1}\left(\frac{\mathcal{F}(I - S^{k+1} + \nabla^T(\gamma^k D^k - J^k))}{1 + \gamma^k(\mathcal{F}(\nabla))^2} \right) \tag{3.10}$$

The D-related sub-problem is followed by

$$\hat{D} = \underset{D}{\operatorname{argmin}}\, \tau \|D\|_1 + \frac{\gamma}{2} \left\| D - \nabla C - \frac{J}{\gamma} \right\|_F^2 \tag{3.11}$$

which can be solved by a soft shrinkage operator

$$\begin{cases} D^{k+1} = \mathrm{shrink}L_1\left(\nabla C^{k+1} + \dfrac{J^k}{\gamma^k}, \dfrac{\tau}{\gamma^k} \right) \\ \mathrm{shrink}L_1(r, \xi) = \dfrac{r}{|r|} * \max(r - \xi, 0). \end{cases} \tag{3.12}$$

Then, we update the Lagrangian multipliers and penalization parameters as follows:

$$\begin{aligned} J^{k+1} &= J^k + \gamma^k(\nabla C^{k+1} - D^{k+1}) \\ \gamma^{k+1} &= \gamma^k \cdot \rho \end{aligned} \tag{3.13}$$

3.2.2 The second stage: segmentation of vessels from stripe denoised images

Active contour models have demonstrated excellent performance in dealing with challenging segmentation problems including vessel segmentation [17, 18]. They can not only provide smooth and closed contours but also achieve subpixel accuracy on vessel boundaries. The Global Minimization of the Active Contour/Snake model (GMAC) [36] is introduced here for vessel segmentation. This model enhances the well-known Chan–Vese (CV) model [15] and provides a convex solution for it. GMAC obtains the segmentation results by detecting large image gradients and homogeneous intensities regions. The GMAC model can be formulated as the energy minimization problem below:

$$\begin{aligned} \min_{u,c_1,c_2} \int_\Omega g(x, y)|\nabla \mathrm{u}(x, y)|dxdy + \beta\int_\Omega u(x, y)(C(x, y) - c_1)^2 dxdy \\ +\beta\int_\Omega (1 - u(x, y))(C(x, y) - c_2)^2 dxdy \end{aligned} \tag{3.14}$$

where $u(x, y)$ is a characteristic function of a closed set Ω_B, B represents the boundaries of Ω, c_1 and c_2 are the average of $C(x, y)$ inside and outside of the segmented region, respectively, and β is a small, positive balancing parameter. The first term is the TV-norm of $u(x, y)$ with weighted edge indicator function $g(x, y)$. A fast minimization method based on a dual formulation of the TV-norm [36] is adopted to solve this problem.

(a) Region average intensity values c_1, c_2

Keeping other arguments fixed and minimizing with respect to c_1 and c_2, respectively. We have the equations for c_1 and c_2 [15]:

$$c_1 = \frac{\int_\Omega u(x, y)C(x, y)dxdy}{\int_\Omega u(x, y)dxdy} \tag{3.15}$$

$$c_2 = \frac{\int_\Omega (1 - u(x, y))C(x, y)dxdy}{\int_\Omega (1 - u(x, y))dxdy} \tag{3.16}$$

which can be evaluated directly by giving the average intensities inside and outside of the segmentation contour.

(b) Segmentation indicator function u

A convex regularization variational model is used according to [37]:

$$\begin{aligned}\{\hat{u}, \hat{v}\} = \underset{u,v}{\operatorname{argmin}}\ \|u\|_{\mathrm{TV},g} + \frac{1}{2\theta}\|u + v - C\|_F^2 \\ + \frac{\beta}{2}(C - v)\|C - c_1\|_F^2 + \frac{\beta}{2}(1 - C + v)\|C - c_2\|_F^2\end{aligned} \tag{3.17}$$

where the parameter θ is small so that we can approximate $C = u + v$. The function u denotes the geometric information, while v represents the texture information in the clean image C. Problem (3.17) can be solved by minimizing u and v iteratively.

The u-related sub-problem is followed by

$$\hat{u} = \underset{u}{\operatorname{argmin}}\ \|u\|_{\mathrm{TV},g} + \frac{1}{2\theta}\|u + v - C\|_F^2 \tag{3.18}$$

The solution of (3.18) can be achieved efficiently by a fast dual projection algorithm [38]. The derived solution is:

$$u^{k+1} = v^k - \theta \mathrm{div} p^{k+1} \tag{3.19}$$

where p is given by a fixed point method as follows:

$$p^{k+1} = \frac{p^k + \delta t \nabla(\mathrm{div}(p^k) - v/\theta)}{1 + \dfrac{\delta t}{g(x, y)}|\nabla(\mathrm{div}(p^k) - v/\theta)|} \tag{3.20}$$

where $\delta t \leqslant 1/8$ is the temporal step.

The v-related sub-problem is followed by

$$\hat{v} = \frac{1}{2\theta}\|u + v - C\|_F^2 + \frac{\beta}{2}(C - v)\|C - c_1\|_F^2 + \frac{\beta}{2}(1 - C + v)\|C - c_2\|_F^2. \tag{3.21}$$

The v-minimization can be achieved through the following update:

$$v^{k+1} = \min\{\max\{u(x, y) - \frac{\beta\theta}{2}[(C(x, y) - c_2)^2 - (C(x, y) - c_1)^2], 0\}, 1\}. \tag{3.22}$$

u in GMAC can be solved in a similar way.

Thus, the vessel segmentation from stripe removed images is achieved by solving problem (3.3) to obtain the stripe removed and denoised image C, followed by segmenting C by solving problem (3.14). The overall algorithm of the two stages strategy is presented in algorithm 3.1.

Algorithm 3.1. Segmenting images corrupted by stripe noises by the two stages strategy (TSS).

Require: Image I with stripe noise, parameters $\tau, \lambda, \beta, \gamma, \theta, \rho$.
Ensure: Clean image C, stripe component S and segmentation indicator function u.
1: Initialize: set $J^1 = 0,\ C^1 = I,\ u^1 = 0,\ v^1 = 0$
Stage 1: stripe noise removal
2: for $k = 1: N$ do
3: Solve (3.5) for S^{k+1};
4: Compute C^{k+1} by solving (3.10);
5: Obtain D^{k+1} by solving (3.12);
6: Update Lagrangian multipliers J^{k+1} and penalization parameters γ^{k+1} via (3.13);
7: end for
Stage 2: segmentation of vessels from stripe denoised images
8: for $k = 1: N$ do
9: Compute ${c_1}^{k+1}$ and ${c_2}^{k+1}$ by solving (3.15) and (3.16);
10: Compute u^{k+1} by solving (3.19)
11: Compute v^{k+1} by solving (3.22)
12: end for

3.3 Joint destriping and segmentation of OCTA images (JDS)

In this section, a joint model [39] is given for simultaneously removing stripe noise and segmenting images. The model tackles the problems of image segmentation and considers the possible presence of stripe noise in a single model. The joint model is formed by replacing the function C in the segmentation model (3.4) with the stripe-removed and denoised image term. Furthermore, the constraints of the decomposition problem (3.3) should be included in the joint model. The illustration of the joint model is shown in figure 3.3. Building the constraints into the segmentation model, the joint minimization model is presented as follows:

$$\begin{aligned}\min_{C,S,u,c_1,c_2} \frac{1}{2} \|C + S - I\|_F^2 + \tau \|C\|_{\mathrm{TV}} + \lambda \|S\|_* \\ + \alpha \|u\|_{\mathrm{TV},g} + \frac{\beta}{2} u \|C - c_1\|_F^2 + \frac{\beta}{2}(1 - u)\|C - c_2\|_F^2\end{aligned} \tag{3.23}$$

where $\|\cdot\|_{\mathrm{TV},g}$ is the TV-norm with weighted function, α, β are small, positive balancing parameters. If α and β equal to 0, then model (3.23) converts to the destriping model (3.3). The segmentation terms (fourth, fifth and sixth terms of (3.23)) obtain the intensity and spatial information from the stripe removed image C. To solve the joint model, each of the arguments is solved in turn. The detailed optimization process is presented as follows.

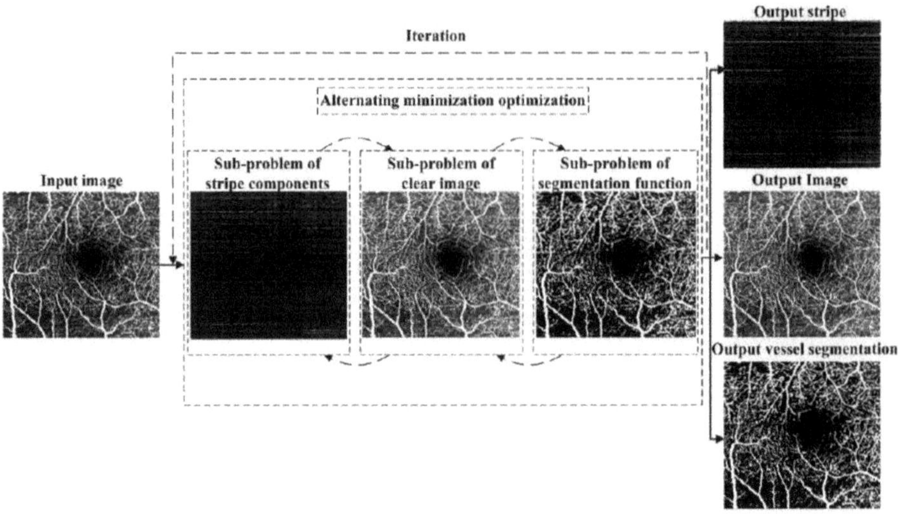

Figure 3.3. Illustration of our joint model for segmentation images with stripe noise.

(a) Stripe component S
Keeping other arguments fixed, S is evaluated by minimizing the following function:

$$\hat{S} = \underset{S}{\operatorname{argmin}} \frac{1}{2} \|C + S - I\|_F^2 + \lambda \|S\|_* \tag{3.24}$$

(3.24) can be solved by the soft-thresholding operation [35]:

$$\begin{cases} S^{k+1} = U\left(\operatorname{shrink}L_*\left(\sum, \lambda\right)\right)V^{\mathrm{T}} \\ \operatorname{shrink}L_*\left(\sum, \lambda\right) = \operatorname{diag}\left\{\max\left(\sum_{ii} - \lambda, 0\right)\right\}_i \end{cases} \tag{3.25}$$

(b) Clean image C
Fixing other arguments and minimizing with respect to C, we have

$$\begin{aligned} \hat{C} = \underset{C}{\operatorname{argmin}} & \frac{1}{2} \|C + S - I\|_F^2 + \tau \|C\|_{\mathrm{TV}} \\ & + \frac{\beta}{2} u \|C - c_1\|_F^2 + \frac{\beta}{2}(1 - u)\|C - c_2\|_F^2 \end{aligned} \tag{3.26}$$

Problem (3.26) is converted into two sub-problems. Let $D_x = \nabla_x C$, $D_y = \nabla_y C$, $D = [D_x, D_y]^{\mathrm{T}}$, $\nabla = [\nabla_x, \nabla_y]^{\mathrm{T}}$, $\tau = [\tau_x, \tau_y]^{\mathrm{T}}$, (3.26) equals the following problem:

$$\{\hat{C}, \hat{D}\} = \underset{C,D}{\operatorname{argmin}} \frac{1}{2} \|C + S - I\|_F^2 + \tau \|D\|_{\text{TV}} + \frac{\beta}{2} u \|C - c_1\|_F^2 + \frac{\beta}{2}(1 - u)\|C - c_2\|_F^2 + \frac{\gamma}{2} \left\| D - \nabla C - \frac{J}{\gamma} \right\|_F^2 \tag{3.27}$$

The C-related sub-problem is followed by

$$\hat{C} = \underset{C}{\operatorname{argmin}} \frac{1}{2} \|C + S - I\|_F^2 + \tau \|C\|_{TV} + \frac{\beta}{2} u \|C - c_1\|_F^2 + \frac{\beta}{2}(1 - u)\|C - c_2\|_F^2 + \frac{\gamma}{2} \left\| D - \nabla C - \frac{J}{\gamma} \right\|_F^2 \tag{3.28}$$

We can obtain a closed form of (3.28) via the fast 2-D Fourier transform (FFT):

$$C^{k+1} = \mathcal{F}^{-1}\left(\frac{\mathcal{F}(I - S^{k+1} + \beta u^k c_1^k + \beta(1 - u^k)c_2^k + \nabla^T(\gamma^k D^k - J^k))}{1 + \beta + \gamma^k(\mathcal{F}(\nabla))^2} \right) \tag{3.29}$$

The D-related sub-problem is followed by

$$\hat{D} = \underset{D}{\operatorname{argmin}}\, \tau \|D\|_{\text{TV}} + \frac{\gamma}{2} \left\| D - \nabla C - \frac{J}{\gamma} \right\|_F^2 \tag{3.30}$$

which can be solved by a soft shrinkage operator

$$\begin{cases} D^{k+1} = \text{shrink}L_1\left(\nabla C^{k+1} + \dfrac{J^k}{\gamma^k}, \dfrac{\tau}{\gamma^k} \right) \\ \text{shrink}L_1(r, \xi) = \dfrac{r}{|r|} * \max(r - \xi, 0). \end{cases} \tag{3.31}$$

(c) Region average intensity values c_1, c_2

Fixing other arguments, c_1 and c_2 are evaluated as follows [15]:

$$c_1 = \frac{\int_\Omega u(x, y)C(x, y)dxdy}{\int_\Omega u(x, y)dxdy} \tag{3.32}$$

$$c_1 = \frac{\int_\Omega (1 - u(x, y))C(x, y)dxdy}{\int_\Omega (1 - u(x, y))dxdy}. \tag{3.33}$$

(d) Segmentation indicator function u

With respect to u and fixing the other arguments, we have

$$\hat{u} = \underset{u}{\operatorname{argmin}}\, \alpha \, \|u\|_{\mathrm{TV},g} + \frac{\beta}{2} u \, \|C - c_1\|_F^2 + \frac{\beta}{2}(1 - u)\|C - c_2\|_F^2 \tag{3.34}$$

Similar to (3.17), a convex regularization variational model is adopted based on [37]:

$$\begin{aligned}\{\hat{u}, \hat{v}\} = \underset{u,v}{\operatorname{argmin}}\, \alpha \, \|u\|_{\mathrm{TV},g} + \frac{1}{2\theta} \, \|u + v - C\|_F^2 \\ + \frac{\beta}{2}(C - v)\|C - c_1\|_F^2 + \frac{\beta}{2}(1 - C + v)\|C - c_2\|_F^2\end{aligned} \tag{3.35}$$

where the parameter θ is small. Problem (3.35) can be solved by minimizing u and v iteratively.

The u-related sub-problem is followed by

$$\hat{u} = \underset{u}{\operatorname{argmin}}\, \alpha \, \|u\|_{\mathrm{TV},g} + \frac{1}{2\theta} \, \|u + v - C\|_F^2 \tag{3.36}$$

The derived solution is given in the following:

$$u^{k+1} = v^k - \alpha\theta \mathrm{div} p^{k+1} \tag{3.37}$$

where p is given by a fixed point method as follows:

$$p^{k+1} = \frac{p^k + \delta t \nabla(\mathrm{div}(p^k) - v/(\alpha\theta))}{1 + \dfrac{\delta t}{g(x, y)} |\nabla(\mathrm{div}(p^k) - v/(\alpha\theta))|}. \tag{3.38}$$

The v-related sub-problem is followed by

$$\hat{v} = \frac{1}{2\theta} \, \|u + v - C\|_F^2 + \frac{\beta}{2}(C - v)\|C - c_1\|_F^2 + \frac{\beta}{2}(1 - C + v)\|C - c_2\|_F^2. \tag{3.39}$$

The solution of (3.39) is given in the following:

$$v^{k+1} = \min\{ \max\{u(x, y) - \frac{\beta\theta}{2}[(C(x, y) - c_2)^2 - (C(x, y) - c_1)^2], 0\}, 1\}. \tag{3.40}$$

Then, the Lagrangian multipliers and penalization parameters are updated as follows:

$$\begin{aligned} J^{k+1} &= J^k + \gamma^k(\nabla C^{k+1} - D^{k+1}) \\ \gamma^{k+1} &= \gamma^k \cdot \rho \end{aligned} \tag{3.41}$$

The overall algorithm of joint model JDS is presented in algorithm 3.2.

Algorithm 3.2. Segmenting images corrupted by stripe noises by the joint model (JDS).

Require: Image I with stripe noise, parameters τ, λ, α, β, γ, θ, ρ.
Ensure: Clean image C, stripe component S and segmentation indicator function u.
1: Initialize: set $J^1 = 0$, $C^1 = I$, $u^1 = 0$, $v^1 = 0$
2: for $k = 1$: N do
3: Solve S^{k+1} by using (3.25);
4: Compute C^{k+1} by solving (3.29);
5: Solve D^{k+1} using (3.31);
6: Obtain ${c_1}^{k+1}$ and ${c_2}^{k+1}$ via (3.32) and (3.33);
7: Compute u^{k+1} by solving (3.37);
8: Solve (3.40) for v^{k+1};
9: Update Lagrangian multipliers J^{k+1} and γ^{k+1} via (3.41);
10: end for

3.4 Results

The proposed algorithms are evaluated in two parts: the effectiveness of destriping and the effectiveness of segmentation. All experiments are run in MATLAB® (R2018a) (Mathworks, MA) on a desktop with 8 GB RAM, Intel (R) Core (TM) i5-7500 CPU @ 3.40 GHz.

3.4.1 Datasets

All the compared methods are extensively evaluated for their effectiveness and efficiency for destriping and segmenting OCTA images. In this section, a brief introduction of the used datasets is provided. All images used in this work were collected with regulatory approvals and patients' consent as appropriate.

(a) **Real dataset**: A real OCTA dataset consists of 85 images in the size of 840*840 that were collected from St Paul's Eye Unit, Royal Liverpool University Hospital, UK. Each image is taken in a 3 mm × 3 mm field of view centered on the fovea from the internal limiting membrane (ILM) to the inner plexiform layer (IPL), i.e., they are taken from superficial vascular plexus (SVP) layer. An example is shown in figure 3.6(a). All images were acquired using SPECTRALIS OCT (Heidelberg Engineering, Heidelberg, Germany). The central lines of vessels in all images were manually annotated using an in-house program by an ophthalmologist after proper training on using the program, as shown in figure 3.4(b).

(b) **Simulated dataset:** All the OCTA images obtained in clinical settings are corrupted by stripe noise. There is no clean image readily for the evaluation of the performance. Following the general practice in image denoising and enhancement, we created a synthetic dataset for comparison. A collection of 10 fluorescein angiography (FA) images from patients with diabetes by

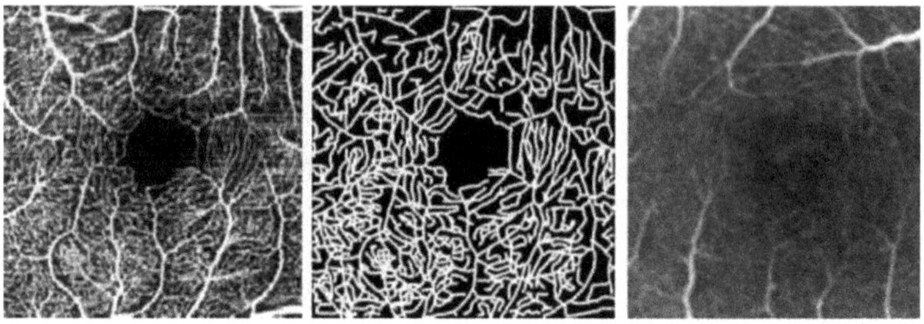

(a) Real dataset (b) Centerline annotation of (a) (c) Simulated dataset

Figure 3.4. Examples of synthetic and real datasets. The central lines in (b) are thickened for better visualization.

HRA2 (Heidelberg Engineering, Heidelberg, Germany) was used as the clean images. All the images have excellent quality as confirmed by visual inspection of clinicians [40]. A region of 256 × 256 pixels centered at the fovea was cropped and used in the experiments, as shown in figure 3.4(c). This dataset is used to validate the effect of destriping.

3.4.2 Effectiveness of destriping

To test the effectiveness of destriping, two classical denoising methods (WMF and WL), two destriping methods of other modalities (FFT and WFT) and two methods described in this chapter (TSS and JDS) are compared and tested. These models are briefly described as follows:

- WMF: The weighted median filter denoising method [41].
- WL: The wavelet denoising method [42].
- FFT: The fast Fourier transform based destriping method [27].
- WFT: The wavelet-Fourier combined transform based stripe artifact removal method [26].
- TSS: The two-stage strategy presented in this chapter by destriping model (3.3) followed by the segmentation model (3.14).
- JDS: The joint model (3.23) given in this chapter for destriping and segmentation simultaneously.

Two evaluation metrics are utilized to test the above methods on simulated dataset and real dataset, i.e., the peak signal-to-noise ratio (PSNR) and the structural similarity (SSIM) [43]. The two metrics are calculated by using equations (3.42) and (3.43), respectively.

$$PSNR = 10 \cdot \log_{10}(\frac{MAX_I^2}{MSE}) \tag{3.42}$$

$$SSIM(x, y) = \frac{(2\mu_x\mu_y + a_1)(2\sigma_{xy} + a_2)}{({\mu_x}^2 + {\mu_y}^2 + a_1)({\sigma_x}^2 + {\sigma_y}^2 + a_2)} \tag{3.43}$$

where MAX_I is the maximum pixel value of the noise-free image and MSE is the mean squared error between the noise-free image and noisy image. x and y are two windows of an image. μ_x, μ_y, σ_x^2, σ_y^2 are the average of x, y and the variance of x, y, respectively. σ_{xy} is the covariance of x and y. a_1, a_2 are two variables.

3.4.2.1 Stripe direction influence

The stripe direction of OCTA image is usually either vertical or horizontal. The effectiveness of stripe noise removal in two stripe directions is tested. We choose an example simulated FA image and add the same stripes in the vertical and horizontal directions separately, as shown in figures 3.5(b) and 3.6(b). Then six destriping methods are used to remove the stripe noise, their visual results are presented in figures 3.5(c)–(h) and figures 3.6(c)–(h). As WMF and WL are mainly designed for random pepper noise or Gaussian noise, they fail to remove the stripe noise in both vertical and horizontal directions. Although FFT and WFT can remove most of the stripe noise, they generate some excessive destriping results with significant blurring artifacts along the stripe direction. TSS and JDS can remove the stripes in both directions and retain stripe-free areas well.

The corresponding quantitative results of all destriping methods are given in tables 3.1 and 3.2 in terms of PSNR and SSIM, in which the highest PSNR and SSIM values in each intensity scenario of simulated experiments are marked in bold. Five different levels of stripes are added to the original image to test the robustness of our methods. The intensity denotes the mean absolute value of the stripe lines. TSS and JDS achieve better performance compared to other methods according to tables 3.1 and 3.2. This demonstrates that the destriping model stated in this chapter is effective to remove the stripe noise in both vertical and horizontal directions.

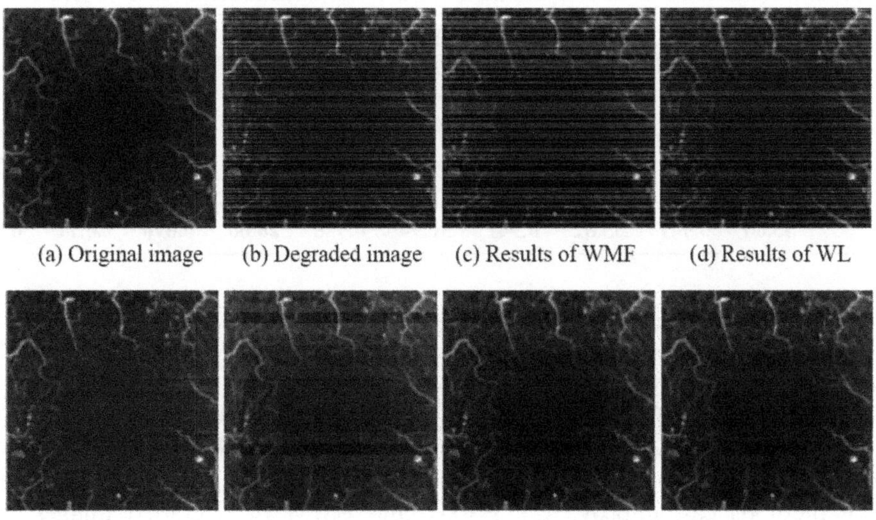

(a) Original image (b) Degraded image (c) Results of WMF (d) Results of WL

(e) Results of FFT (f) Results of WFT (g) Results of TSS (h) Results of JDS

Figure 3.5. Simulated results of horizontal stripe noise removal. The intensity of the added stripe noise is 40 in figure 3.5(b).

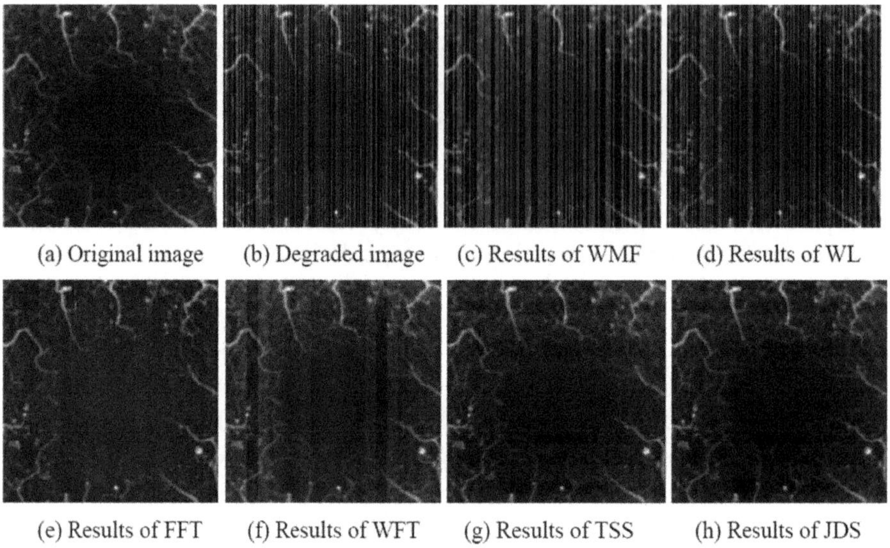

Figure 3.6. Simulated results of vertical stripe noise removal. The intensity of added stripe noise is 40 in figure 3.6(b).

Table 3.1. Quantitative simulated results of horizontal stripe noise removal.

	Stripe noise in the horizontal direction									
	Intensity = 10		Intensity = 20		Intensity = 30		Intensity = 40		Intensity = 50	
Method	PSNR	SSIM	PSNR	SSIM	PSNR	SSIM	PSNR	SSIM	PSNR	SSIM
Degraded	26.65	0.636	21.24	0.360	18.29	0.230	16.33	0.161	14.90	0.120
WMF	26.78	0.624	21.27	0.377	18.15	0.254	16.09	0.187	14.60	0.147
WL	29.81	0.763	22.99	0.447	19.35	0.266	16.99	0.177	15.42	0.128
FFT	24.48	0.821	25.53	0.821	24.25	0.771	22.99	0.717	22.28	0.677
WFT	31.77	0.941	27.33	0.884	24.40	0.828	22.28	0.779	20.63	0.735
TSS	34.99	0.945	33.81	**0.942**	31.31	0.923	28.61	0.896	26.03	0.858
JDS	**35.00**	**0.947**	**33.82**	**0.942**	**31.94**	**0.928**	**28.66**	**0.896**	**26.11**	**0.859**

3.4.2.2 Simulated experiments

To simulate stripe noise in OCTA images, we will add some synthetic stripes to the clean images in the simulated dataset. We only test horizontal stripes in this section and the intensities of added stripes are the same as section 3.4.2.1. Table 3.3 reports the destriping results in terms of PSNR and SSIM on the simulated dataset. Almost all the methods can improve the PSNR and SSIM of degraded images after destriping, while WMF and WL decrease the image quality when the added stripe noise is weak (intensity equals 10). It can also be observed from table 3.3 that the

Table 3.2. Quantitative simulated results of vertical stripe noise removal.

	Stripe noise in the vertical direction									
	Intensity = 10		Intensity = 20		Intensity = 30		Intensity = 40		Intensity = 50	
Method	PSNR	SSIM	PSNR	SSIM	PSNR	SSIM	PSNR	SSIM	PSNR	SSIM
Degraded	26.63	0.635	21.22	0.360	18.27	0.230	16.31	0.162	14.88	0.121
WMF	26.51	0.620	21.07	0.371	17.98	0.250	15.94	0.185	14.47	0.145
WL	29.65	0.744	23.11	0.425	19.36	0.249	16.96	0.164	15.29	0.118
FFT	22.85	0.793	23.67	0.796	23.60	0.765	23.09	0.724	22.98	0.697
WFT	32.76	0.947	28.55	0.898	25.23	0.840	22.91	0.788	21.14	0.743
TSS	34.69	**0.955**	33.45	0.946	31.77	0.930	28.87	0.901	26.36	**0.862**
JDS	**34.70**	**0.955**	**33.57**	**0.949**	**31.78**	**0.931**	**29.04**	**0.902**	**26.40**	**0.862**

Table 3.3. Quantitative results (PSNR (dB) and SSIM values) on the simulated dataset.

	Stripe noise									
	Intensity = 10		Intensity = 20		Intensity = 30		Intensity = 40		Intensity = 50	
Method	PSNR	SSIM	PSNR	SSIM	PSNR	SSIM	PSNR	SSIM	PSNR	SSIM
Degraded	29.36	0.848	23.74	0.634	20.72	0.480	18.73	0.376	17.29	0.304
WMF	26.59	0.701	23.93	0.592	21.51	0.500	19.61	0.422	18.10	0.366
WL	26.57	0.668	25.91	0.638	23.62	0.543	21.24	0.439	19.37	0.353
FFT	34.71	0.939	27.97	0.870	25.86	0.821	26.20	0.804	23.66	0.747
WFT	32.01	0.961	30.72	0.952	29.06	0.934	27.25	0.909	25.96	0.910
TSS	32.49	**0.942**	31.67	0.939	30.47	0.934	28.91	0.923	27.33	**0.911**
JDS	**32.50**	**0.942**	**31.76**	**0.940**	**30.66**	**0.935**	**28.92**	**0.924**	**27.36**	**0.911**

results of JDS outperform the competitors for all of the experiments under five different levels of stripes.

3.4.2.3 Real experiments

The real OCTA stripe images are tested in this subsection. We choose a representative image (see figure 3.7) and zoom in four regions of this image, i.e. R1, R2, R3 and R4. It is shown that TSS and JDS can remove most stripe noise, while WMF and WL can just remove some random noise, and FFT and WFT remove a little stripe noise, but also add some noises in the image.

3.4.3 Effectiveness of segmentation

The effectiveness of segmentation is tested on two aspects: the destriping influence on vessel segmentation and the vessel segmentation results. All the segmentation models to be compared and tested in this section are introduced as follows:

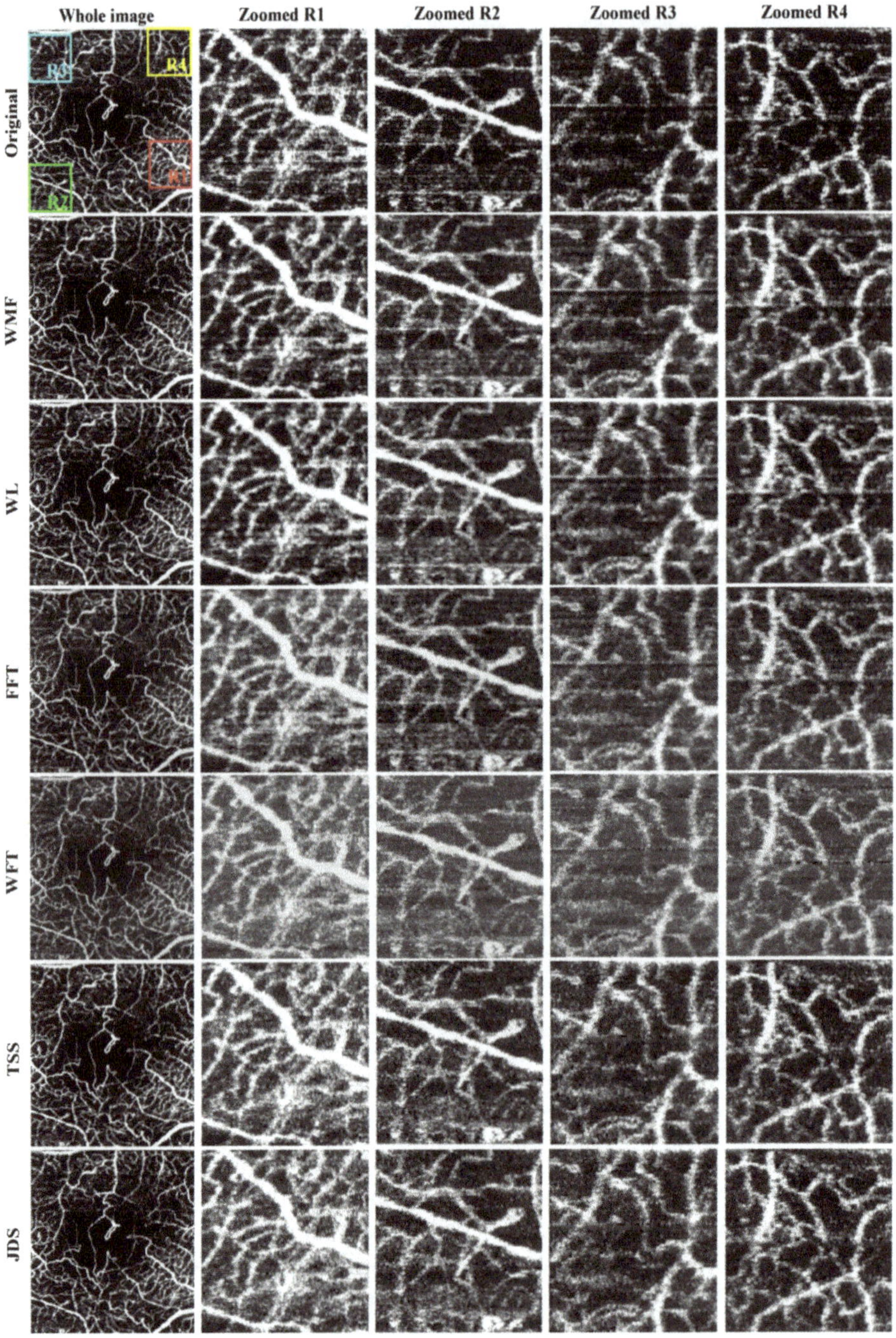

Figure 3.7. Illustration of real OCTA destriping/denoising results. The first row shows the original OCTA image and its four zoomed regions. The destriping results of WMF, WL, FFT, WFT, TSS and JDS are shown from the second to seventh rows, respectively. The second to fifth columns show the zoomed regions R1, R2, R3 and R4 of the whole image and their corresponding destriping results, respectively.

- CV: The Chan–Vese (CV) model of active contours without edges [15]
- GMAC: The Global Minimization of the Active Contour/Snake model (GMAC) [36]
- TSS: The two-stage strategy presented in this chapter by the destriping model (3.3) followed by the segmentation model (3.14).
- JDS: The joint model (3.23) given in this chapter for destriping and segmentation simultaneously.

All of the vessel segmentation results are quantitatively evaluated on two metrics: accuracy (Acc) and the area under the receiver operating characteristic curve (AUC) [18]. In particular, AUC is a better metric for the overall performance measurement for the imbalanced problem, i.e., in the vessel segmentation problem, the vessel pixels are typically much fewer than the background pixels. The computations of Acc and AUC are based on the centerline of the segmentation results and the centerline annotation (see figure 3.4(b)). They are calculated as follows:

$$Acc = \frac{TP + TN}{TP + FP + TN + FN} \tag{3.44}$$

$$AUC = \frac{1}{2}\left(\frac{TN}{TN + FP} + \frac{TP}{TP + FN}\right) \tag{3.45}$$

where TP, FP, TN, FN indicate the true positive, false positive, true negative and false negative, respectively.

3.4.3.1 Destriping influence on vessel segmentation

The destriping influence on vessel segmentation for the real OCTA dataset is investigated in the following. This is an indirect method to show the effectiveness of different destriping methods. Here we adopt two basic segmentation methods: CV [7] and GMAC [14]. The source codes of these segmentation algorithms are acquired from the authors or freely available on the Internet. These two segmentation methods were utilized to segment the original noisy OCTA images and the destriping images produced by different destriping methods (FFT, WFT, TSS and JDS). WMF and WL are denoising methods and will not be tested in this section. It should be noticed that we just utilize the destriping results for both TSS and JDS, the segmentation results of these two methods will be compared in section 3.4.3.2. The centerlines of the vessels from the segmentation were detected and the central lines of the segmented SVP images are compared to the centerlines from manual annotations. Quantitative results are given in table 3.4 in terms of Acc and AUC. After destriping, almost all the Acc and AUC values improve, which demonstrates the usefulness of destriping for vessel segmentation. JDS performs best in terms of Acc and AUC. It improves Acc from 0.7851, 0.7705 to 0.7879, 0.7753 and AUC from 0.6367, 0.6221 to 0.6403, 0.6289, respectively. Visual comparison results are illustrated in figure 3.8. It can be observed that the segmentation performance is improved for all the segmentation methods after removing the stripe noise by JDS.

Table 3.4. Destriping influence on vessel segmentation (CV) in terms of Acc and AUC.

	CV		GMAC	
Method	Acc	AUC	Acc	AUC
Original	0.7851	0.6367	0.7705	0.6221
FFT	0.7825	0.6352	0.7706	0.6223
WFT	0.7819	0.6349	0.7705	0.6222
SD	0.7874	0.6389	0.7746	0.6287
JDS	**0.7879**	**0.6403**	**0.7753**	**0.6289**

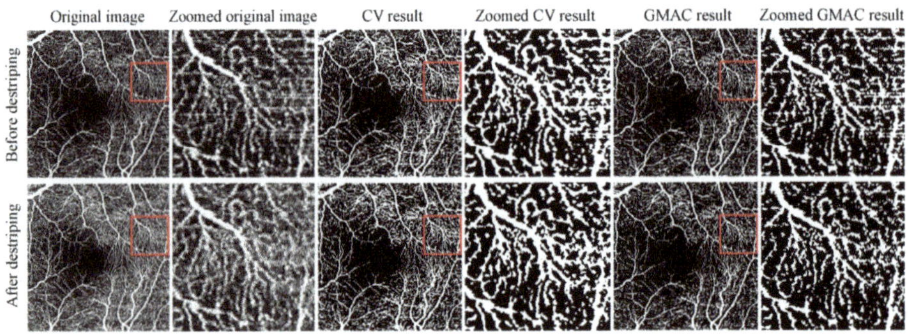

Figure 3.8. Examples of the influence of destriping on vessel segmentation in real OCTA images by different segmentation methods. From left to right: original image, a zoomed region of the original image, vessel segmentation results of CV, a zoomed region of CV result, vessel segmentation results of GMAC, a zoomed region of GMAC result. From top to bottom: before destriping, after destriping by JDS.

Table 3.5. Average accuracy values and CPU times of OCTA image segmentation.

Method	Acc	AUC	Times
CV	0.7851	0.6367	177.82
GMAC	0.7705	0.6221	12.01
TSS	0.7743	0.6285	96.68
JDS	**0.7877**	**0.6369**	**65.23**

3.4.3.2 Vessel segmentation results

The average accuracy values and CPU times of TSS and JDS are listed in table 3.5. The joint model JDS can achieve higher Acc and AUC compared to the other three models, and spend less time than CV and TSS. We also give two segmentation examples in figure 3.9. JDS can generate competitive segmentation results to TSS, GMAC and CV efficiently.

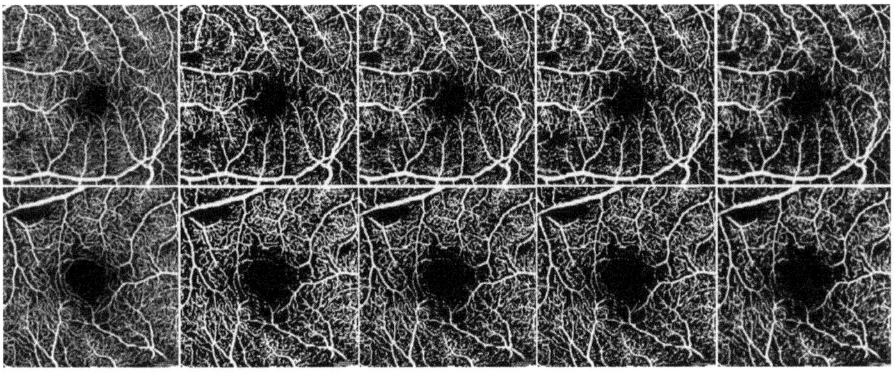

(a) Image (b) CV result (c) GMAC result (d) TSS result (e) JDS result

Figure 3.9. Illustration of OCTA segmentation results.

3.5 Discussion and conclusions

OCTA is a new, non-invasive imaging technology that can provide blood vessel maps at different depths. As an important element of retinal vasculature detection, vessel segmentation separates a retinal vessel from its background and can be applied in other vascular analysis. During the OCTA vessel segmentation, the unavoidable stripe noise arose from the OCTA image acquisition affects the accuracy of results. To address this issue, a two-stage strategy TSS and a joint model JDS are presented in this chapter to segment retinal blood vessels in OCTA images corrupted by stripe noise.

Both of TSS and JDS are based on the image decomposition and the active contour theory. The TSS removed the stripe noise by a low-rank representation model utilizing the low-rank constraints for stripes and TV regularization for preserving vessel edges, and then segmented the retinal blood vessels by an active contour model called GMAC. The JDS combined the models of stripe noise removal and vessel segmentation. The performance of two models is tested on a simulated FA dataset and a real OCTA dataset from the Royal Liverpool University Hospital. The quantitative, visual and efficiency comparison results of destriping and segmentation showed that both two models presented in this chapter can solve the two problems efficiently, with JDS solving the problems simultaneously, faster and more accurately.

The stripe direction of the real OCTA image depends on its scan direction and is usually either vertical or horizontal. TSS and JDS are able to remove stripe noise in both two directions. The experimental results in section 3.4.2.1 showed their effectiveness and practical value of both TSS and JDS with preferable PSNR and SSIM values. To further prove the effectiveness of the destriping results, the results of destriping influence on vessel segmentation are compared in section 3.4.3.1. The results from four vessel segmentation methods proved that destriping is useful to improve the segmentation performance and JDS can provide the highest improvement compared to other methods in terms of Acc and AUC.

All parameters used in TSS and JDS are empirically chosen in the experiments. The performance could be further improved if these parameters are fine 'tuned'. On the other hand, we have implemented the algorithms by using MATLAB®, which is by no means optimal in terms of speed. Improvement can still be made by parallel processing and GPU if needed.

In conclusion, this chapter presented a two-stage strategy TSS and a joint model JDS for segmenting retinal blood vessels in OCTA images corrupted by stripe noise. The experimental results demonstrate the effectiveness and efficiency of TSS and JDS in comparison to classic denoising/destriping and segmentation models. Besides, the destriping influence on vessel segmentation is also proved in this chapter. It is believed that our work can inspire a way to consider the inclusion of stripe removal in clinical vessel analysis models.

References

[1] Sheng B, Li P, Mo S, Li H, Hou X, Wu Q, Qin J, Fang R and Feng D D 2018 Retinal vessel segmentation using minimum spanning superpixel tree detector *IEEE Trans. Cybern.* **99** 1–13

[2] De-Carlo T E, Romano A, Waheed N K and Duker J S 2015 A review of optical coherence tomography angiography (OCTA) *Int. J. Retin. Vitr.* **1** 5

[3] Spaide R F, Fujimoto J G, Waheed N K, Sadda S R and Staurenghi G 2018 Optical coherence tomography angiography *Prog. Retin. Eye Res.* **64** 1–55

[4] Heimann H and Kellner U 2011 *Atlas of Fundus Angiography* (New York: Thieme)

[5] Venugopal J *et al* 2018 Repeatability of vessel density measurements of optical coherence tomography angiography in normal and glaucoma eyes *Br. J. Ophthalmol.* **102** 352–7

[6] Fingler J, Readhead C, Schwartz D and Fraser S E 2008 Phase-contrast OCT imaging of transverse flows in the mouse retina and choroid *Investig. Ophthalmol. Vis. Sci.* **49** 5055–9

[7] Jia Y, Tan O, Tokayer J, Potsaid B, Wang Y, Liu J J and Huang D 2012 Split-spectrum amplitude-decorrelation angiography with optical coherence tomography *Opt. Express* **20** 4710–25

[8] Tan A C S, Tan G S, Denniston A K, Keane P A, Ang M, Milea D, Chakravarthy U and Cheung C M G 2018 An overview of the clinical applications of optical coherence tomography angiography *Eye* **32** 262–86

[9] Zhu C, Zou B, Zhao R, Cui J, Duan X, Chen Z and Liang Y 2017 Retinal vessel segmentation in colour fundus images using extreme learning machine *Comput. Med. Imaging Graph.* **55** 68–77

[10] Martinez-Perez M E, Highes A, Stanton A V, Thorn S A, Chapman N, Bharath A A and Parker K H 2002 Retinal vascular tree morphology: a semi-automatic quantication *IEEE Trans. Biomed. Eng.* **49** 912–7

[11] Niemeijer M, Van Ginneken B, Staal J, Suttorp-Schulten M S and Abramoff M D 2005 Automatic detection of red lesions in digital color fundus photographs *IEEE Trans. Med. Imaging* **24** 584–92

[12] Zana F and Klein J C 1999 A multimodal registration algorithm of eye fundus images using vessels detection and Hough transform *IEEE Trans. Med. Imaging* **18** 419–28

[13] Eladawi N, Elmogy M, Helmy O, Aboelfetouh A, Riad A, Sandhu H, Schaal S and El-Baz A 2017 Automatic blood vessels segmentation based on different retinal maps from OCTA scans. *Comput. Biol. Med.* **89** 150–61

[14] Eladawi N, Elmogy M, Khalifa F, Ghazal M, Ghazi N, Aboelfetouh A, Riad A, Sandhu H, Schaal S and El-Baz A 2018 Early diabetic retinopathy diagnosis based on local retinal blood vessel analysis in optical coherence tomography angiography (OCTA) images *Med. Phys.* **45** 4582–99

[15] Chan T F and Vese L A 2001 Active contours without edges *IEEE Trans. Image Process.* **10** 266–77

[16] Sum K and Cheung P Y 2008 Vessel extraction under non-uniform illumination: a level set approach *IEEE Trans. Biomed. Eng.* **55** 358–60

[17] Al-Diri B, Hunter A and Steel D 2009 An active contour model for segmenting and measuring retinal vessels *IEEE Trans. Med. Imaging* **28** 1488–97

[18] Zhao Y, Rada L, Chen K, Harding S P and Zheng Y 2015 Automated vessel segmentation using infinite perimeter active contour model with hybrid region information with application to retinal images *IEEE Trans. Med. Imaging* **34** 1797–807

[19] Moult E *et al* 2014 Ultrahigh-speed swept-source OCT angiography in exudative AMD *Ophthalmic Surg. Lasers Imaging Retin.* **45** 496–505

[20] Zang P *et al* 2016 Automated motion correction using parallel-strip registration for wide-field en face OCT angiogram *Biomed. Opt. Express* **7** 2823–36

[21] Spaide R F, Fujimoto J G and Waheed N K 2015 Image artifacts in optical coherence angiography *Retina* **35** 2163

[22] Camino A, Zhang M, Gao S S, Hwang T S, Sharma U, Wilson D J, Huang D and Jia Y 2016 Evaluation of artifact reduction in optical coherence tomography angiography with real-time tracking and motion correction technology *Biomed. Opt. Express* **7** 3905–15

[23] Braaf B, Vienola K V, Sheehy C K K, Yang Q, Vermeer K A, Tiruveedhula P, Arathorn D W, Roorda A and Boer J F D 2013 Real-time eye motion correction in phase-resolved OCT angiography with tracking SLO *Biomed. Opt. Express* **4** 51–65

[24] Zhang Q, Huang Y, Zhang T, Kubach S, An L, Laron M, Sharma U and Wang R K 2015 Wide-field imaging of retinal vasculature using optical coherence tomography-based micro-angiography provided by motion tracking *J. Biomed. Opt.* **20** 066008

[25] Kraus M F, Potsaid B, Mayer M A, Bock R, Baumann B, Liu J J, Hornegger J and Fujimoto J G 2012 Motion correction in optical coherence tomography volumes on a per A-scan basis using orthogonal scan patterns *Biomed. Opt. Express* **3** 1182–99

[26] Münch B, Trtik P, Marone F and Stampanoni M 2009 Stripe and ring artifact removal with combined wavelet-Fourier filtering *Opt. Express* **17** 8567–91

[27] Shu-wen W C and Pellequer J L 2011 DeStripe: frequency-based algorithm for removing stripe noises from AFM images *BMC Struct. Biol.* **11** 7

[28] Liang X, Zang Y, Dong D, Zhang L, Fang M, Yang X, Arranz A, Ripoll J, Hui H and Tian J 2016 Stripe artifact elimination based on nonsubsampled contourlet transform for light sheet fluorescence microscopy *J. Biomed. Opt.* **21** 106005

[29] Chang Y, Chen M, Yan L, Zhao X L, Li Y and Zhong S 2019 Toward universal stripe removal via wavelet-based deep convolutional neural network *IEEE Trans. Geosci. Remote Sens.* **58** 2880–97

[30] Fehrenbach J, Weiss P and Lorenzo C 2012 Variational algorithms to remove stationary noise: applications to microscopy imaging *IEEE Trans. Image Process.* **21** 4420–30

[31] Sun L, Jeon B, Zheng Y and Wu Z 2017 Hyperspectral image restoration using low-rank representation on spectral difference image *IEEE Geosci. Remote Sens. Lett.* **14** 1151–55

[32] Chang Y, Yan L, Wu T and Zhong S 2016 Remote sensing image stripe noise removal: From image decomposition perspective *IEEE Trans. Geosci. Remote Sens.* **54** 7018–31
[33] Kamilov U S 2016 A parallel proximal algorithm for anisotropic total variation minimization *IEEE Trans. Image Process.* **26** 539–48
[34] Fazel M 2002 Matrix rank minimization with applications *PhD Thesis* Stanford University, Stanford, CA
[35] Cai J F, Candes E J and Shen Z 2010 A singular value thresholding algorithm for matrix completion *SIAM J. Optim.* **20** 1956–82
[36] Bresson X, Esedolu S, Vandergheynst P, Thiran J P and Osher S 2007 Fast global minimization of the active contour/snake model *J. Math. Imaging Vis.* **28** 151–67
[37] Aujol J F, Gilboa G, Chan T and Osher S 2006 Structure-texture image decomposition modeling, algorithms, and parameter selection *Int. J. Comput. Vision* **67** 111–36
[38] Chambolle A 2004 An algorithm for total variation minimization and applications *J. Math. Imaging Vis.* **20**(1-2) 89–97
[39] Wu X, Gao D, Williams B M, Stylianides A and Zheng Y 2019 Joint destriping and segmentation of OCTA images *Annual Conf. on Medical Image Understanding and Analysis* pp 423–35
[40] Zheng Y, Gandhi J S, Stangos A N, Campa C, Broadbent D and Harding S P 2010 Automated segmentation of foveal avascular zone in fundus fluorescein angiography *Investig. Ophthalmol. Vis. Sci.* **51** 3653–9
[41] Brownrigg D 1984 The weighted median filter *Commun. ACM* **27** 807–18
[42] Chang S G, Yu B and Vetterli M 2000 Adaptive wavelet thresholding for image denoising and compression *IEEE Trans. Image Process.* **9** 1532–46
[43] Wang Z, Bovik A C, Sheikh H R and Simoncelli E P 2004 Image quality assessment: from error visibility to structural similarity *IEEE Trans. Image Process.* **13** 600–12

IOP Publishing

Photo Acoustic and Optical Coherence Tomography Imaging, Volume 3
Angiography: an application in vessel imaging
Ayman El-Baz and Jasjit S Suri

Chapter 4

Optical coherence tomography angiography changes in early type 3 neovascularization after anti-vascular endothelial growth factor treatment

Riccardo Sacconi, Carlo Di Biase, Enrico Borrelli, Lea Querques, Francesco Prascina, Ilaria Zucchiatt, Francesco Bandello and Giuseppe Querques

Type 3 macular neovascularization (MNV) is a peculiar form of neovascular age-related macular degeneration (AMD) characterized by a predilection for the neurosensory retina. In the early stages, type 3 MNV is characterized by the presence of neovascular capillaries detected mainly in the deep vascular complex (DVC) and the avascular slabs, without connection with the retinal pigment epithelium (RPE). At this stage, type 3 MNV is characterized by the absence of intraretinal fluid. Over time, typically, nascent type 3 MNV displays a down-growth progression from the DVC to the RPE. Only when the neovascular complex reaches the RPE and sub-RPE space does the type 3 neovascularization become 'active' and it is complicated by intraretinal fluid. Due to the 'aggressive' nature of type 3 neovascularization, it is very important to treat this kind of neovessel in the early stages of this lesion, offering thus hope for more favorable results. The treatment with anti-vascular endothelial growth factor (VEGF) injections should be started when the complete progression of nascent type 3 MNV (i.e. from the DVC to the RPE) is detected.

After anti-VEGF treatment, the detectable flow of type 3 MNV is no longer continuous with the RPE and sub-RPE space, and thus remains isolated in the retinal layers, featuring absence typical exudative signs of type 3 MNV. However, at the time of recurrence, type 3 MNVs show the presence of intra/sub-retinal exudation with the restoration of the flow deepening from the DVC to the RPE/sub-RPE space. This restoration is therefore featuring the early phases of recurrent exudation in type 3 MNVs. Thus, the communication of the neovascular complex

doi:10.1088/978-0-7503-2060-3ch4

with the RPE and sub-RPE space from the DVC seems necessary to develop an 'active' form of type 3 MNV in both nascent type 3 MNV and recurrent type 3 MNV after treatment.

4.1 Introduction

Type 3 macular neovascularization (MNV) is a particular kind of neovascular age-related macular degeneration (AMD) with neurosensory retina [1] predilection. It has been initially described by Hanett *et al* [2] as an 'Abnormal deep retinal vascular complex' in patients with AMD. Subsequently, several study groups have speculated about the genesis of this type of MNV. A choroidal origin has been suggested by Gass *et al* [3] who defined these MNVs as 'occult chorio-retinal anastomosis'. Contrarily, Yannuzzi *et al* [4] hypothesized an intraretinal origin, proposing the term 'retinal angiomatous proliferation' (RAP). On the other hand, Freund *et al* [1] coined the term 'type 3 MNV' referring to any sign of intraretinal MNV in AMD, without considering its retinal or choroidal origin. The term has been proposed so as to expand the Gass classification of MNV on the basis of the anatomical site. The evolution of type 3 MNV has been studied by different groups with the use of high-resolution imaging techniques. Their evaluations have supported the intraretinal origin of this common form of MNV in patients with AMD [5–11]. Since no histopathological evidence of retinal neovessels and choriocapillaris anastomosis has been found, novel imaging techniques suggested a crucial role of the deep vascular complex (DVC) in the pathogenesis of the disease. Optical coherence tomography angiography (OCT-A), has been of central importance defining the structure of type 3 MNV. In this latter, OCT-A shows an intraretinal vascular complex deriving from the DVC which could be frequently associated with abutting telangiectatic vessels [12–14]. Early identification of type 3 is of primary importance in order to perform an immediate therapy and limit late stage disease progression.

4.2 Nascent type 3 MNV

Nascent type 3 macular neovascularization has been recently identified by our study group. It is characterized by hyperreflective foci in the retinal layers (HRF) at structural OCT with a noticeable flow at OCT-A [15]. Therefore, OCT-A is capable of neovascular flow detection earlier than the clinical outbreak of type 3 MNV and prior to the observation of exudation (e.g. intraretinal fluid at structural OCT) [15]. In our series, the presence of neovascular flow within the HRF was observed at structural OCT in all cases. Moreover, every case evolved to an exudative form of type 3 MNV. In a past publication, Su *et al* [16] explained the appearance of HRF using structural OCT as a first sign of type 3 MNV ascribing intraretinal foci to migration of retinal pigment epithelial (RPE) cells. In another study, the presence of HRF at spectral domain OCT corresponds to a small capillary bunch if evaluated using OCT-A, although only posterior to the detection of intraretinal edema [17–19]. Even if HRF could correspond to migrated RPE cells, our group disclosed the existence of detectable flow inside these foci even in the lack of exudate, thus an 'active' type 3 MNV forms. The landmark of the nascent type 3 MNV is HRF downgrowth advancement, with flow starting from the DVC toward the RPE. For this reason, we described that type 3 MNVs become 'active' (i.e. presence of intraretinal fluid) only at the time of detectable flow between the DVC to the RPE/ sub-RPE space.

In the early phases of nascent type 3, neovascularizations are seen as HRF with detectable flow noticeable exclusively in the DVC and the avascular slabs, and not consecutive to the RPE. At this stage, nascent type 3 MNV displays the lack of intraretinal fluid. Over time, commonly, nascent type 3 MNV shows a downgrowth advancement from the DVC to the RPE. Type 3 MNV becomes 'active' and it is characterized by intraretinal fluid, solely when the MNV reaches the RPE and sub-RPE space. OCT-A is of predominant significance, in this process, to recognize the early stages of the disease, changing our treatment and follow-up approach.

The retinal derivation theory of type 3 MNV is supported by the OCT-A findings of nascent type 3 neovascularization. In fact, prior to leakage, nascent type 3 undeniably originates from the DVC over a drusenoid pigment epithelial detachment (PED) and subsequently advances into the RPE and sub-RPE space generating intraretinal exudation.

Still, more rarely, the intraretinal neovascularization has been related to simultaneous type 1 MNV below the drusenoid PED, and no evident anastomosis amid these findings.

Miere *et al* [28] interestingly theorized that the existence of a drusenoid PED related to the loss of photoreceptors and to the atrophy of the outer retinal layers and of the RPE may stimulate the formation of a DVC neovascularization with latter down-growth towards the RPE/sub-RPE space.

Our results likewise support this hypothesis.

In our series, several HRFs with detectable flow using OCT-A did not evolve to 'active' type 3 MNV during follow-up, displaying only an isolated flow to the deep capillary plexus (DCP) and avascular slab.

Furthermore, we also published an HRF case showing signs of flow that displayed a disappearance of the lesion during the follow-up, using fluorescein angiography, indocyanine green angiography and OCT-A.

For this matter, it is noteworthy to stress that not every HRF with flow advance to an exudative form of type 3 MNV, but exclusively HRFs with flow that display a progressive growth starting from the deep capillary plexus and reaching the RPE/ sub-RPE space.

It is strategic to treat these types of neovessels in the initial phases of this lesion, owing to the 'aggressive' nature of type 3 MNV, hence offering hope for more favorable results [22, 28].

Even so, we need to know the correct timing of when to start treatment.

We should underline that timely diagnosis of HFR with flow owing to migrating of the RPE cells inside the retinal layers is consequentially more frequently met than HRF by nascent type 3 MNV. The cardinal dissimilarity among these two types is that, in the case of migrating RPE cells, hyper-reflective foci may seldom show down-growth advancement.

A higher possibility of downward advancement was suggested with OCT-A signal of flow matching with HRF, while nascent type 3 neovascularization infrequently naturally reverted without treatment unless there was progress into the RPE/sub-RPE space.

Owing to this possibility of a nascent form of type 3 MNV regression, our suggestion is no treatment until the neovascularization progresses as far as the RPE/sub-RPE space.

The diagnosis of nascent type 3 MNV should accordingly authorize an earlier follow-up using OCT-A in order to record the downgrowth towards the RPE/sub-RPE space.

Our suggestion is that nascent form of type 3 MNV complete progression (i.e. starting from the DVC reaching the RPE) may point to the necessity of prompt treatment.

4.3 Treatment of active type 3 MNV

In various MNV types, multiple clinical studies have evaluated the outcomes of anti-VEGF treatment [20–25].

Ebenezer *et al* engaged in a two-year clinical outcome trial, 'Comparison of AMD Treatments Trials (CATT)' [22]. The study assessed 126 eyes with subtype 3 MNV. Daniel *et al* disclosed results that type 3 MNV eyes treated using anti-VEGF injections required fewer injections in relationship to the type 1 and 2 MNV subtypes (6.1 versus 7.4 during the first year and 5.4 versus 6.6 during the second year).

Moreover, at one-year follow-up, the average BCVA improvement was better in type 3 NV patients than in other types of MNV (10.6 versus 6.9 letters) [22].

Even so, at follow-up finish, type 3 lesions displayed exhibited an augmented incidence of RPE atrophy [22].

In a real-life practice series completed in seven Italian centers, Parodi *et al* [23] confirmed that fewer anti-VEGF treatments (bevacizumab or ranibizumab) are needed in type 3 MNV than in other MNV subtypes.

Specifically, 95 patients had undergone an average of 4.4 injections during the first year of treatment recording a considerable upgrading of the BCVA (from 0.66 to 0.53 logMAR). [23]

Different papers have presented the value of ranibizumab or bevacizumab in the treatment of type 3 NV, while less is specified about the efficacy of aflibercept.

Cho *et al* [26] studied the efficiency of aflibercept in relationship to ranibizumab analyzing 63 type 3 MNV naïve eyes treatments. Their conclusions showed similar BCVA improvement in the two groups at one year follow-up.

Nevertheless, they showed an increased RPE atrophy occurrence in the aflibercept group [26].

In fact, Matsumoto *et al* [27] presented comparable improvements in BCVA in type 3 MNV treated with a treat-and-extend protocol based on aflibercept.

Mean BCVA by ETDRS chart was 9.5 letters, while central macular thickness fell from 340 μm to 133 μm. In a recently published paper [28], the authors observed improvements after three sequential aflibercept injections in 47% and vision stability in 42% of the examined eyes.

The therapy was less effective presumably owing to a higher incidence of type 3 lesions in the advanced forms (73% of the examined eyes showed PED).

Summarizing, both aflibercept and ranibizumab injections, in terms of BCVA and central macular thickness improvement, obtained good results in type 3 NV treatment.

The treatment goal is early stage disease patient management to reduce treatment during follow-up obtaining a better VA.

Nevertheless, to minimize atrophy development, we must be cautious with the number of injections.

4.4 OCT-A features of recurrent type 3 MNV after anti-VEGF injections

In a current study, our group examined OCT-A early changes characterizing post treatment recurrent type 3 MNV.

This is especially practical to finding implicit imaging bio-markers for an immediate re-treatment that may likely preclude advancement to advanced stages of the disease.

After visiting 12 post anti-VEGF treated patients, we saw that at the post treatment non-exudative stage, all lesions incompletely reversed, in fact, detectable flow was observed exclusively in the Deep Vascular Complex and/or avascular slab (no detectable flow at or beneath the RPE).

Hence, in all patients, the appreciable flow was then found disjointed with the RPE/sub-RPE space, and thus lasted apart within the retinal layers, underlining lack of type 3 MNV characteristic exudative signs (figures 4.1 and 4.2).

While, during relapse, all type 3 MNVs displayed the appearance of intraretinal and or subretinal exudation with the regeneration of the flow out of the DVC towards the RPE space. This regeneration, present in each case, was thus an early phase characteristic of type 3 MNV recurrent exudation (figures 4.1 and 4.2).

Therefore, the neovascular complex relationship between the deep vascular complex and the RPE/sub-RPE space appears essential to evolve to an 'active' pattern of type 3 MNV in both nascent type 3 MNV and post treatment recurrent type 3 MNV.

Outcomes about naïve eyes treatment by anti-VEGF shots in type 3 MNVs secondary to AMD were previously published by Miere *et al* [29].

Two different OCT-A scenes were described. The initial one presents eyes with exudation at one-year examination showing duration of a well-defined lesion in the external retinal layers and choriocapillaris.

However, the latter one is characterized by eyes showing a complete exudate resolution at 12-month follow-up with a total loss of the early tuft-shaped and claw-like flow signals, without outer retina and choriocapillaris level flow, regardless of the DVC feeder vessels permanence.

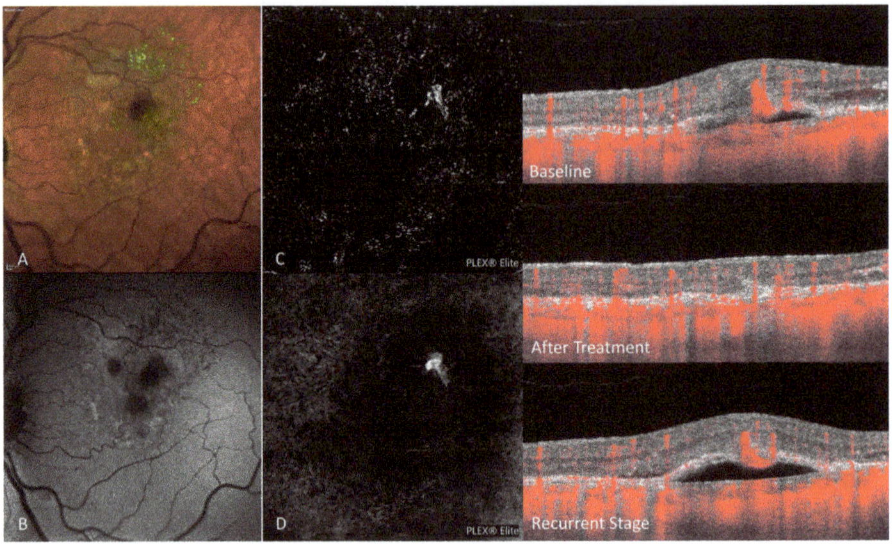

Figure 4.1. Multimodal imaging evaluation including (a) multicolor image, (b) fundus autofluorescence, and (c) optical coherence tomography angiography (OCT-A) of the left eye of a patient affected by type 3 macular neovascularization. In detail, enface OCT-A of the (c) avascular slab and (D) RPE-RPEfit at the baseline showed the neovascular network. Cross-sectional B-scan with flow showed the presence of a lesion growing from the DVC to the RPE space at the baseline, with a regression after the anti-VEGF treatment, and a restoration at the time of the recurrence.

Our study results were in agreement with Miere *et al.* In our study, naïve treatment of type 3 neovascularization displayed, at baseline, neovessels rising from the DVC to the RPE/sub-RPE. This OCT-A detected linkage was lost in all cases post intra-subretinal exudation resolution with the neovascular tuft persistency at the DVC and/or avascular slab level. However, regeneration of the flow starting from the deep vascular plexus to the RPE/sub-RPE defined the recurrence of type 3 MNV related exudative signs.

In a previous histological study, Skalet *et al* [25] presented the eye globe findings of a 93 year-old male patient with type 3 MNV treated by several anti-VEGF shots. They reported that ranibizumab had been unsuccessful in reducing the type 3 lesion entirely.

Actually, regardless of no exudative signs, the type 3 lesion showed up as thick-walled neovessels in the retinal layers. Remarkably at the RPE level, with no signs of neovascular tissue, a thin band of hyperreflective tissue underlying the type 3 lesion was found [25]. The previous histopathologic findings were confirmed by our *in vivo* OCT-A findings.

Our group reported that type 3 neovascularization did not completely dissolve after anti-VEGF injections. Lesion regression was obtained with the vanishing of recognizable flow at the outer retina level and without evidence of a link with the RPE, even if a neovascular flow was still visible at the DVC. At recrudescence, an OCT-A detectable flow remerged at the outer retina level with a neovascularization

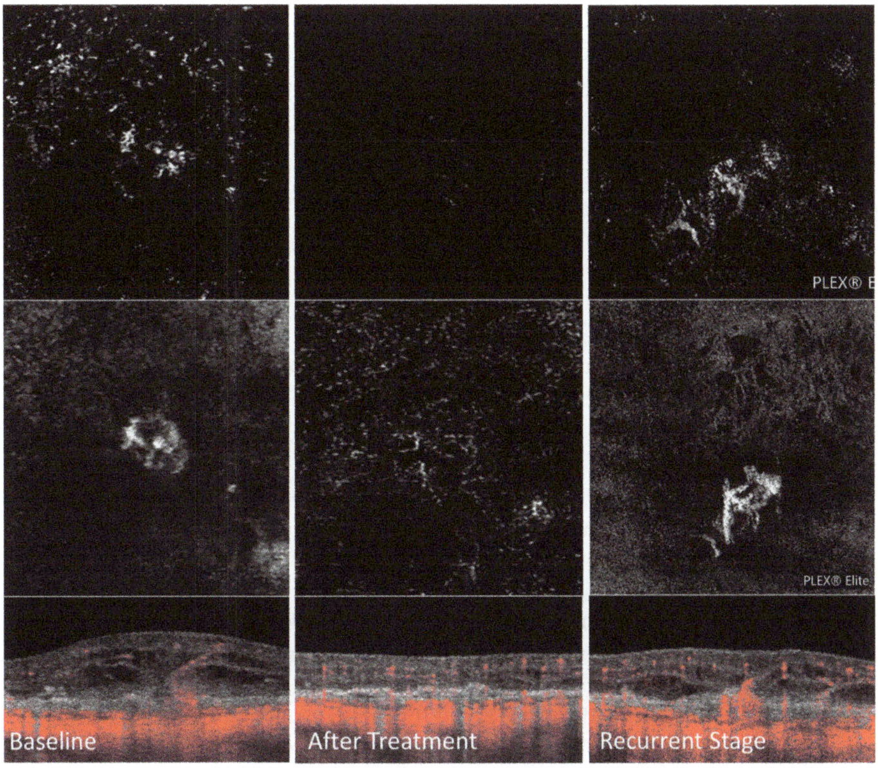

Figure 4.2. Optical coherence tomography angiography (OCT-A) evaluation of the right eye of a patient with a type 3 macular neovascularization. *En face* OCT-A of the avascular slab (first row) and of the RPE-RPEfit slab (second row), and cross-sectional B-scan with flow (third row) showed the neovascular network with the presence of flow growing from the DVC to the RPE space at the baseline, with a regression after the anti-VEGF treatment, and a restoration at the time of the recurrence.

down-growth advancement from the DVP to the RPE. In concert with former histopathological records and up to date *in vivo* OCT-A results, we embrace the thesis that type 3 lesions do not dissolve after anti-VEGF shots, since the lesion between the DVC and the RPE/sub-RPE space is still present as a fibrous tissue under the DVC neovascular complex. OCT-A image lesion disappearance at the RPE and sub-RPE space level may be associated both with the lack of or with a very sluggish flow (i.e. no flow observable at OCT-A) within the hyperreflective tissue of the type 3 neovascularization post anti-VEGF injections. Nonetheless, at recrudescence, there might be an incrementing flow inside the type 3 MNV that defines the recovered flow in the entire neovascularization from the DVC to the RPE.

Post treatment, type 3 MNVs show no exudative signs before the retinal neovascular complex flow does not advance *de novo* from the DVC into the RPE. When this happens, type 3 MNV recrudescence is characterized by exudation (i.e. intraretinal fluid), prefiguring the 'active' form of type 3 neovascularization. This linkage amidst the DVC and the RPE appears to be necessary for new exudation production secondary to recrudescence type 3 MNV. Consequently, the discovery of appreciable flow down-growth advancement from the DVC to the RPE, indicates impending recrudescence and may encourage prompt retreatment. Amidst these lines, OCT-A could be intended as a new useful imaging technique to determine initial signs of recrudescence, in reality prior to exudative signs found by structural OCT.

References

[1] Freund K B, Ho I V and Barbazetto I A *et al* 2008 Type 3 neovascularization: the expanded spectrum of retinal angiomatous proliferation *Retina* **28** 201–11

[2] Hartnett M E, Weiter J J, Garsd A and Jalkh A E 1992 Classification of retinal pigment epithelial detachments associated with drusen *Graefes. Arch. Clin. Exp. Ophthalmol* **230** 11–9

[3] Gass J D, Agarwal A, Lavina A M and Tawansy K A 2003 Focal inner retinal hemorrhages in patients with drusen: an early sign of occult choroidal neovascularization and chorioretinal anastomosis *Retina* **23** 741–51

[4] Yannuzzi L A, Negrão S and Iida T *et al* 2001 Retinal angiomatous proliferation in age-related macular degeneration *Retina* **21** 416–34

[5] Miere A, Querques G, Semoun O, El Ameen A, Capuano V and Souied E H 2015 Optical coherence tomography angiography in early type 3 neovascularization *Retina* **35** 2236–41

[6] Miere A, Querques G and Semoun O *et al* 2017 Optical coherence tomography angiography changes in early type 3 neovascularization after anti-vascular endothelial growth factor treatment *Retina* **37** 1873–9

[7] Querques G, Souied E H and Freund K B 2013 Multimodal imaging of early stage 1 type 3 neovascularization with simultaneous eye-tracked spectral-domain optical coherence tomography and high-speed real-time angiography *Retina* **33** 1881–7

[8] Querques G, Souied E H and Freund K B 2015 How has high-resolution multimodal imaging refined our understanding of the vasogenic process in type 3 neovascularization? *Retina* **35** 603–13

[9] Su D, Lin S and Phasukkijwatana N *et al* 2016 An updated staging system of type 3 neovascularization using spectral domain optical coherence tomography *Retina* **36** S40–9

[10] Tan A C, Dansingani K K, Yannuzzi L A, Sarraf D and Freund K B 2017 Type 3 neovascularization imaged with cross-sectional and en face optical coherence tomography angiography *Retina* **37** 234–46

[11] Freund K B, Zweifel S A and Engelbert M 2010 Do we need a new classification for choroidal neovascularization in age-related macular degeneration? *Retina* **30** 1333–49

[12] Miere A, Querques G and Semoun O *et al* 2015 Optical coherence tomography angiography in early type 3 neovascularization *Retina* **35** 2236–41

[13] Kuehlewein L, Dansingani K K and de Carlo T E *et al* 2015 Optical coherence tomography angiography of type 3 neovascularization secondary to age-related macular degeneration *Retina* **35** 2229–35

[14] Miere A, Querques G and Semoun O *et al* 2017 Optical coherence tomography angiography changes in early type 3 neovascularization after anti-vascular endothelial growth factor treatment *Retina* **37** 1873–9

[15] Sacconi R, Sarraf D and Garrity S *et al* 2018 Nascent type 3 neovascularization in age-related macular degeneration *Ophthalmol. Retina* **2** 1097–106

[16] Su D, Lin S and Phasukkijwatana N *et al* 2016 An updated staging system of type 3 neovascularization using spectral domain optical coherence tomography *Retina* **36** S40–9

[17] Kuehlewein L, Dansingani K K and de Carlo T E *et al* 2015 Optical coherence tomography angiography of type 3 neovascularization secondary to age-related macular degeneration *Retina* **35** 2229–35

[18] Miere A, Querques G, Semoun O, El Ameen A, Capuano V and Souied E H 2015 Optical coherence tomography angiography in early type 3 neovascularization *Retina* **35** 2236–41

[19] Dansingani K K, Naysan J and Freund K B 2015 En face OCT angiography demonstrates flow in early type 3 neovascularization (retinal angiomatous proliferation) *Eye* **29** 703–6

[20] Park Y G and Roh Y J 2015 One year results of intravitreal ranibizumab monotherapy for retinal angiomatous proliferation: a comparative analysis based on disease stages *BMC Ophthalmol* **15** 182

[21] Tsaousis K T, Konidaris V E and Banerjee S *et al* 2015 Intravitreal aflibercept treatment of retinal angiomatous proliferation: a pilot study and short-term efficacy *Graefes. Arch. Clin. Exp. Ophthalmol* **253** 663–5

[22] Daniel E, Shaffer J and Ying G S *et al* 2016 Comparison of age-related macular degeneration treatments trials (CATT) research group. Outcomes in eyes with retinal angiomatous proliferation in the comparison of age-related macular degeneration treatments trials (CATT) *Ophthalmology* **123** 609–16

[23] Parodi M B, Donati S and Semeraro F *et al* 2017 Intravitreal anti-vascular endothelial growth factor drugs for retinal angiomatous proliferation in real-life practice *J. Ocul. Pharmacol. Ther* **33** 123–7

[24] Inoue M, Arakawa A, Yamane S and Kadonosono K 2014 Long-term results of intravitreal ranibizumab for the treatment of retinal angiomatous proliferation and utility of an advanced RPE analysis performed using spectral-domain optical coherence tomography *Br. J. Ophthalmol* **98** 956–60

[25] Skalet A H, Miller A K, Klein M L, Lauer A K and Wilson D J 2017 Clinicopathologic correlation of retinal angiomatous proliferation treated with ranibizumab *Retina* **37** 1620–4

[26] Cho H J, Hwang H J, Kim H S, Han J I, Lee D W and Kim J W 2018 Intravitreal aflibercept and ranibizumab injections for type 3 neovascularization *Retina* **38** 2150–8

[27] Matsumoto H, Sato T and Morimoto M *et al* 2016 Treat-and-extend regimen with aflibercept for retinal angiomatous proliferation *Retina* **36** 2282–9

[28] Chou H D, Wu W C, Wang N K, Chuang L H, Chen K J and Lai C C 2017 Short-term efficacy of intravitreal aflibercept injections for retinal angiomatous proliferation *BMC Ophthalmol* **17** 104 Erratum in: BMC Ophthalmol 2017;17 (1):144

[29] Miere A, Querques G and Semoun O *et al* 2017 Optical coherence tomography angiography changes in early type 3 neovascularization after anti-vascular endothelial growth factor treatment *Retina* **37** 1873–9

Chapter 5

Optical coherence tomography angiography in multiple sclerosis and neuromyelitis optica

Olwen C Murphy and Shiv Saidha

The anterior visual pathway is a key site of disease in both multiple sclerosis (MS) and neuromyelitis optica spectrum disorder (NMOSD). The retina is considered a 'window into the brain' and optical coherence tomography (OCT) is now a widely used imaging tool for evaluation of the anterior visual pathway in MS and NMOSD. Thinning of the peripapillary retinal nerve fiber layer (pRNFL) and ganglion cell + inner plexiform layer (GCIPL) reflecting neuroaxonal loss are key retinal changes known to occur in MS and NMOSD as a result of acute optic neuritis (ON), subclinical optic neuropathy, or a combination of both. In MS, retinal layer thinning is considered an imaging biomarker of neurodegeneration that can be assessed in a rapid, reproducible manner using OCT. In recent years the advent of OCT angiography (OCTA) has offered the opportunity to explore metabolic demand and vascular integrity in the retina in these diseases using a non-invasive depth-resolved technique. Studies using OCTA have demonstrated that retinal vascular plexus densities are reduced in people with MS (in ON eyes more than non-ON eyes), and that greater reductions in vascular plexus densities correlate with longer MS disease duration, higher levels of disability and poorer visual acuity. Reductions in retinal vascular plexus densities have also been identified with OCTA in NMOSD—particularly in ON eyes. We review and interpret the growing body of research utilizing OCTA in MS and NMOSD, and discuss some limitations of OCTA imaging in the context of these diseases.

5.1 Introduction

The anterior visual pathway is an important site of disease in demyelinating disorders of the central nervous system (CNS). Around half of patients with multiple sclerosis (MS) will experience a clinical episode of acute optic neuritis (ON) at some

doi:10.1088/978-0-7503-2060-3ch5

point in their disease course, and subclinical optic neuropathy is almost ubiquitous, with demyelinated plaques identified in the optic nerves of up to 99% of people with MS at *post-mortem* [1, 2]. Neuromyelitis optica spectrum disorder (NMOSD) also frequently affects the anterior visual pathway, with ON being the inaugural manifestation of disease in around 40% of aquaporin-4 antibody seropositive patients [3, 4]. ON and subclinical optic neuropathy cause axonal injury and degeneration within the optic nerve, which may proceed anterogradely (towards the thalamus and posterior visual pathway), and retrogradely (towards the retina). Ultimately, retrograde neuroaxonal degeneration can result in thinning of the retinal nerve fiber layer (RNFL), and loss of retinal ganglion cell bodies—which is seen histopathologically in approximately 80% of MS eyes at *post-mortem* [5, 6]. Research using optical coherence tomography (OCT) has offered a useful tool for measuring these dynamic retinal processes *in vivo*.

Retinal tissue is highly metabolically active and highly vascularized (with a distinctly-structured vascular network), and represents a useful accessible location for examining these processes. Neuroaxonal degeneration in the anterior visual pathway in MS and NMOSD is likely to result in reduced metabolic demand in retinal tissue, and exploring retinal vascular metrics can offer useful information regarding metabolic demand and vascular supply. Moreover, aberrant vascular processes are known to be extensive in patients with MS, including excess platelet activation, increased thrombophilic markers, altered blood–brain-barrier (BBB) permeability, endothelial cell dysfunction, hypoxia-like tissue injury, perivascular iron deposition, and vascular occlusion within lesions [7–12]. While pathobiologic processes differ in NMOSD as compared to MS, vascular processes are also thought to be at play in the NMOSD disease state; altered BBB permeability is a pathological hallmark of the disease, and NMOSD lesions are characterized by perivascular immunoglobulin deposition, perivascular complement activation, and blood vessel thickening [13]. Brain imaging studies have suggested that reductions in cerebral perfusion are present in patients with MS and NMOSD [14–18], and that MS lesions have a predilection for peri-venular locations [19, 20]. Many of these pathobiological processes occurring in the brain may be capitulated in the retina. Indeed, *post-mortem* studies of MS eyes have shown that venular thickening, peri-vascular inflammation, peri-venular gliosis, and alterations in endothelial tight junctions are present in the retina and optic nerve head [5, 21]. While similar histopathological studies of the retina are lacking in NMOSD, aquaporin-4 (the target of the pathogenic antibody identified in most patients with NMOSD) is highly expressed in the retina by astrocytes and Müller glial cells [22], and retinitis has been demonstrated in animal models of NMOSD [23]. *In vivo* imaging studies using fluorescein angiography or fundus photography have suggested the presence of retinal peri-phlebitis in some MS eyes [24–26], and the presence of retinal focal arteriolar narrowing and attenuation of the peripapillary vascular tree in NMOSD eyes [27].

The advent of OCT angiography (OCTA) in recent years has offered a new technique to explore the retinal vasculature *in vivo*. The combination of OCT and OCTA may allow the concomitant evaluation of both neurodegenerative and vascular processes operative in MS and NMOSD. Furthermore, insights gained from the retina may reflect global disease processes. For these reasons, there has been substantial interest in employing OCTA for research in MS and NMOSD. In this chapter, we will focus on the background provided by extensive studies of OCT-derived retinal layer thicknesses in MS and NMOSD, the expanding body of research employing OCTA in these disorders, and the potential limitations of OCTA as an imaging technique.

5.2 Optical coherence tomography (OCT) of the retina

Retinal layer thicknesses can be measured *in vivo* using optical coherence tomography (OCT)—a rapid, non-invasive, well-tolerated, inexpensive, reproducible and high-resolution imaging technique that uses near-infrared light to generate 2D or 3D images of tissues. Spectral-domain OCT (SD-OCT) is the fourth generation iteration of this technology, which offers higher-definition and more rapid imaging, as compared to previously available time-domain OCT (TD-OCT). SD-OCT allows segmentation of discrete retinal layers within the macula, with each layer potentially having different biological and clinical relevance in the context of neurological diseases (figure 5.1). As the retina is unmyelinated, OCT offers a unique avenue to examine neurodegenerative and inflammatory processes occurring outside of the context of demyelination. Insights into these pathobiological processes are of paramount importance, since neuroaxonal dysfunction and degeneration (rather than demyelination) are now believed to be the principal susbstrates of clinical disability in MS and related diseases [28–30]. Thus, OCT has become an extremely useful imaging tool in these disease states, with a multitude of research establishing the role of OCT-derived retinal layer thicknesses in the evaluation and monitoring of MS and related disorders over the last two decades. Insights into these diseases gained through OCT form an important context for more recent studies using OCTA.

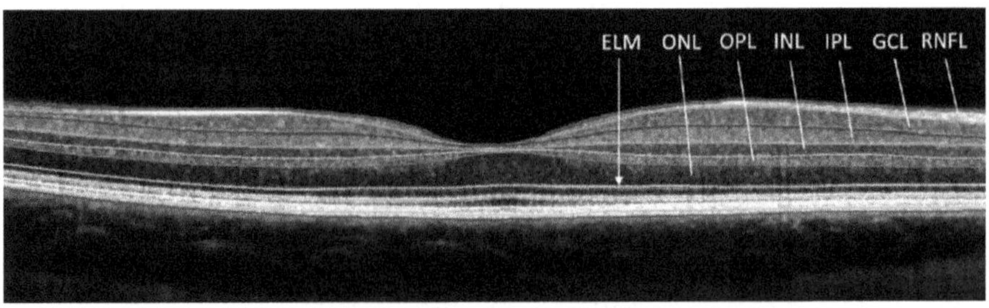

Figure 5.1. Example of a macular SD-OCT image. This macular image was acquired from a consenting healthy research participant using a Cirrus SD-OCT device (model 5000, software version 8.1, Carl Zeiss Meditec, California, USA), with automated segmentation of discrete retinal layers. RNFL = retinal nerve fiber layer, GCL = ganglion cell layer, IPL = inner plexiform layer, INL = inner nuclear layer, OPL = outer plexiform layer, ONL = outer nuclear layer, ELM = external limiting membrane. (Reprinted from [46], with permission.)

5.2.1 Retinal OCT in multiple sclerosis (MS)

Extensive studies using OCT in people with MS have been undertaken in recent years. Peripapillary RNFL (pRNFL) and ganglion cell + inner plexiform layer (GCIPL) thicknesses have been repeatedly shown to be reduced in the eyes of people with MS as compared to healthy controls, and reductions are most marked in MS eyes with a history of ON (MS-ON eyes) [31, 32]. GCIPL thickness measures are more reliable and reproducible (with an inter-visit intra-class correlation coefficient of 0.99 in both MS patients and healthy controls using Cirrus HD-OCT) [33] than pRNFL thickness measures, and are less susceptible to confounding by edema in the context of ON [33, 34]. In people with MS, GCIPL thickness has been demonstrated to correlate with brain volume (particularly gray matter volumes), quantitative spinal cord MRI measures, high and low-contrast visual function, and global disability [33, 35–39]. Longitudinal studies have demonstrated that progressive GCIPL atrophy occurs in MS, and correlates with worsening visual function, global disability accumulation, and brain (and brain substructure) atrophy [40, 41]. Furthermore, accelerated GCIPL atrophy is seen in patients with clinical and/or radiological inflammatory activity, and atrophy is differentially modulated by various disease modifying therapies (DMT) [41, 42]. Based on these abundant data, GCIPL thickness is thought to be a useful biomarker of neurodegeneration in MS. In addition, other qualitative and quantitative OCT findings may also be informative in MS; microcystic macular edema is associated with disability progression, inner nuclear layer (INL) thickening is associated with inflammatory disease activity, and thinning of the deeper retinal layers (the INL and the outer nuclear layer—ONL) may be a hallmark of progressive MS [37, 43–45]. Overall, OCT is considered to offer a 'window into the brain' in people with MS, and has clinical applications as a biomarker for prognostication, monitoring of disease progression, and evaluation of treatment response [46–48].

5.2.2 Retinal OCT in neuromyelitis optica spectrum disorder (NMOSD)

ON behaves somewhat differently in NMOSD as compared to MS. In patients with aquaporin-4 seropositive NMOSD, ON is typically more severe, more frequently bilateral, associated with poorer visual function outcomes, and more likely to be recurrent [49–51]. This difference in severity is reflected in OCT measures, as reductions in pRNFL and GCIPL thicknesses are substantially greater in NMOSD-ON eyes as compared to MS-ON eyes, or idiopathic-ON eyes [49, 50, 52]. Additionally, microcystic macular edema and INL thickening may also be present in NMOSD-ON eyes—perhaps more frequently after severe ON episodes associated with worse structural and functional outcomes [53, 54]. Similar to findings in MS, pRNFL and GCIPL thickness correlate well with visual function measures in NMOSD-ON [50, 55, 56]. However, in contrast with MS, correlations between retinal layer thicknesses and global disability measures such as the expanded disability status scale (EDSS) have not been robustly and reproducibly demonstrated in NMOSD [49]. Similarly, the evidence for subclinical optic neuropathy in NMOSD is not as conclusive as it is in MS, although a number of studies suggest that reductions in pRNFL and GCIPL thicknesses may be present in NMOSD eyes without a history of ON [52, 54, 57]. In addition, foveal thinning has been observed in NMOSD eyes with and without a history of AON, as compared to control eyes, and thus has been suggested as an alternative putative marker of subclinical optic neuropathy in these patients [58]. The lack of conclusive evidence for subclinical optic neuropathy or for a clear relationship between retinal layer thicknesses and global disability measures in NMOSD may be explained by the pathophysiology of the disorder. In NMOSD, accumulation of disability is thought to be primarily attack-related, rather than driven by subclinical pathophysiological processes [59]. By contrast, subclinical inflammation and neuroaxonal loss are important components of MS disease pathology (that are capitulated in the anterior visual pathway) [60], and much disability accumulation in MS is not associated with overt attacks [61]. Furthermore, large volume retinal tissue loss occurring after severe ON attacks in NMOSD may mask potential relationships between retinal layer thicknesses and global disease parameters in those eyes, and may also confound longitudinal changes through a ‘floor effect’ whereby further neuroaxonal loss may not be detectable.

5.3 Optical coherence tomography angiography (OCTA) of the retina

OCTA is a relatively recent advancement of OCT technology, which detects inherent motion of predominantly red blood cells in retinal vessels to generate *en face* images of the retinal vasculature. OCTA imaging is a rapid and well-tolerated technique, and does not require the injection of dye. Furthermore, the ability of OCTA to produce depth-resolved images allowing segmentation of discrete vascular plexuses represents a distinct advantage as compared to other retinal vascular imaging techniques (such as fluorescein angiography).

The anatomy of the retinal vasculature is highly organized, with certain vascular plexuses consistently lying within certain histological layers, and minimal variation in this anatomy occurring between individuals [62]. Thus, retinal layer boundaries detected using OCT can be used to segment OCTA images into distinct vascular plexuses (figure 5.2). Arterial supply to the inner retina is derived from the central retinal artery—a branch of the ophthalmic artery. The vascular plexuses supplying the inner retinal layers include the radial peripapillary capillary plexus (RPCP), the superficial vascular plexus (SVP), and the deep vascular plexus (DVP). Each of these vascular plexuses comprises a network of arteries, arterioles, capillaries, venules and veins. The SVP lies primarily within the ganglion cell layer and supplies the RNFL and the GCIPL. The DVP lies primarily within the INL, and supplies the INL and the outer plexiform layer (OPL). Arterial supply and venous drainage of the DVP is through the SVP. The choriocapillaris lies within Bruch’s membrane and provides vascular supply to the outer retina. Arterial supply to the choriocapillaris is derived from the posterior ciliary arteries—a branch of the ophthalmic artery. Visualization of the outer retinal vasculature with OCTA is considered by some experts to be substantially more challenging than the inner retinal circulation, due to (1) projection

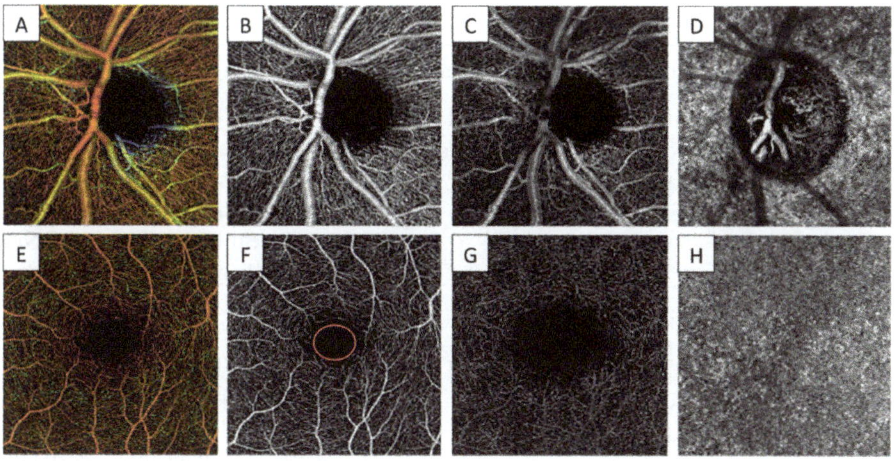

Figure 5.2. Examples of retinal OCTA images of the optic nerve head and macula. 3 mm × 3 mm angiography images acquired from the left eye of a healthy control acquired using Cirrus AngioPlex SD-OCT (Carl Zeiss) are demonstrated here, centered over the optic nerve head or fovea. Depth-encoded images are used to generate images of the discrete vascular plexuses. A = optic nerve head depth-encoded image, B = optic nerve head superficial vascular plexus, C = optic nerve head deep vascular plexus, D = optic nerve head choriocapillaris, E = macular depth-encoded image, F = macular superficial vascular plexus with the foveal avascular zone illustrated by the red circle, G = macular deep vascular plexus, H = macular choriocapillaris. The foveal avascular zone is the capillary-free area at the center of macular images E–G.

artifact from the inner retinal vasculature, and (2) the hyper-reflective retinal pigment epithelium (RPE) acting as a barrier to light beam penetration [63].

OCTA images can be reviewed qualitatively, which is particularly useful in the clinical evaluation of certain ophthalmologic disorders associated with focal or regional retinal abnormalities e.g. retinal artery or vein occlusions, and diabetic retinopathy. In the case of neurological disorders that are thought to be associated with a more diffuse effect on the retinal vasculature, quantitative OCTA measurements are likely to be more useful than qualitative analyses. A range of different quantitative measures have been derived using OCTA to date, including vascular plexus density (sometimes referred to as vessel density), vessel length density, area of the foveal avascular zone, flow index, flow velocity, skeleton density, fractal dimension and non-perfused area [64]. In addition, a range of different techniques (using various algorithms) for the derivation of these measurements has been reported. The choice of quantitative metrics analyzed may vary according to the device utilized, the disease-state being examined, and the study hypothesis.

5.3.1 OCTA in MS

Retinal OCTA has become a research tool of interest in people with MS in recent years. A number of studies have explored retinal OCTA measures in people with MS, and compared these to healthy controls (table 5.1).

Macular SVP density has repeatedly been demonstrated to be lower in people with MS than in healthy controls [67, 70, 71]. Correlations between vascular plexus

Table 5.1. Selection of published studies exploring OCTA measures in people with MS.

Study authors	Study design	OCTA measures	Key findings
Wang *et al* 2014 [65]	Cross-sectional study of 35 people with MS and 21 healthy controls	ONH-FI and macular parafoveal FI	• ONH-FI was lower in MS-ON eyes as compared to healthy control eyes and MS-NON eyes • No significant differences in parafoveal FI between MS-ON/MS-NON eyes and healthy control eyes
Spain *et al* 2018 [66]	Cross-sectional study of 45 people with MS and 32 healthy controls	ONH-FI	• ONH-FI was lower in MS-ON and MS-NON eyes as compared to healthy control eyes • No significant correlations between ONH-FI and retinal layer thicknesses
Lanzillo *et al* 2017 [67]	Cross-sectional study of 50 people with MS and 46 healthy controls	Macular SVP density	• SVP density was lower in MS-ON eyes and MS-NON eyes as compared to healthy control eyes • SVP density significantly correlated with RNFL and GCC thicknesses • Lower SVP density significantly correlated with higher disability (EDSS and MSSS scores)
Feucht *et al* 2018 [68]	Cross-sectional study of 42 people with MS and 50 healthy controls	Macular SVP, DVP and choriocapillaris densities	• SVP and DVP densities were lower in MS-ON eyes as compared to MS-NON eyes and healthy control eyes • SVP and DVP densities significantly correlated with GCIPL volumes, INL volumes and TMV
Lanzillo *et al* 2018 [69]	Longitudinal study (1 year) of 50 people with MS	Macular SVP density	• Across the whole group, SVP density in the parafoveal region showed a small but significant increase over 1 year • Reductions in SVP density in the parafoveal region correlated significantly with increasing disability (EDSS scores)

(*Continued*)

Table 5.1. (*Continued*)

Study authors	Study design	OCTA measures	Key findings
Murphy *et al* 2019 [70]	Cross-sectional study of 111 people with MS and 50 healthy controls	Macular SVP and DVP densities	• SVP density was lower in MS-ON and MS-NON eyes as compared to healthy control eyes • Lower SVP density correlated significantly with longer disease duration, higher disability (EDSS and MSFC scores), and poorer high- and low-contrast letter acuity scores • SVP density correlated significantly with RNFL and GCIPL thicknesses, DVP density correlated significantly with GCIPL and INL thicknesses
Yilmaz *et al* 2020 [71]	Cross-sectional study of 47 people with MS and 61 healthy controls	Macular SVP and DVP densities, RPCP density, FAZ area	• SVP and DVP densities were lower in people with MS as compared to healthy controls, across all EDTRS sectors of the macula • No significant differences in FAZ and RPCP density between people with MS and healthy controls • SVP and DVP densities were inversely correlated with visual evoked potential (VEP) latency

ONH-FI = optic nerve head flow index, FI = flow index, MS-ON = MS optic neuritis, MS-NON = MS non-optic neuritis, SVP = superficial vascular plexus, DVP = deep vascular plexus, RNFL = retinal nerve fiber layer, GCC = ganglion cell complex, GCIPL = ganglion cell + inner plexiform layer, INL = inner nuclear layer, TMV = total macular volume, EDSS = expanded disability status scale, MSSS = multiple sclerosis severity scale, MSFC = multiple sclerosis functional composite, RPCP = radial peripapillary capillary plexus, FAZ = foveal avascular zone.

densities and retinal layer thicknesses have been observed in both MS eyes and healthy control eyes, suggesting a strong relationship between structural and vascular measures in the macula [67, 68, 70, 71]. Additionally, SVP density reductions in MS eyes as compared to healthy control eyes have been demonstrated to be present in all sectors of the EDTRS grid, demonstrating that alterations in vascular measures occur diffusely within the macula in MS [67, 71]. Reductions in SVP density appear to be more marked in MS-ON eyes than in MS non-ON (MS-NON) eyes, however significant differences have been detected even in MS-NON eyes alone as compared to healthy control eyes [67, 70]. In some cases, the rarefication of the macular SVP in MS-ON eyes may be so evident as to be visible to the naked eye with qualitative image evaluation (figure 5.3). It is notable that one

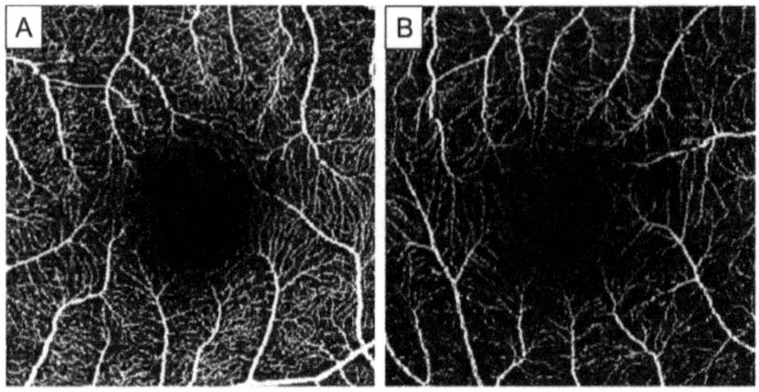

Figure 5.3. Macular SVP OCTA image in the MS-NON eye and MS-ON eye of one individual with MS. These images show the qualitative differences in the SVP visible on macular Spectralis OCTA (Heidelberg Engineering, Germany) images from the eyes of an individual with MS who has a history of unilateral ON. The patient's left eye (A) was unaffected by ON and had an SVP density of 26.9% calculated using an image binarization method, whereas the patient's right eye (B) was affected by ON 5 years previously and had an SVP density of 11.7% calculated using the same method. The capillary network is noticeably less dense in the eye with a history of ON.

study failed to identify significant differences in macular SVP density of MS-NON eyes versus healthy control eyes, although that study was relatively small and average MS disease duration was short [68]. As the magnitude of SVP reductions in MS-NON appears to be lower than in MS-ON eyes, it is likely that examining larger patient cohorts with longer disease durations is necessary to detect macular SVP density reductions in MS-NON eyes.

Differences in OCTA measures between people with MS and healthy controls have also been identified at the optic nerve head. Specifically, optic nerve head flow index (ONH-FI) has been shown to be lower in MS-ON eyes as compared to healthy control eyes [65, 66]. Additionally, one study suggested that ONH-FI may be lower in MS-NON eyes as compared to healthy control eyes [66]. Two studies have also compared OCTA measures at the optic disc in ON eyes to non-arteritic ischemic optic neuropathy (NAION) eyes, with both studies finding no difference in RPCP density between these groups at either the acute or chronic stages [72, 73].

Based on data published to date, it appears that OCTA-derived macular vascular measures may demonstrate relationships with certain clinical parameters in MS. Studies have shown that lower SVP density is associated with (1) longer disease duration, (2) higher levels of global disability measured by the expanded disability status scale (EDSS), multiple sclerosis severity scale (MSSS) and multiple sclerosis functional composite (MSFC) scores, (3) poorer visual function measured by high- and low-contrast letter acuity scores, and (4) longer visual evoked potential (VEP) latency [67, 70, 71]. Furthermore, a short longitudinal study (1 year of follow-up) identified that progressive reductions in SVP density were significantly associated with increasing levels of disability [69]. Studies have noted the importance of considering ON history in study methodology and statistical analyses, since the magnitude of retinal tissue loss that occurs after ON is known to confound the relationships between retinal layer thicknesses and clinical characteristics in MS [74], and it is assumed that similar confounding occurs with vessel density reductions after ON. Once this confounding is controlled for, there is certainly accumulating evidence that OCTA-derived vascular measures have the potential to be a useful biomarker of disease severity in MS.

5.3.2 OCTA in NMOSD

OCTA measures have not been as well-explored in NMOSD to date as in MS, although a number of studies have already provided interesting results. Lower RPCP and SVP densities have been identified in people with NMOSD as compared to healthy controls, with differences compared to healthy controls being more marked in NMOSD eyes with a history of ON (NMOSD-ON eyes) than in NMOSD eyes without a history of ON (NMOSD-NON eyes) [75–77]. Lower DVP density has also been detected in some people with NMOSD as compared to healthy controls [77]. People with NMOSD often experience multiple attacks of ON, and a higher number of prior ON attacks is associated with lower SVP and DVP densities in those eyes [77].

Similar to relationships identified in MS eyes, vascular plexus densities in NMOSD eyes have been found to correlate well with retinal layer thicknesses [76, 77]. However, in one study where RPCP and SVP densities were found to be lower in NMOSD-NON eyes than healthy control eyes, pRNFL thickness was not significantly different between the same groups, suggesting that vascular densities can be altered in these patients even where substantial axonal loss is not detected [76]. RPCP, SVP and DVP have all been suggested to correlate with visual function in patients with NMOSD [76, 77].

5.4 Limitations of OCTA

OCTA images can be acquired using a range of commercially available devices, which employ different algorithms to segment the retinal vascular plexuses according to retinal layer boundaries. The quantitative vascular measurements which can be acquired (vascular plexus densities, flow indices, foveal avascular zone area) may differ across devices. Accurate segmentation of the vascular plexus layers may be impaired by the presence of pathology distorting the retinal tissue (such as cysts) [78]. The number of vascular structures which can be segmented and the boundaries used for these may also differ, e.g. the macular DVP can be subdivided into the intermediate capillary plexus and the deep capillary plexus, and the RPCP can be combined with the SVP [62]. In addition, there is a lack of accepted standardized nomenclature in the field of OCTA, with the terms 'capillary plexus', 'vascular plexus' and 'vascular complex' often used interchangeably to describe vascular structures, and the terms 'vessel density', 'vascular plexus density' and 'vessel density index' variably used to report derived measurements. Furthermore, some devices provide images of the deeper vascular structures (such as the choriocapillaris and choroid) while other devices do not. Visualizing the vasculature beyond the highly reflective RPE using OCTA is technically challenging, and where devices do provide images of these deeper vascular structures, there remains some uncertainty regarding the reliability of these images and the quantitative measures gleaned from them [63, 79]. In addition to differences across devices, many research studies employing OCTA utilize different image processing tools or novel research algorithms, adding further complexity to this field. For the reasons outlined here, it is not clear yet whether quantitative data acquired from different devices or processed using difference algorithms can be directly compared, and accordingly calls into question the generalizability of study findings.

Artifacts are known to be a significant issue impacting the acquisition and interpretation of OCTA images, across healthy controls and various disease states [80–82]. Possible artifacts include projection artifact (from inner vascular structures onto deeper structures), motion artifact, loss of focus artifact (figure 5.4), and blink lines, amongst others [82]. It has been demonstrated that imaging artifacts are more frequent in people with MS as compared to healthy controls, and that these artifacts may lead to underestimation of vascular plexus densities in people with MS [82]. The greater frequency of imaging artifact in people with MS may be due to nystagmus, poor visual acuity, fatigue, and physical disability—factors which may be also be present in people with NMOSD or other disabling neurological diseases. However, many research studies employing OCTA fail to describe clear quality control mechanisms to counteract the difficulties associated with either image acquisition or artifact-degradation.

There is still much to be understood regarding the interpretation of OCTA measurements. Even in healthy controls, OCTA measures appear to be influenced by demographic factors, including age and ethnicity [83, 84], yet there is a lack of accepted age and ethnicity-adjusted normative data available for clinical, or research application. OCTA images are typically generated and quantitatively analyzed in *en face* format, and thus are usually not normalized or adjusted for tissue volume,

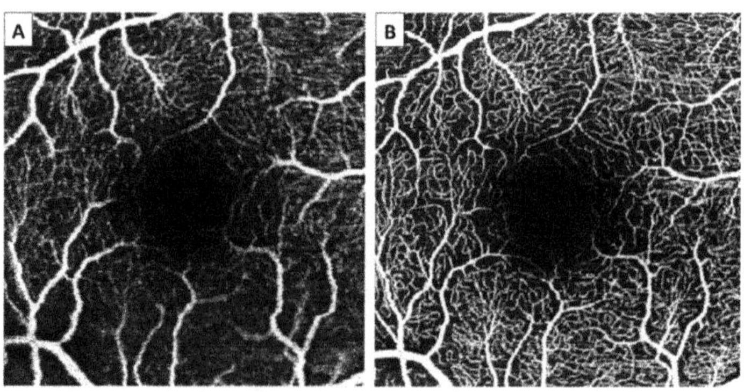

Figure 5.4. Impact of artifact on visualization of the SVP with OCTA, in a single eye. These SVP images were acquired from the same eye of the same subject during a single sitting, using Spectralis OCTA (Heidelberg Engineering, Germany). Image A is degraded by loss of focus artifact, resulting in a lack of definition of the capillary architecture, and apparent 'dropout' of the smallest vessels. By contrast, image B is a high-quality image with minimal artifact. The obvious difference between these images illustrates how artifacts can impair the qualitative and quantitative interpretation of OCTA images.

which may affect the interpretation of the vascular parameters. Moreover, the histological correlate of altered vascular plexuses is uncertain in MS and NMOSD. Since OCTA depends on vascular flow to generate images, reduced vascular plexus densities can represent loss of capillaries, shrinking of capillaries, or simply slow flow below the limits of motion-based detection. The pathobiological processes driving vascular density alterations in demyelinating CNS disease could include reduced metabolic demand due to retinal atrophy, altered metabolism due to neuronal dysfunction, primary vascular processes, or a combination of these factors. Retinal periphlebitis is also well-documented in MS [5, 21], and may impair retinal vascular flow. Furthermore, metabolic and vascular diseases such as hypertension and diabetes mellitus can impact the retinal vasculature, making interpretation of findings complex in patients with co-morbidities.

5.5 Conclusions

OCTA has become a useful non-invasive tool to examine the retinal vasculature *in vivo*. MS and NMOSD are neuroimmune disorders with prominent anterior visual pathway involvement, and there is a growing body of research employing OCTA in these disorders. Retinal vascular measurements appear to be reduced across MS eyes, and may correlate with visual function and global disability, thus representing a potential biomarker of disease. In NMOSD, research is more limited but also suggests reduced retinal vascular measurements, particularly in eyes previously affected by ON. OCTA technology continues to improve, yet there are a number of limitations which should be considered—including technical limitations and challenges in the interpretation of findings. Further research should focus on a range of issues including the impact of demographic characteristics and co-morbidities on OCTA measurements in individuals; the standardization of quality control processes across research studies; and longitudinal studies examining changes in retinal vascular measures over time in MS and NMOSD.

References

[1] Ikuta F and Zimmerman H M 1976 Distribution of plaques in seventy autopsy cases of multiple sclerosis in the united states *Neurology* **26** 26–8

[2] Toussaint D, Périer O, Verstappen A and Bervoets S 1983 Clinicopathological study of the visual pathways, eyes, and cerebral hemispheres in 32 cases of disseminated sclerosis *J. Clin. Neuroophthalmol.* **3** 211–20

[3] Kitley J, Leite M I and Nakashima I *et al* 2012 Prognostic factors and disease course in aquaporin-4 antibody-positive patients with neuromyelitis optica spectrum disorder from the United Kingdom and Japan *Brain: A J. Neurol.* **135** 1834–49

[4] Kim S, Kim W, Li X F, Jung I and Kim H J 2012 Clinical spectrum of CNS aquaporin-4 autoimmunity *Neurology* **78** 1179–85

[5] Green A J, McQuaid S, Hauser S L, Allen I V and Lyness R 2010 Ocular pathology in multiple sclerosis: retinal atrophy and inflammation irrespective of disease duration *Brain* **133** 1591–601

[6] Kerrison J B, Flynn T and Green W R 1994 Retinal pathologic changes in multiple sclerosis *Retina (Philadelphia, Pa)* **14** 445–51

[7] Minagar A, Jy W and Jimenez J J *et al* 2001 Elevated plasma endothelial microparticles in multiple sclerosis *Neurology* **56** 1319–24

[8] Aksungar F B, Topkaya A E, Yildiz Z, Sahin S and Turk U 2008 Coagulation status and biochemical and inflammatory markers in multiple sclerosis *J. Clin. Neurosci.* **15** 393–7

[9] Dardiotis E, Arseniou S and Sokratous M *et al* 2017 Vitamin B12, folate, and homocysteine levels and multiple sclerosis: a meta-analysis *Mult. Scler. Relat. Disord.* **17** 190–7

[10] Cramer S P, Modvig S, Simonsen H J, Frederiksen J L and Larsson H B W 2015 Permeability of the blood–brain barrier predicts conversion from optic neuritis to multiple sclerosis *Brain* **138** 2571–83

[11] Wakefield A J, More L J, Difford J and McLaughlin J E 1994 Immunohistochemical study of vascular injury in acute multiple sclerosis *J. Clin. Pathol.* **47** 129–33

[12] Adams C W 1988 Perivascular iron deposition and other vascular damage in multiple sclerosis *J. Neurol. Neurosurg. Psychiatry* **51** 260–5

[13] Popescu B F G and Lucchinetti C F 2016 Immunopathology: autoimmune glial diseases and differentiation from multiple sclerosis *Handb. Clin. Neurol.* **133** 95–106

[14] Saindane A M, Law M, Ge Y, Johnson G, Babb J S and Grossman R I 2007 Correlation of diffusion tensor and dynamic perfusion MR imaging metrics in normal-appearing corpus callosum: support for primary hypoperfusion in multiple sclerosis *AJNR Am. J. Neuroradiol.* **28** 767–72

[15] Law M, Saindane A M and Ge Y *et al* 2004 Microvascular abnormality in relapsing-remitting multiple sclerosis: perfusion MR imaging findings in normal-appearing white matter *Radiology* **231** 645–52

[16] Varga A W, Johnson G, Babb J S, Herbert J, Grossman R I and Inglese M 2009 White matter hemodynamic abnormalities precede sub-cortical gray matter changes in multiple sclerosis *J. Neurol. Sci.* **282** 28–33

[17] Adhya S, Johnson G, Herbert J, Jaggi H, Babb J S, Grossman R I and Inglese M 2006 Pattern of hemodynamic impairment in multiple sclerosis: dynamic susceptibility contrast perfusion MR imaging at 3.0 T *Neuroimage* **33** 1029–35

[18] Zhang X, Guo X and Zhang N *et al* 2018 Cerebral blood flow changes in multiple sclerosis and neuromyelitis optica and their correlations with clinical disability *Front. Neurol.* **9** 305

[19] Maggi P, Absinta M and Grammatico M *et al* 2018 Central vein sign differentiates multiple sclerosis from central nervous system inflammatory vasculopathies *Ann. Neurol.* **83** 283–94

[20] Absinta M, Nair G, Sati P, Cortese I C M, Filippi M and Reich D S 2015 Direct MRI detection of impending plaque development in multiple sclerosis *Neurol. Neuroimmunol. Neuroinflamm.* **2** e145

[21] Rucker C W 1944 Sheathing of the retinal veins in multiple sclerosis *Mayo. Clin. Proc.* **19** 176–8

[22] Hamann S, Zeuthen T, Cour M L, Nagelhus E A, Ottersen O P, Agre P and Nielsen S 1998 Aquaporins in complex tissues: distribution of aquaporins 1–5 in human and rat eye *Am. J. Physiol.: Cell Physiol.* **274** C1332

[23] Zeka B, Hastermann M and Kaufmann N *et al* 2016 Aquaporin 4-specific T cells and NMO-IgG cause primary retinal damage in experimental NMO/SD *Acta Neuropathol. Commun.* **4** 82

[24] Younge B R 1976 Fluorescein angiography and retinal venous sheathing in multiple sclerosis *Can. J. Ophthalmol.* **11** 31–6

[25] Birch M K, Barbosa S, Blumhardt L D, O'Brien C and Harding S P 1996 Retinal venous sheathing and the blood-retinal barrier in multiple sclerosis *Arch. Ophthalmol.* **114** 34–9

[26] Lightman S, McDonald W I, Bird A C, Francis D A, Hoskins A, Batchelor J R and Halliday A M 1987 Retinal venous sheathing in optic neuritis. its significance for the pathogenesis of multiple sclerosis *Brain* **110** 405–14

[27] Green A J and Cree B A C 2009 Distinctive retinal nerve fibre layer and vascular changes in neuromyelitis optica following optic neuritis *J. Neurol. Neurosurg. Psychiatry* **80** 1002–5

[28] Dutta R and Trapp B D 2011 Mechanisms of neuronal dysfunction and degeneration in multiple sclerosis *Prog. Neurobiol.* **93** 1–12

[29] Correale J, Gaitán M I, Ysrraelit M C and Fiol M P 2017 Progressive multiple sclerosis: from pathogenic mechanisms to treatment *Brain* **140** 527–46

[30] Trapp B D, Peterson J, Ransohoff R M, Rudick R, Mörk S and Bö L 1998 Axonal transection in the lesions of multiple sclerosis *N. Engl. J. Med.* **338** 278–85

[31] Petzold A, Balcer L J and Calabresi P A *et al* 2017 Retinal layer segmentation in multiple sclerosis: a systematic review and meta-analysis *Lancet Neurol.* **16** 797–812

[32] Petzold A, de Boer J F and Schippling S *et al* 2010 Optical coherence tomography in multiple sclerosis: a systematic review and meta-analysis *The Lancet Neurol.* **9** 921–32

[33] Saidha S, Syc S B and Ibrahim M A *et al* 2011 Primary retinal pathology in multiple sclerosis as detected by optical coherence tomography *Brain* **134** 518–33

[34] Syc S B, Saidha S and Newsome S D *et al* 2012 Optical coherence tomography segmentation reveals ganglion cell layer pathology after optic neuritis *Brain* **135** 521–33

[35] Nguyen J, Rothman A and Gonzalez N *et al* 2019 Macular ganglion cell and inner plexiform layer thickness is more strongly associated with visual function in multiple sclerosis than bruch membrane opening-minimum rim width or peripapillary retinal nerve fiber layer thicknesses *J. Neuroophthalmol.*

[36] Nguyen J, Rothman A and Fitzgerald K *et al* 2018 Visual pathway measures are associated with neuropsychological function in multiple sclerosis *Curr. Eye Res.* **43** 941–8

[37] Saidha S, Sotirchos E S and Oh J *et al* 2013 Relationships between retinal axonal and neuronal measures and global central nervous system pathology in multiple sclerosis *JAMA Neurol* **70** 34–43

[38] Saidha S, Syc S B and Durbin M K *et al* 2011 Visual dysfunction in multiple sclerosis correlates better with optical coherence tomography derived estimates of macular ganglion cell layer thickness than peripapillary retinal nerve fiber layer thickness *Mult. Scler.* **17** 1449–63

[39] Oh J, Sotirchos E S and Saidha S *et al* 2015 Relationships between quantitative spinal cord MRI and retinal layers in multiple sclerosis *Neurology* **84** 720–8

[40] Saidha S, Al-Louzi O and Ratchford J N *et al* 2015 Optical coherence tomography reflects brain atrophy in multiple sclerosis: a four-year study *Ann. Neurol.* **78** 801–13

[41] Ratchford J N, Saidha S and Sotirchos E S *et al* 2013 Active MS is associated with accelerated retinal ganglion cell/inner plexiform layer thinning *Neurology* **80** 47–54

[42] Button J, Al-Louzi O and Lang A *et al* 2017 Disease-modifying therapies modulate retinal atrophy in multiple sclerosis: a retrospective study *Neurology* **88** 525–32

[43] Gelfand J M, Nolan R, Schwartz D M, Graves J and Green A J 2012 Microcystic macular oedema in multiple sclerosis is associated with disease severity *Brain: A J. Neurol.* **135** 1786–93

[44] Saidha S, Sotirchos E S and Ibrahim M A *et al* 2012 Microcystic macular oedema, thickness of the inner nuclear layer of the retina, and disease characteristics in multiple sclerosis: a retrospective study *Lancet Neurol.* **11** 963–72

[45] Sotirchos E S, Caldito N G and Filippatou A *et al* 2020 Progressive multiple sclerosis is associated with faster and specific retinal layer atrophy *Ann. Neurol.*

[46] Lambe J, Murphy O and Saidha S 2018 Can optical coherence tomography be used to guide treatment decisions in adult or pediatric multiple sclerosis? *Curr. Treat. Options Neurol.* **20** 1–15

[47] Sotirchos E S and Saidha S 2018 OCT is an alternative to MRI for monitoring MS—YES *Mult. Scler.* **24** 701–3

[48] Lambe J, Saidha S and Bermel R A 2020 Optical coherence tomography and multiple sclerosis: update on clinical application and role in clinical trials *Mult. Scler.* **26** 624–39

[49] Green A, Bennett J and Lana-Peixoto M *et al* 2015 Neuromyelitis optica and multiple sclerosis: seeing differences through optical coherence tomography *Multiple Sclerosis J.* **21** 678–88

[50] Sotirchos E S, Filippatou A and Fitzgerald K C *et al* 2019 Aquaporin-4 IgG seropositivity is associated with worse visual outcomes after optic neuritis than MOG-IgG seropositivity and multiple sclerosis, independent of macular ganglion cell layer thinning *Mult. Scler. J.* **26** 1360–71

[51] Fernandes D, Ramos R I, Falcochio C, Apóstolos-Pereira S, Callegaro D and Monteiro M L 2012 Comparison of visual acuity and automated perimetry findings in patients with neuromyelitis optica or multiple sclerosis after single or multiple attacks of optic neuritis *J. Neuro-Ophthalmol.* **32** 102–6

[52] Peng A, Qiu X, Zhang L, Zhu X, He S, Lai W and Chen L 2017 Evaluation of the retinal nerve fiber layer in neuromyelitis optica spectrum disorders: a systematic review and meta-analysis *J. Neurol. Sci.* **383** 108–13

[53] Kaufhold F, Zimmermann H, Schneider E, Ruprecht K, Paul F, Oberwahrenbrock T and Brandt A U 2013 Optic neuritis is associated with inner nuclear layer thickening and microcystic macular edema independently of multiple sclerosis *PLoS One* **8** e71145

[54] Sotirchos E S, Saidha S and Byraiah G *et al* 2013 In vivo identification of morphologic retinal abnormalities in neuromyelitis optica *Neurology* **80** 1406–14

[55] Cheng L, Wang J, He X, Xu X and Ling Z F 2016 Macular changes of neuromyelitis optica through spectral-domain optical coherence tomography *Int. J. Ophthalmol.* **9** 1638–45

[56] Peng C, Wang W and Xu Q *et al* 2016 Structural alterations of segmented macular inner layers in Aquaporin4-antibody-positive optic neuritis patients in a chinese population *PLoS One* **11** e0157645

[57] Hu S J and Lu P R 2018 Retinal ganglion cell-inner plexiform and nerve fiber layers in neuromyelitis optica *Int. J. Ophthalmol.* **11** 89–93

[58] Jeong I, Kim H, Kim N, Jeong K and Park C 2016 Subclinical primary retinal pathology in neuromyelitis optica spectrum disorder *J. Neurol.* **263** 1343–8

[59] Pittock S J and Lucchinetti C F 2016 Neuromyelitis optica and the evolving spectrum of autoimmune aquaporin-4 channelopathies: a decade later *Ann. N. Y. Acad. Sci.* **1366** 20–39

[60] Reich D S, Lucchinetti C F and Calabresi P A 2018 Multiple sclerosis *N. Engl. J. Med.* **378** 169–80

[61] Kappos L, Wolinsky J S and Giovannoni G *et al* 2020 Contribution of relapse-independent progression vs relapse-associated worsening to overall confirmed disability accumulation in typical relapsing multiple sclerosis in a pooled analysis of 2 randomized clinical trials *JAMA Neurol.* **77** 1132–40

[62] Campbell J P, Zhang M, Hwang T S, Bailey S T, Wilson D J, Jia Y and Huang D 2017 Detailed vascular anatomy of the human retina by projection-resolved optical coherence tomography angiography *Sci. Rep.* **7** 42201

[63] Wang J C, Laíns I, Silverman R F, Sobrin L, Vavvas D G, Miller J W and Miller J B 2018 Visualization of choriocapillaris and choroidal vasculature in healthy eyes with en face swept-source optical coherence tomography versus angiography *Transl. Vis. Sci. Technol.* **7** 25
[64] Coffey A M, Hutton E K, Combe L, Bhindi P, Gertig D and Constable P A 2020 Optical coherence tomography angiography in primary eye care *Clin. Exp. Optom.* **104** 3–13
[65] Wang X, Jia Y and Spain R *et al* 2014 Optical coherence tomography angiography of optic nerve head and parafovea in multiple sclerosis *Br. J. Ophthalmol.* **98** 1368–73
[66] Spain R I, Liu L, Zhang X, Jia Y, Tan O, Bourdette D and Huang D 2018 Optical coherence tomography angiography enhances the detection of optic nerve damage in multiple sclerosis *Br. J. Ophthalmol.* **102** 520–4
[67] Lanzillo R, Cennamo G and Criscuolo C *et al* 2017 Optical coherence tomography angiography retinal vascular network assessment in multiple sclerosis *Mult. Scler.* **24** 1706–14
[68] Feucht N, Maier M and Lepennetier G *et al* 2018 Optical coherence tomography angiography indicates associations of the retinal vascular network and disease activity in multiple sclerosis *Mult. Scler.* **25** 224–34
[69] Lanzillo R, Cennamo G and Moccia M *et al* 2018 Retinal vascular density in multiple sclerosis: a one year follow up *Eur. J. Neurol.* **26** 198–201
[70] Murphy O C, Kwakyi O and Iftikhar M *et al* 2019 Alterations in the retinal vasculature occur in multiple sclerosis and exhibit novel correlations with disability and visual function measures *Mult. Scler.* **26** 815–28
[71] Yilmaz H, Ersoy A and Icel E 2020 Assessments of vessel density and foveal avascular zone metrics in multiple sclerosis: an optical coherence tomography angiography study *Eye (Lond.)* **34** 771–8
[72] Fard M A, Jalili J, Sahraiyan A, Khojasteh H, Hejazi M, Ritch R and Subramanian P S 2018 Optical coherence tomography angiography in optic disc swelling *Am. J. Ophthalmol.* **191** 116–23
[73] Fard M, Yadegari S, Ghahvechian H, Moghimi S, Soltani-Moghaddam R and Subramanian P 2019 Optical coherence tomography angiography of a pale optic disc in demyelinating optic neuritis and ischemic optic neuropathy *J. Neuro-Ophthalmol.* **39** 339–44
[74] Martinez-Lapiscina E H, Arnow S and Wilson J A *et al* 2016 Retinal thickness measured with optical coherence tomography and risk of disability worsening in multiple sclerosis: a cohort study *Lancet Neurol* **15** 574–84
[75] Chen Y, Shi C, Zhou L, Huang S, Shen M and He Z 2020 The detection of retina microvascular density in subclinical aquaporin-4 antibody seropositive neuromyelitis optica spectrum disorders *Front. Neurol.* **11** 35
[76] Huang Y, Zhou L and ZhangBao J *et al* 2019 Peripapillary and parafoveal vascular network assessment by optical coherence tomography angiography in aquaporin-4 antibody-positive neuromyelitis optica spectrum disorders *Br. J. Ophthalmol.* **103** 789–96
[77] Kwapong W R, Peng C, He Z, Zhuang X, Shen M and Lu F 2018 Altered macular microvasculature in neuromyelitis optica spectrum disorders *Am. J. Ophthalmol.* **192** 47–55
[78] Wang J C and Miller J B 2019 Optical coherence tomography angiography: review of current technical aspects and applications in chorioretinal disease *Semin. Ophthalmol.* **34** 211–7
[79] Diaz J D, Wang J C and Oellers P *et al* 2018 Imaging the deep choroidal vasculature using spectral domain and swept source optical coherence tomography angiography *J. Vitreoretin. Dis.* **2** 146–54

[80] Spaide R F, Fujimoto J G and Waheed N K 2015 Image artifacts in optical coherence tomography *Retina* **35** 2163–80

[81] Enders C, Lang G E, Dreyhaupt J, Loidl M, Lang G K and Werner J U 2019 Quantity and quality of image artifacts in optical coherence tomography angiography *PLoS One* **14** e0210505

[82] Iftikhar M, Zafar S and Gonzalez N *et al* 2019 Image artifacts in optical coherence tomography angiography among patients with multiple sclerosis *Curr. Eye Res.* **44** 558–63

[83] Zheng F, Zhang Q and Shi Y *et al* 2019 Age-dependent changes in the macular choriocapillaris of normal eyes imaged with swept-source optical coherence tomography angiography *Am. J. Ophthalmol.* **200** 110–22

[84] Chun L Y, Silas M R, Dimitroyannis R C, Ho K and Skondra D 2019 Differences in macular capillary parameters between healthy black and white subjects with optical coherence tomography angiography (OCTA) *PLoS One* **14** e0223142

Chapter 6

Optical coherence tomography angiography for the diagnosis of polypoidal choroidal vasculopathy

Talisa De Carlo and Gregg T Kokame

Optical coherence tomography angiography (OCTA) is a novel non-invasive technology that can be useful for detecting and monitoring polypoidal choroidal vasculopathy (PCV) lesions. It is important to distinguish this subtype from other forms of age-related macular degeneration because the treatment algorithm may be different and it predicts anti-vascular endothelial growth factor resistance in all ethnic populations. We describe how to detect PCV using both the OCT angiograms and corresponding b-scans.

6.1 Introduction

Optical coherence tomography angiography (OCTA) is a relatively novel approach to visualizing the retinal and choroidal vasculature *in vivo*. The technique uses non-invasive motion-contrast to generate volumetric angiographic images in less than one minute. OCTA compares the differences in backscattered OCT signal intensity, or decorrelation signal, between sequential OCT b-scans obtained at each given cross-section of the retina. The decorrelation signal between repeated OCT b-scans are assumed to correlate to erythrocyte movement within the retinal vasculature and can therefore be utilized to construct a blood flow map of the retina. OCTA images correlate well with fluorescein angiography (FA) and indocyanine green angiography (ICGA), providing even more detailed and high-resolution images of the retinal and choroidal microvasculature. Furthermore, the OCTA images can be segmented to show the individual vascular plexuses.

Each OCTA image set contains *en face* OCT angiograms, OCT b-scans, and structural *en face* OCT images that are co-registered in order to visualize both

doi:10.1088/978-0-7503-2060-3ch6

structural and blood flow information in tandem. This allows for colocalization of microvascular pathology to the structural images. Using the entire OCTA image set or multiple components offers more information than just the OCT angiogram or OCT b-scans alone. This can be particularly helpful in diseases such as polypoidal choroidal vasculopathy (PCV), as we will see in this chapter.

PCV is a subtype of exudative or wet age-related macular degeneration (AMD), in which polypoidal-shaped dilated lesions are a prominent part of the choroidal neovascular network. This subtype has a different prevalence in different ethnic populations [1, 2]. In Asian populations PCV can often make up over 50% of patients presenting with exudative AMD. Caucasian patients were initially felt to have a low incidence of PCV less than 10%, but with better screening with ICGA using the scanning laser ophthalmoscope, prevalence has now been shown to be 20%–31% in Caucasian populations. Although wet AMD is infrequent in the black population, when it occurs, it is usually associated with PCV.

PCV is the most important subtype of exudative AMD to identify, because it predicts for anti-VEGF resistance [3]. PCV eyes are more anti-VEGF resistant than typical wet AMD eyes, and this is critical, as anti-VEGF medications, such as bevacizumab, ranibizumab, aflibercept, and brolucizumab, are the first line treatment for exudative AMD. In the recent EVEREST II study, it was shown that combination therapy with verteporfin photodynamic therapy (vPDT) not only resulted in better vision than anti-VEGF monotherapy, but this was accomplished with one half the number of intravitreal injections compared to anti-VEGF monotherapy [4]. In addition, PCV responds different to different anti-VEGF agents with a better response anatomically with brolucizumab, which is better than aflibercept, which is better than ranibizumab [5].

6.2 OCTA for the diagnosis of PCV

OCTA is a promising and continually developing new technology for noninvasive imaging of choroidal neovascularization related to exudative AMD. For the diagnosis of PCV OCTA is not as sensitive as the gold standard of ICGA, as the diagnostic polypoidal lesions are often not seen on OCTA, possibly due to slower blood flow through the polypoidal lesions. However, the branching vascular network (BVN) is often very well imaged with OCTA, again possibly due to faster blood flow through the BVN. Because of the non-invasive nature of OCTA, it may be very useful in following the response to therapy of PCV anatomically.

6.2.1 OCT angiograms

The PCV complex can be visualized on the OCT angiogram as a branching vascular network (BVN) and circular flow lesions representing polyps [6]. The polyps can be seen as either high-flow vascular dilations or faint low-flow lesions (figure 6.1). PCV is a type 1 choroidal neovascular complex, located between Bruch's membrane and the retinal pigment epithelium. Therefore, segmentation of the OCT angiogram to this region or to the choriocapillaris can give a higher yield for detection of the PCV complex.

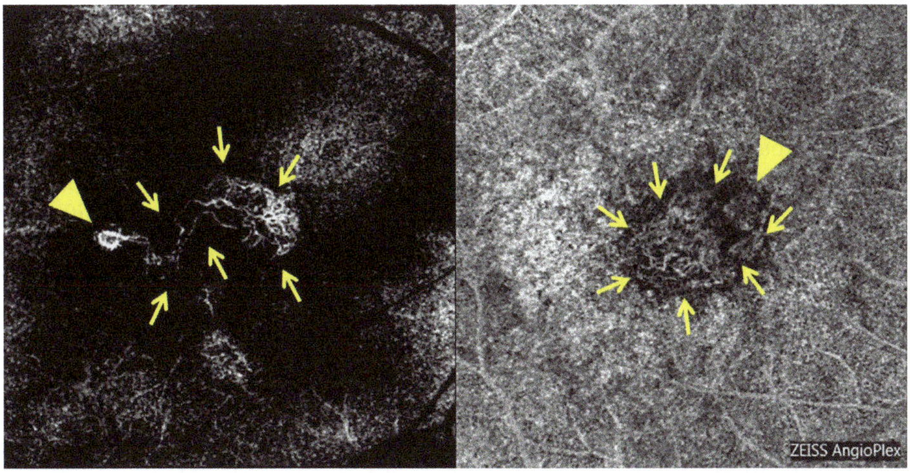

Figure 6.1. Two examples of optical coherence tomography angiograms of polypoidal choroidal vasculopathy complexes. The polyps (arrowheads) can be seen as either vascular dilations (left) or faint low-flow lesions (right) adjacent to branching vascular networks (arrows). This is seen best on the outer retina (left) or choriocapillaris (right) segmentations.

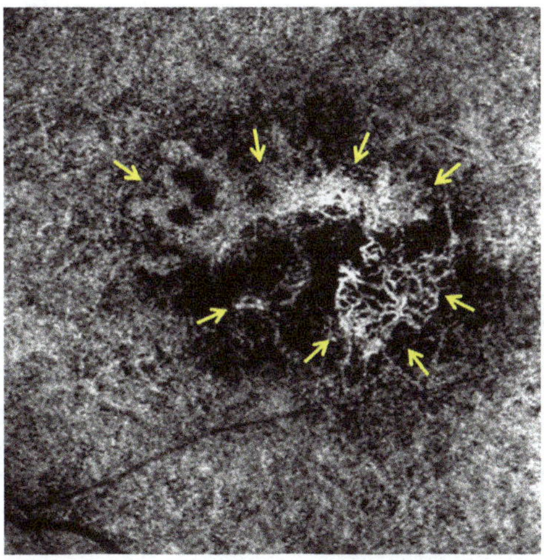

Figure 6.2. The branching vascular network (arrows) of the polypoidal choroidal vasculopathy complex is readily seen on the optical coherence tomography angiograms but no polyps are definitively visualized although the rounded branches at the upper left aspect of the complex are suspicious for possible polyps. The corresponding optical coherence tomography b-scans are invaluable in equivocal cases such as these.

The BVN is readily delineated on the OCT angiogram but the polyps are visualized less consistently ranging from approximately 75%–92% of the time (figure 6.2) [6–10]. Detection of the polyps is limited by a couple of factors. Polyps that are too small or have too slow flow can be missed on the OCT angiogram because of the principle of slowest detectable flow determined by the time between repeated OCT b-scans [11]. If the blood flow within a polyp is slower than the time between repeated OCT b-scans, then no decorrelation signal will be produced, resulting in no blood flow on the OCT angiogram. One study compared characteristics of visualized versus non-visualized polyps on OCTA noting that polyps with a higher height, that have a pulsatile nature, or that have overlying hemorrhage may be less likely to be visualized on OCTA [12].

Furthermore, in eyes that have undergone treatment the polyps may regress while the BVN remains. Therefore, detecting the entire PCV complex can be difficult in previously treated eyes. Studies comparing OCTA to ICGA in treatment naïve eyes demonstrated that the imaging modalities were comparable for detecting of the PCV complex including the polyps [13].

6.2.2 OCT b-scans

As previously mentioned, the OCTA image set contains multiple OCT b-scans co-registered with the OCT angiogram. It is useful to review these b-scans in concert with the OCT angiograms. PCV is seen on OCT b-scans as elongated inverted u-shaped pigment epithelial detachments (PED) with hyperreflective polyps adherent to the underside of the PED, and an adjacent low-lying PED dubbed a 'double layer sign' that represents the BVN (figure 6.3). Using the OCT b-scans alone has a reasonably high sensitivity and specificity for detection of PCV [14, 15].

6.2.3 OCTA image set

Viewing the OCT angiogram and co-registered OCT b-scans in concert can improve the sensitivity of detection of the PCV complex with minimal to no compromise to

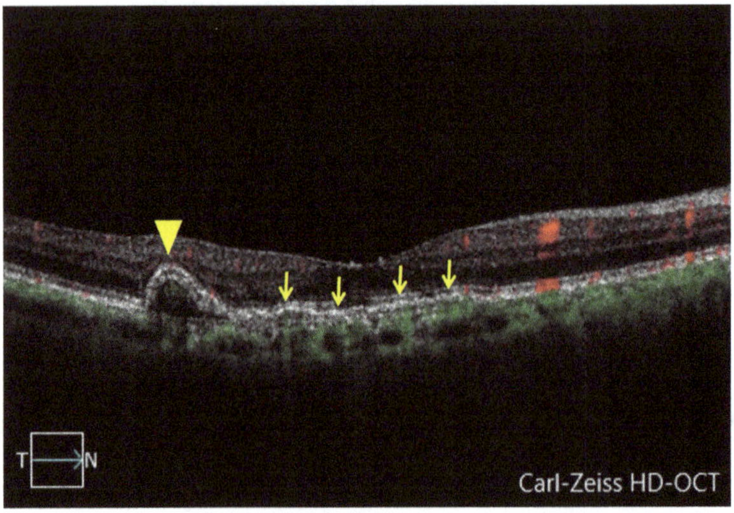

Figure 6.3. The polypoidal choroidal vasculopathy complex is shown on the corresponding optical coherence tomography b-scans as an inverted u-shaped polyp (arrowhead) adjacent to a double layer sign or low level pigment epithelial detachment representing the branching vascular network (arrows).

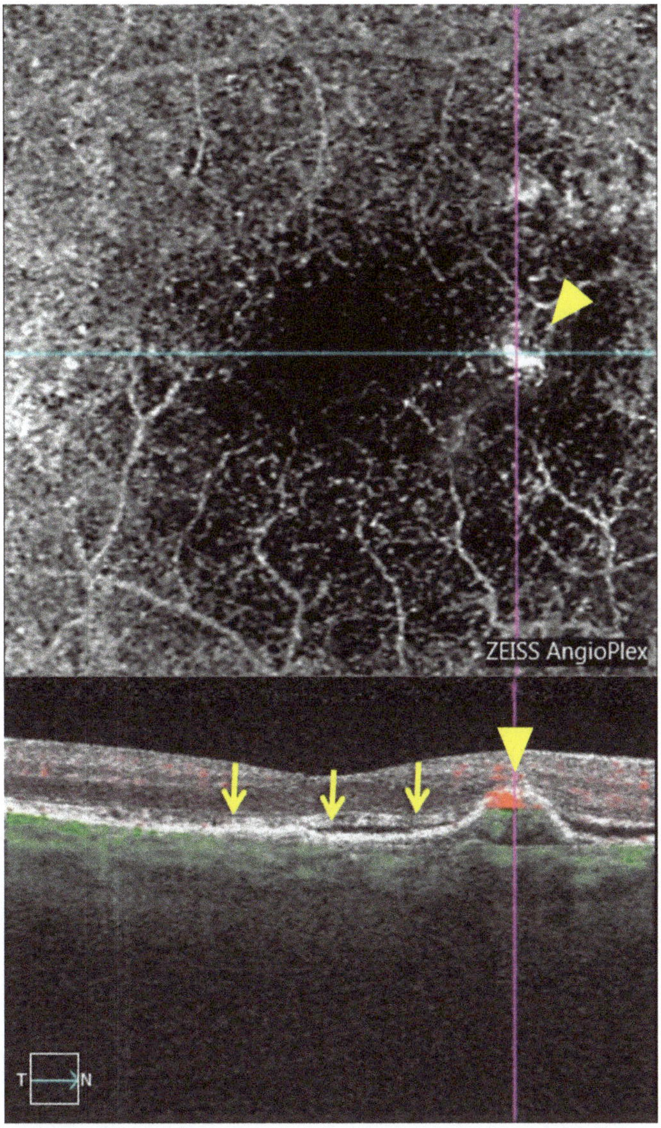

Figure 6.4. There is a circular area of flow noted on the optical coherence tomography angiogram that could represent a polyp (arrowhead, top). The corresponding optical coherence tomography shows an inverted u-shaped lesion (arrowhead) adjacent to a double layer sign or low-level pigment epithelial detachment representing the branching vascular network (arrows). The cross-hairs colocalize the circular area of flow on the angiogram with the inverted u-shaped lesion on the b-scan increasing our suspicion that this lesion is a polyp.

specificity (figure 6.4) [16, 17]. Oftentimes clinicians use only the OCT angiogram to evaluate PCV, but the co-registered OCT b-scans can be an asset by providing additional clues and signs that a PCV complex is present. The blood flow and structural changes can be cross-referenced in order to better characterize the lesions.

Interpreting the OCTA is limited by the presence of artifacts and inaccuracies in the automated segmentation. These artifacts and segmentation discrepancies are important to keep in mind as they can both falsely appear as or mask a PCV complex. Poor signal quality due to cataracts or other media opacity can further confuse the picture. Therefore, there is a learning curve for accurately interpreting OCTA images and differentiating artifact from true pathology.

6.3 Discussion

In this chapter, we presented an introduction to the OCTA technology and how it can be used to facilitate the diagnosis of PCV. As OCTA is a non-invasive and time-efficient addition to the standard OCT machines frequently utilized to image PCV, understanding how to use OCTA to visualize the PCV complex can be a useful clinical skill. PCV is the most important subtype of exudative AMD, because it predicts for anti-VEGF resistance and also can guide therapy, such as choice of anti-VEGF therapy, and whether or not to use combination vPDT, which has been shown to have better vision results than anti-VEGF monotherapy with one half the number of injections for anti-VEGF monotherapy. We demonstrated that while OCTA is limited by artifacts and the principle of slowest detectable flow, the technology can accurately detect the PCV complex in many cases. This is particularly true in treatment of naïve eyes and when viewing the OCTA dataset as a whole—by using the OCT angiogram and OCT b-scans together. We described the appearance of the polyps and the BVN on the OCTA dataset.

6.4 Conclusion

In conclusion, OCTA is a novel non-invasive technology that can be a useful tool in the clinician's arsenal for detecting and monitoring PCV lesions. It is important to distinguish the PCV subtype from other forms of AMD because the treatment algorithm may be different and it predicts anti-VEGF resistance in all ethnic populations.

References

[1] Kokame G T, Liu K, Kokame K A, Kaneko K N and Omizo J N 2020 Clinical characteristics of polypoidal choroidal vasculopathy and anti-vascular endothelial growth factor treatment response in caucasians *Ophthalmologica* **243** 178–86

[2] Pereira F B, Veloso C E, Nehemy M B and Kokame G T 2015 Characterization of neovascular macular degeneration in Brazilian patients *Ophthalmologica* **234** 233–42

[3] Kokame G T, deCarlo T E, Kaneko K N, Omizo J N and Lian R 2019 Anti-vascular endothelial growth factor resistance in exudative macular degeneration and polypoidal choroidal vasculopathy *Ophthalmol. Retin.* **3** 744–52

[4] Lim T H, Lai T Y Y and Takahshi K *et al* 2020 Comparison of ranibizumab with or without verteporfin photodynamic therapy for polypoidal choroidal vasculopathy: the EVEREST II randomized clinical trial *JAMA Ophthalmol.* **138** 935–42

[5] Kokame G T, Lai J C, Wee R, Yanagihara R, Shantha J G, Ayabe J and Hirai K 2016 Prospective clinical trial of intravitreal aflibercept treatment for polyploidal choroidal vasculopathy with hemorrhage or exudation (EPIC study): 6 month results *BMC Ophthalmol.* **16** 1–10
[6] de Carlo T E, Kokame G T, Shantha J G, Lai J C and Wee R *et al* 2018 Spectral-domain optical coherence tomography angiography for the diagnosis and evaluation of polypoidal choroidal vasculopathy *Ophthalmologica.* **239** 103–9
[7] Wang M, Zhou Y and Gao S S *et al* 2016 Evaluating polypoidal choroidal vasculopathy with optical coherence tomography angiography *Invest. Ophthalmol. Vis. Sci.* **57** 526–32
[8] Takayama K, Ito Y and Kaneko H *et al* 2017 Comparison of indocyanine green angiography and optical coherence tomographic angiography in polypoidal choroidal vasculopathy *Eye* **31** 45–52
[9] Tomiyasu T, Nozaki M, Yoshida M and Ogura Y 2016 Characteristics of polypoidal choroidal vasculopathy evaluated by optical coherence tomography angiography *Invest. Ophthalmol. Vis. Sci.* **57** 324–30
[10] Kim J Y, Kwon O W and Oh H S *et al* 2016 Optical coherence tomography angiography in patients with polypoidal choroidal vasculopathy *Graefes. Arch. Clin. Exp. Ophthalmol.* **254** 1505–10
[11] Fukuyama H, Iwami H, Araki T, Ishikawa H, Ikeda N and Gomi F 2018 Indocyanine green dye filling time for polypoidal lesions in polypoidal choroidal vasculopathy affects the visibility of the lesions on OCT angiography *Ophthalmol. Retin.* **2** 803–7
[12] Zhan Z, Sun L and Jin C *et al* 2019 Comparison between non-visualized polyps and visualized polyps on optical coherence tomography angiography in polypoidal choroidal vasculopathy *Graefes. Arch. Clin. Exp. Ophthalmol.* **257** 2349–56
[13] Kim K, Yang J and Feuer W *et al* 2020 A comparison study of polypoidal choroidal vasculopathy imaged with indocyanine green angiography and swept source OCT angiography *Am. J. Ophthalmol.* **S0002–9394** 30247–6
[14] De Salvo G, Vaz-Pereira S and Keane P A *et al* 2014 Sensitivity and specificity of spectral-domain optical coherence tomography in detecting idiopathic polypoidal choroidal vasculopathy *Am. J. Ophthalmol.* **158** 1228–38 E1
[15] Liu R, Li J and Li Z *et al* 2016 Distinguishing polypoidal choroidal vasculopathy from typical neovascular age-related macular degeneration based on spectral domain optical coherence tomography *Retina* **36** 778–86
[16] de Carlo T E, Kokame G T, Kaneko K N, Lian R, Lai J C and Wee R 2019 Determining the sensitivity and specificity of detecting PCV with en-face OCT and OCTA *Retina.* **39** 1343–52
[17] Cheung C M G, Yanagi Y and Akiba M *et al* 2019 Improved detection and diagnosis of polypoidal choroidal vasculopathy using a combination of optical coherence tomography and optical coherence tomography angiography *Retina* **39** 1655–63

IOP Publishing

Photo Acoustic and Optical Coherence Tomography Imaging, Volume 3

Angiography: an application in vessel imaging

Ayman El-Baz and Jasjit S Suri

Chapter 7

Quantitative features for objective assessment of OCT angiography

Minhaj Nur Alam, David Le, Taeyoon Son, Jennifer I Lim and Xincheng Yao

As a new quantitative imaging modality, optical coherence tomography (OCT) angiography (OCTA) provides a non-invasive platform for detecting microvascular abnormalities caused by different eye diseases. Since its inception, OCTA has generated widespread interest because of its unprecedented capability to visualize high resolution capillary structures in superficial and deep retina plexuses. In recent years, researchers have demonstrated several quantitative imaging biomarkers for describing retinal distortions caused by diseases such as diabetic retinopathy (DR), diabetic macular edema (DME), age-related macular degeneration (AMD), glaucoma, vein occlusion, sickle cell retinopathy (SCR), etc. However, it is also essential to standardize the quantitative OCTA features for integrating them in clinical workflow and to establish objective interpretations of clinical outcomes. In this chapter, we summarize the state of the art of quantitative OCTA features with technical rationale and describe potential clinical applications of quantitative OCTA analysis. In addition, we also discuss the artificial intelligence (AI) based applications demonstrated for OCTA disease classification, feature segmentation and image quality improvement along with their future prospect in retinal disease management.

7.1 Introduction

The retina is regularly targeted by eye diseases. As an extension of the central nervous system (CNS), the retinal neurovascular complex can serve as a window for assessing functionality of the brain or cardiovascular conditions. Systemic health conditions such as diabetes can also cause retinal abnormalities such as diabetic retinopathy (DR) and macular edema (DME). Long- term existence of diabetes can cause hyperglycemia-induced vascular damage [1], hyperlipidemia and

doi:10.1088/978-0-7503-2060-3ch7

hypertension. Moreover, blood clots and atherosclerosis can further develop into pathologies such as retinal vein occlusions (RVO) [2]. Therefore, in ophthalmology practices, retinal imaging has been extensively used for screening and diagnosing systematic diseases. Different imaging modalities include color fundus photography, scanning laser ophthalmoscope (SLO), fluorescein angiography (FA), etc. Color fundus photography is a widely used modality which provides valuable information for disease detection and treatment assessment. However, the spatial resolution, field of view and image contrast are limited to identify subtle distortions in early stages of retinal diseases. Furthermore, many diseases manifest within the distinct retinal layers, which cannot be observed just by visualizing the fundus of the eye. SLO [3, 4] and adaptive optics (AO) [5–7] can provide enhanced image resolution, while FA [8, 9] allows identification of vascular leakage and hemorrhage within the retina.

However, these imaging modalities lack the capability to differentiate individual retinal layers and vascular plexuses. It has been established that different diseases and their stages can target retinal vasculatures in different ways, within different layers of the retina. Given the unprecedented capability to non-invasively differentiate individual retinal layers, optical coherence tomography (OCT) [10] has been widely employed for depth-resolved investigation of morphological abnormalities caused by different retinal diseases [11–13]. OCT angiography (OCTA) is a recent OCT imaging modality that can provide a non-invasive method to visualize blood flow information in individual capillary plexus layers in the outer and inner retina [14, 15] with high resolution. Since its first commercial launch in 2014, OCTA has rapidly demonstrated its excellence as a quantitative imaging technique in clinical management of diabetic retinopathy (DR) [16, 17], glaucoma [18, 19], sickle cell retinopathy (SCR) [20], age-related macular degeneration (AMD) [21] and other eye diseases. Standardized quantitative OCTA analysis is essential for objective interpretation of clinical outcomes and designing treatment strategies. Researchers have developed multiple OCTA features in recent years for quantitative analysis of vascular distortions caused by a myriad of eye conditions and systematic diseases [22]. In this chapter, we discuss, summarize and provide technical rationales of these quantitative OCTA features for a better understanding of current trend and state of the art. Additionally, current status and future prospects of using OCTA features for artificial intelligence (AI) based objective detection and classification of eye diseases or specific features will be discussed in brief.

7.2 Quantitative OCTA features

We will briefly introduce and provide brief technical rationale of different OCTA features developed in literature, including blood vessel density (BVD), blood vessel caliber (BVC), vessel perimeter index (VPI), blood vessel tortuosity (BVT), fovea avascular zone features (area-FAZ-A, contour irregularity-FAZ-CI), vessel complexity features, branchpoint features (BPF), differential artery and vein (A-V) features, and features based retinal flow analysis and choroidal vasculature quantification.

7.2.1 Blood vessel density

BVD is one of the most commonly analyzed OCTA features, and has been demonstrated to be associated with abnormalities associated with diseases such as DR [23, 24], SCR [25, 26], AMD [27, 28], glaucoma [29, 30] and vein occlusion (VO) [31, 32] which induce vessel abnormalities such as ischemia and drop-out areas in retinal vasculatures. Most of these diseases initially manifest at the capillary level, which can be detected in OCTA. Similar features that have been demonstrated in literature with different terminology include vessel density (VD) [33], percent area of nonperfusion (PAN) [34], capillary density (CD) [35], vessel area density (VAD) [36], etc. All together these density features reflect the ratio of the image area that is occupied by blood vessels (figure 7.1(b)). BVD can be quantified with the following equation [37]:

$$BVD = \frac{\sum_{x=1,y=1}^{n} A(x, y)}{\sum_{x=1,y=1}^{n} I(x, y)} \tag{7.1}$$

where $A(x,y)$ is all the pixels occupied by the blood vessels, and I(x,y) is all the pixels in the OCTA image. Similar metrics can be measured for vessel skeletons density as well, where A can be replaced with S (skeleton pixels).

7.2.2 Blood vessel caliber

BVC is a metric that is used to quantify the dilation or shrinkage of the blood vessels. Similar studies have also referred to BVC as vessel diameter, vessel width or vessel diameter index (VDI) [38]. BVC distortions are prevalent in patients with hypertension, DR [39] and SCR [37, 40]. BVC is defined as the ratio of the vascular area to the total vascular length [37]:

$$BVC = \frac{\sum_{x=1,y=1}^{n} A(x, y)}{\sum_{x=1,y=1}^{n} S(x, y)} \tag{7.2}$$

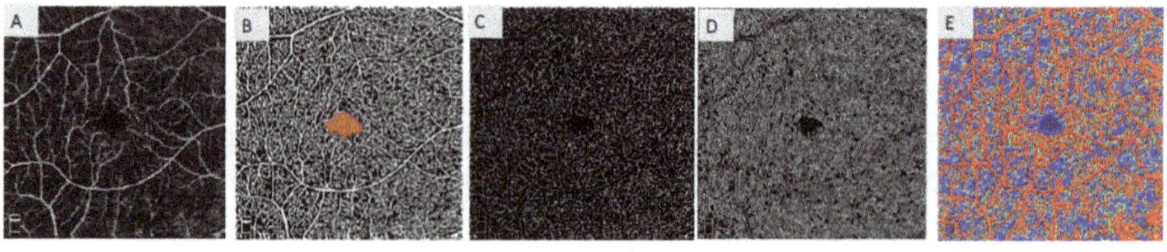

Figure 7.1. OCTA Feature extraction. (a) A representative 6 mm × 6 mm OCTA superficial scan from a DR patient. (b) Extracted vessel map showing segmented fovea in orange. (c) Skeletonized vessel map. (d) Vessel perimeter map. (e) Fractal contour map.

where $A(x,y)$ is the pixels in the segmented vessel map (figure 7.1(b)) and $S(x,y)$ is the pixels in the skeletonized vessel map (figure 7.1(c)). So, $\sum_{x=1,y=1}^{n} A(x, y)$ represents the total vascular area, and $\sum_{x=1,y=1}^{n} S(x, y)$ represents the total vessel length.

7.2.3 Blood vessel tortuosity

BVT [37, 40] is a measure of the geometric tortuosity change in blood vessels. It can be correlated to the degree of individual vessel distortion. Normally, apart from the regular patterning of the blood vessel network, for efficient blood/oxygen transport, the blood vessels have a relatively smooth structure. In diseased conditions [37, 39–42], the epithelial tissue surrounding the blood vessel walls are affected or in some cases, distorted shape of blood cells cause occlusion and distortions in the structure of the blood vessels, affecting the blood/oxygen transportation efficiency of blood vessel network. This structural distortion can be correlated to the vessel tortuosity change, which can be quantified by taking the ratio of the geodesic and Euclidean distance [37] between two vessel endpoints:

$$BVT = \frac{1}{n} \sum_{i=1}^{n} \left(\frac{\text{Geodesic distance between two endpoints of a vessel branch } i}{\text{Euclidean distance between two endpoints of a vessel branch } i} \right) \tag{7.3}$$

where i is equal to the ith branch and n is equal to the total number of branches; Euclidian distance here represent the straight-line distance between two vessel end points, whereas Geodesic distance represents the total curve length between two vessel end points (as shown in figure 7.1(c)).

7.2.4 Vessel perimeter index

VPI [43] is a metric that measures the ratio between overall vessel boundary (figure 7.1(d)) and total blood vessel area in segmented vessel maps (figure 7.1(b)). VPI has been demonstrated to reflect vessel dropout and early ischemia [39, 44], in DR [44] and SCR [40]. It can be measured using the following equation [37]:

$$VPI = \frac{\sum_{x=1,y=1}^{n} P(x, y)}{\sum_{x=1,y=1}^{n} I(x, y)} \tag{7.4}$$

where P(x,y) is the vessel perimeter pixels and I(x,y) is all the vessel pixels.

7.2.5 Foveal avascular zone features

The fovea is one of the most affected areas in the retina by retinal diseases such as DR [45], VO [46], and SCR [40] due to foveal ischemia and parafoveal vessel drop-outs. FAZ-A is a sensitive feature that can quantify the shape change of the fovea and has been demonstrated to be able to differentiate the severities of different non-proliferative diabetic retinopathy (NPDR) stages [39]. In general, FAZ-A is determined by first segmenting the FAZ (blue region in figure 7.1(c)) and then calculating the total segmented FAZ area with the following equation [37]:

$$FAZ - A = [\text{Area of single pixel(in } \mu\text{m}^2) \times \sum_{x=1,y=1}^{n} A(x, y)] \tag{7.5}$$

where A(x,y) is the segmented FAZ pixels.

FAZ-CI, is another foveal feature, that measures the structural irregularity of the fovea [40]. Recent studies has been also termed it as FAZ-circularity [40] or FAZ acircularity index (FAZ-AI) [47] and demonstrated FAZ-CI to be sensitive to vascular contortions caused by DR [47], AMD [48], SCR [37], and glaucoma [49]. The FAZ-CI can be measured as the ratio of the foveal perimeter and the perimeter of a reference circle that has an area identical to the fovea [37]:

$$FAZ - CI = \frac{\sum_{x=1,y=1}^{n1} O(x, y)}{\sum_{x=1,y=1}^{n2} R(x, y)} \tag{7.6}$$

where $O(x,y)$ is the foveal perimeter pixel (green demarcation—figure 7.1(c)), $R(x,y)$ is the reference circle pixel.

7.2.6 Vessel complexity features

Vessel complexity features have been shown to be sensitive in DR detection, in particularly for monitoring the progression of DR from the non-proliferative to the proliferative stage [50]. Vessel complexity index (VCI) is one of the complexity features and can be quantified from the perimeter map and vessel map [25]:

$$VCI = \frac{(\text{Pixels enclosed by perimeter map})^2}{4\pi \times (\text{Pixels enclosed by vessel map})} \tag{7.7}$$

VCI uniquely enables the localization of complex vascular morphologies within the retina.

Similarly, fractal dimension (FD) has also been used to quantify the complexity of the vasculature (figure 7.1(e)). FD is generally calculated using the box-counting method, which establishes a relationship between the number of boxes (of a certain size/scale) that can enclose the pattern within an image (figure 7.1(e)) [51].

This box-count can be iteratively measured for different scales. The box-counting method is equated as [51]:

$$FD = \frac{\log N_r}{\log r^{-1}} \tag{7.8}$$

where N_r is the number of boxes which enclose the image pattern by the scale, r. Another complimentary parameter to FD is Lacunarity (LAC) which is a measure of rotational inhomogeneity between vessel structures [52], specifically LAC provides information regarding the size of the gaps and their distribution around objects of interest in binary images.

7.2.7 Branchpoint analysis

Decreased efficiency in blood/oxygen transport in the retina can affect bifurcation in vascular structures [53]. Quantitative OCTA branchpoint analysis with BPFs, including both angle- and width-based features (figure 7.2), has been recently explored for objective detection of DR [54]. Angle parameters including the vessel branching angle (VBA) and the child branching angles (CBA1 and CBA2), quantify in vessel bifurcation changes [54]. To measure the branching angles, first the branchpoint and branch endpoints must be determined (green dots and red dots in figures 7.2(b) and (c), respectively). Then the branchpoint angles are measured. Analogously, two width-based parameters were also quantified, namely the vessel branching coefficient (VBC) and child width ratios (CWRs).

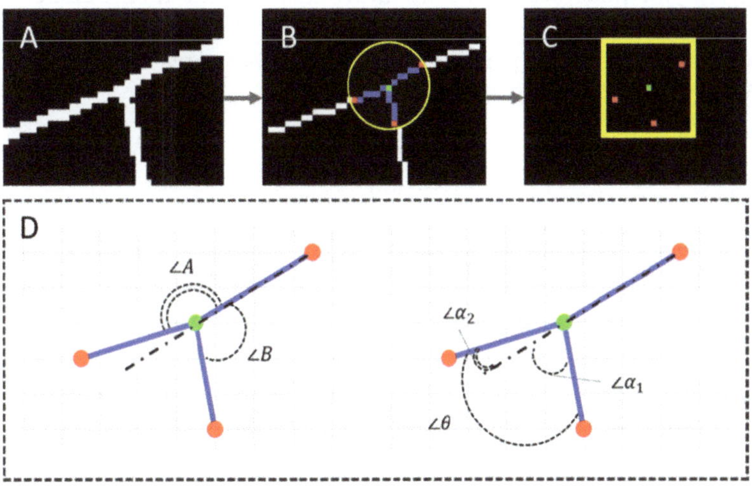

Figure 7.2. (a) Sample branchpoint; (b) Branchpoint showed in a vessel skeleton. Here the green dot represents a branchpoint and the red dots represent three end points, the blue pixels represent the parent and children vessels of interest, and the yellow circle shows the dilated area. (c) Identified branchpoint (green) and endpoints (red), the yellow square is the window area. (d) Branch angle measurement. Modified from [54].

Width-based features examine the structural change of the blood vessels as a result of bifurcation [54]:

$$VBC = \frac{d_1^2 + d_2^2}{d_0^2} \tag{7.9}$$

$$CWR_i = \frac{d_i}{d_0} \tag{7.10}$$

where d_0 is parent vessel width, d_1 and d_2 are the child vessel widths, and i represents the ith child vessel. As a result of ischemia and neovascularization, changes in branching angles and widths could be correlated to vascular remodeling.

7.2.8 Flow analysis

Several parameters have been developed to quantify the blood flow in retinal tissues, namely the flow index (FI) [55], flow void (FV) [56], adjusted flow index (AFI) [34], vascular connectivity [57], and total retinal blood flow (TRBF). FI is defined as the average OCTA decorrelation value within a region in the retina [58]. The FI can be measured using the following equation,

$$FI = \frac{\sum_{x=1,y=1}^{n} A(x, y)}{N} \tag{7.11}$$

where A(x,y) is the pixel decorrelation value, and N is the total number of OCTA pixels.

The FV is measured as the percentage of the non-flow area over the total scanned region [56]

$$FV = \frac{\text{Area}_{\text{Flowvoid}}}{\text{Area}_{\text{whole}}} \times 100\% \tag{7.12}$$

TRBF is a parameter demonstrated using Doppler OCT and can be used to estimate the flow in all the vessel segments [59]:

$$TRBF = -\iint_{xy-\text{plane}} v_z(x, y)dxdy \tag{7.13}$$

where the flow in a blood vessel is computed by integrating the axial flow velocity, $v_z(x, y)$ measured from the different pixels in the OCTA image over the surface normal to $v_z(x, y)$ [59]. This method is typically applied in OCTA systems with high acquisition speeds (e.g. 100 kHz).

Another OCTA parameter related to flow is named variable interscan time analysis (VISTA) [60, 61], which can be performed to assess the changes in choriocapillaris for differentiating varying degrees of flow impairment. In different analysis, VISTA has been demonstrated to quantify changes in AMD, geographic atrophy (GA) and NPDR/PDR eyes.

7.2.9 Choroidal vasculature

Aside from the retinal vasculature, OCTA features have also been demonstrated for analyzing choroidal vasculature. For example, qualitative observations of CNV using OCTA have revealed distinct morphological distortions such as medusa, seafan, tangled, and dead-tree type patterns within the choroidal vasculature [28]. CNV area is a standardized parameter and can be measured as follows [62]:

$$\text{CNV area}(\text{mm}^2) = \text{CNV area}(\text{pixel}) \times (\text{Field of view in mm}/304\ \text{pixel})^2 \quad (7.14)$$

The first step for CNV analysis is a reliable segmentation (figure 7.3). In literature, the segmentation has typically been performed using manual or semi-manual methods [14]. Recently, projection-resolved (PR) OCTA techniques have been explored to determine the CNV area and skeleton automatically [57]. The binarized CNV vessel or skeleton map can be used for FD analysis [28], and measurement of BVD or LAC. Other CNV features include CNV maturity, location, presence of core vessels and loops [63], etc.

7.2.10 Differential artery–vein (A–V) analysis

Differential A–V analysis in OCTA compares the changes due to diseases that affect artery and veins differently. There have been only few studies that demonstrated

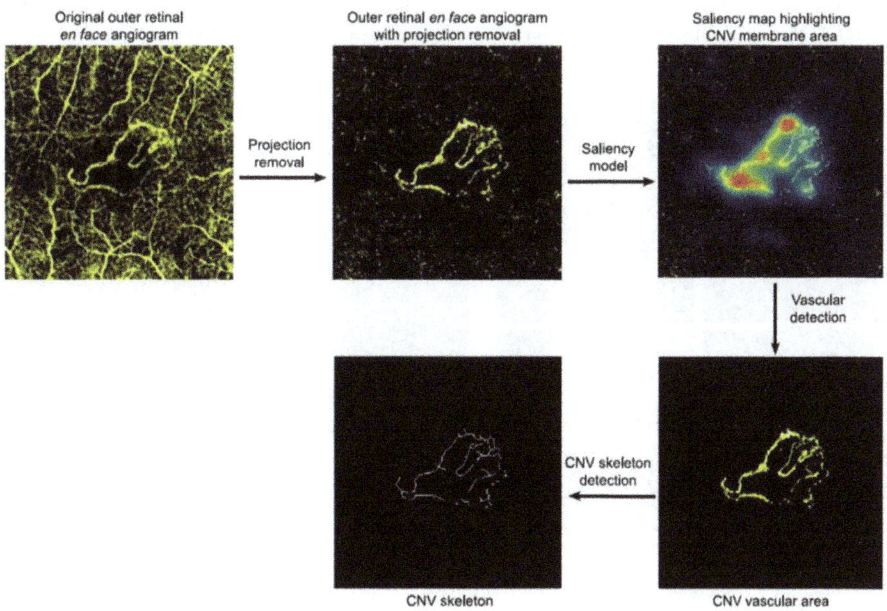

Figure 7.3. Illustration of different image processing steps for generating the CNV area and its skeleton from an original outer retinal *en face* OCTA choroidal image using saliency model. Reprinted from [57].

A–V classification in OCTA. The first study employed color fundus image guided A–V classification in OCTA (figure 7.4). Differential A–V analysis revealed improved sensitivity to identify DR [42] and SCR [37, 40] associated changes. One limitation of these methods is the use of two imaging modalities and image registration. This limitation is addressed in following studies by replacing color fundus image with *en face* OCT images, which retain some intensity profile information to identify A–V near the optic disks (figure 7.5) [64]. Using *en face* OCT for A–V classification in OCTA could improve the efficiency for clinical deployment of differential A–V analysis. Since OCT and OCTA images are intrinsically reconstructed from the same raw spectrogram, it essentially removes any requirement for image registration, and enables single instrument deplyment of OCTA classification. Apart from this, another method for A–V classification in OCTA was demonstrated using near infrared oximetry from OCT images [65] (figure 7.6). Son *et al* [66] reported an automated method for OCTA A–V classification by employing near-infrared OCT oximetry to guide the A–V classification in macular OCTA scans. This study developed a custom

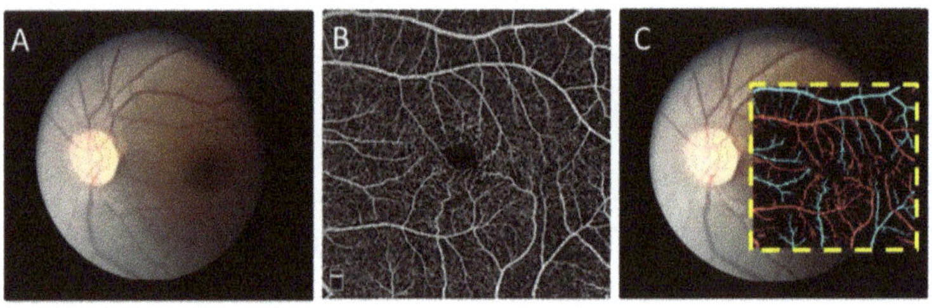

Figure 7.4. Fundus guided A–V classification in OCTA [42]. (a) Fundus image. (b) Corresponding OCTA image; (c) A–V map of an OCTA image overlaid on the fundus image. Modified from [41].

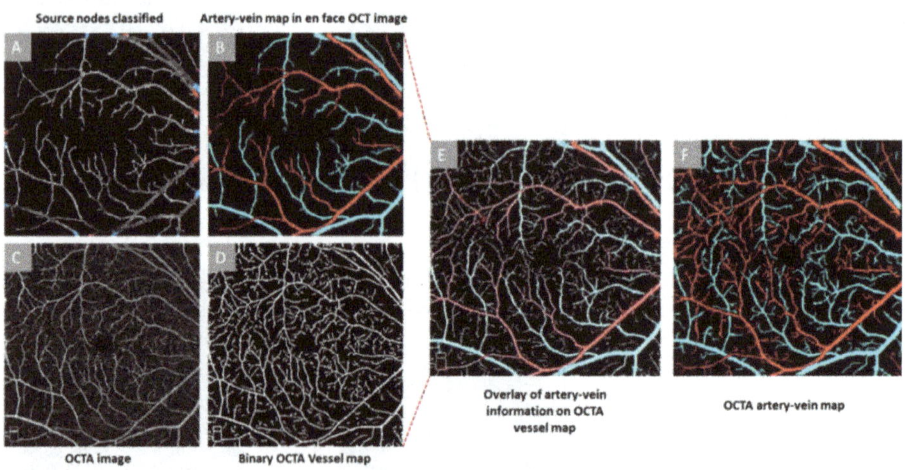

Figure 7.5. *En face* OCT guided A–V classification in OCTA [64]. (a) An *en face* OCT vessel map in which source nodes have been classified into artery–vein (b) *en face* OCT A–V map. (c) OCTA image, (d) OCTA binary map, (e) *en face* OCT A–V map overlaid on the OCTA binary map, (f) final OCTA A–V map (Modified from [64]).

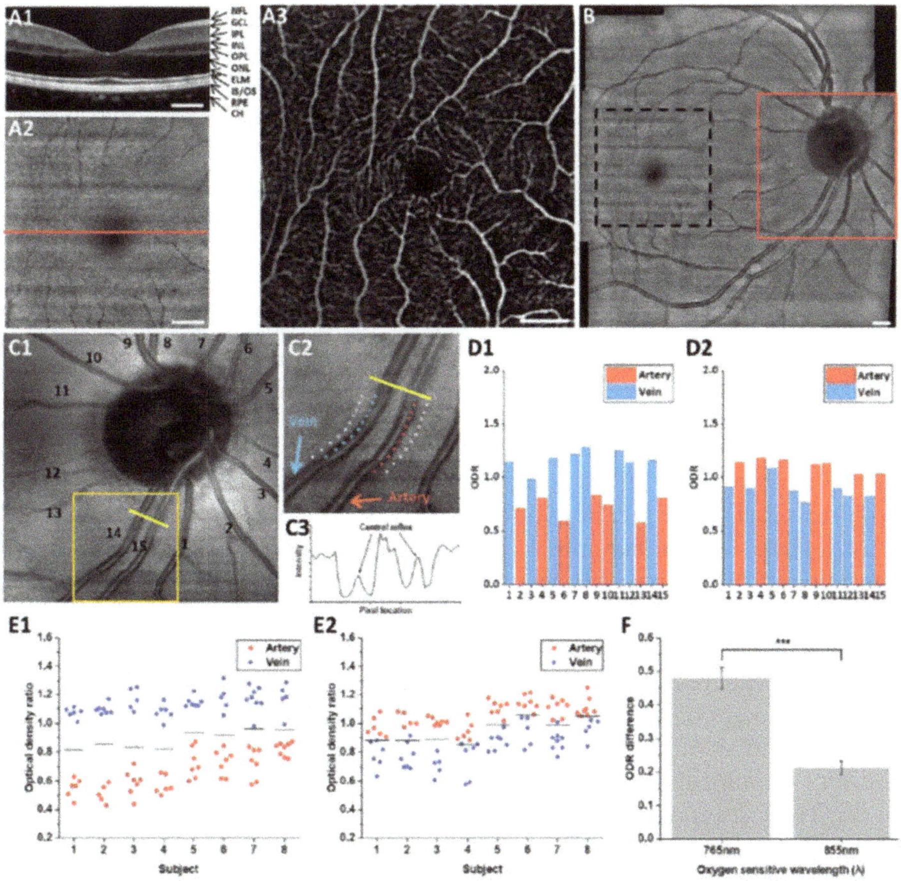

Figure 7.6. Oxymetry guided A–V classification. (a1) OCT B-scan. (a2) *En face* OCT. (a3) Macular OCTA scan. (b) *En face* OCT large field of view image covering optic disc and foveal region. (c1) *En face* OCT image of optic disc. (c2) Sampling locations of tissue and vessel area. (c3) Intensity profile of the vessels and tissue. (d1–d2) ODRs from blood vessels in (c1) at 765 nm and 855 nm wavelength. (e1–e2) ODRs from each subject with 765/805 nm and 805/855 nm analysis. Solid black lines show averaged ODR of all arteries' and veins' ODRs for each subject. (f) Averaged ODR difference between artery and vein. (Reprinted from [66].)

OCT/OCTA device which used an oxygen sensitive 765 nm wavelength that provided 2.8× higher oxygen absorption coefficient between artery and vein (figure 7.6).

7.3 Machine learning techniques in objective OCTA analysis

Machine learning techniques have garnered huge attention in the field of ophthalmology in recent years. This includes supervised techniques that rely on quantitative biomarkers or features that correlate to pathophysiological changes in the retina, and more recently unsupervised 'deep' machine learning methods uses representation-learning to identify intricate structures in high-dimensional data. However, the use of machine learning has been primarily limited to fundus images because fundus data is widely available due to its regular use in clinical ophthalmic practices.

Some studies have applied machine learning techniques in OCT and achieved excellent results for layer segmentation [67–69]. Only recently have we seen some machine learning applications in OCTA [37, 39]. The limitation of database size is a major issue for OCTA. However, because the OCTA quantitative features have shown remarkable results in identifying disease onset and progression of stages, a supervised machine learning technique can be implemented for preliminary studies using OCTA. In recent studies, Alam *et al* used a support vector machine (SVM) classifier for objective classification of SCR and DR. In the first study, six OCTA features, i.e., BVD, BVC, BVT, VPI, FAZ-A, and FAZ-CI, were used to train SVM classifiers for identifying control, mild and severe SCR subjects. The SVM was able to identify control versus disease and mild versus severe SCR with 100% and 97% accuracies, respectively. In a subsequent study, they adapted a similar approach for classification of NPDR and observed 94.41% accuracy for control versus disease (i.e. NPDR) and 92.96% accuracy for control versus mild NPDR classification. The control versus mild NPDR was significantly important because it is crucial to identify the onset of NPDR in patients with DM. In a similar study, Sandhu *et al* [70] used three OCTA features, i.e., BVD, BVC and DAZ-A, for automated classification of NPDR with 94.3% accuracy.

In both studies, the best results were obtained using all OCTA features as a combined set to train SVM classifiers. However, the accuracy value for a combined set was only compared with individual features. A more optimal solution would be to choose the best combination of OCTA feature-sets for classification tasks. For example, Ashraf *et al* conducted a study with multiple OCTA features to find a statistically significant combination of features to distinguish DR subjects from controls. Their study employed a regression model to identify superficial plexus FAZ-A, deep plexus VD and FAZ-CI as the best combination for objective classification. Similarly, in a recent study, Alam *et al* proposed a multi-task AI classifier (sample workflow in figure 7.7), which used a logistic regression based backward elimination technique to identify an optimized feature set for multiple tasks, such as control versus disease, inter-disease and disease staging classification. For a proof of method study, only DR and SCR OCTAs from a retinal clinic were used. The classifier identified perifoveal BVD (superficial plexus), FAZ-A (superficial plexus) and FAZ-CI (deep plexus) for control versus disease classification;

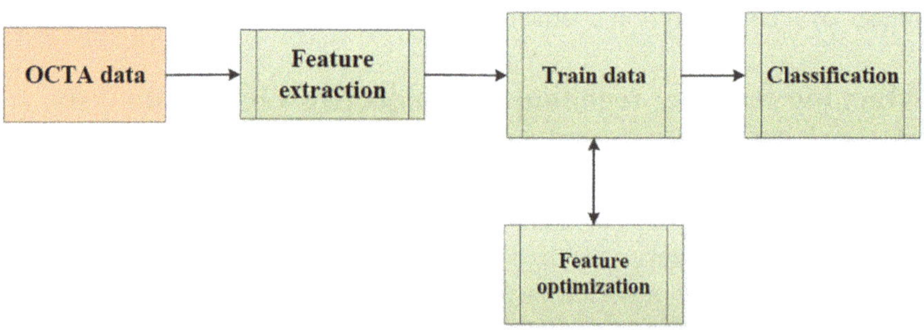

Figure 7.7. Supervised machine learning for objective classification of OCTA.

BVT (superficial plexus), BVD (superficial plexus), FAZ-A (superficial plexus), and FAZ-CI (deep plexus) for DR versus SCR classification; BVD (superficial plexus) and FAZ-A (superficial plexus) for NPDR staging; and BVT (superficial plexus), BVD (superficial plexus), and FAZ-CI (superficial plexus) for SCR staging.

This study demonstrated the importance of identifying the most sensitive features for different retinopathies. Furthermore, by choosing the optimized features from both superficial and deep plexuses, the AI-based classifier ensured representation of layer specific abnormalities due to retinopathies. Although machine learning based OCTA studies showed excellent preliminary results, they were not validated with independent clinical data. Therefore, the generalized performance of OCTA classification on a population of different races and ethnicities could not be analyzed. Further studies with multi-center OCTA data and multi-device OCTA images are required for clinical validation of AI-based OCTA classification.

The use of unsupervised and deep machine learning on OCTA has recently been demonstrated, however, it remains not fully explored due to limited availability of large datasets. Initial studies have used convolutional neural networks (CNNs) for DR detection [71]. Le and Alam *et al* demonstrated a transfer learning based approach for identifying control and three NPDR stages [72]. Aside from disease diagnosis, few studies have implemented deep machine learning for image processing-based applications in OCTA. For example, Prentasic *et al* [73] demonstrated a convolutional neural network (CNN) based technique to segment microvasculature in a foveal area. They use 80 OCTA images and a cross validation technique to test their CNN on identifying vessel and non-vessel pixels in a parafovea OCTA scan. Compared to manual segmentation results, this study showed 83% accuracy for vessel segmentation. When the result was compared with inter- and intra-rater accuracies, it was observed that the CNN was equivalent to a second manual rater. A similar application was adapted by Guo *et al* [74] to segment retinal non-prefusion areas in ultra-wide-field angiograms. In a follow-up study, they further demonstrated a CNN that was able to identify non-perfusion from signal reduction artifacts [75]. In a separate study, Kasaragod *et al* [76] used machine learning based techniques for segmentation of the optical nerve head (ONH). However, they used Jones matrix OCT (JM-OCT), which provides OCT, OCTA, birefringence tomography and attenuation coefficient tomography at the same time. The semi-automatic unsupervised classification framework was specially designed for ONH tissue segmentation. Deep learning has been also widely used for image processing applications in OCT images, as previously mentioned. OCT layer segmentation is a major application. Although not directly a machine learning application to OCTA, OCT retinal layer segmentation can be correlated to OCTA superficial and deep plexus segmentation, which in turn provides better image quality from OCTA. Fang *et al* demonstrated a deep learning based strategy to identify nine retinal layers in OCT B-scans [68]. This information could be transferred to OCTA B-scan volumes for better segmentation of inner and outer plexiform layers, which include retinal information. The clinical OCTA devices can integrate robust segmentation for removing any shadow artifacts in OCTA *en face* projection images.

7.4 Discussion

Compared to imaging modalities like fundus photography, SLO, and FA, OCTA provides a label-free solution for high resolution visualization and quantification of retinal vasculatures. With the capability to image depth-resolved retinal vasculatures at capillary level resolution, OCTA is being rapidly adopted for the clinical management of different eye diseases. We have summarized the use of quantitative OCTA features in table 7.1, organizing literature which utilized commercial devices and software. Different quantitative features have been demonstrated and widely explored to foster the standardization of objective OCTA interpretation.

It has been observed that pathological mechanisms of different eye diseases can affect the sensitivity of the OCTA features. For example, SCR produces sickle shaped blood cells, that induces tortuous and dilated vessels, which can be quantified using BVT and BVC, respectively. From OCTA analysis of SCR patients, BVT has been demonstrated as the most sensitive feature for SCR classification [77]. Analogously, DR patients frequently have associated hypertension which can cause arterial narrowing [42], which can be measured using BVC. BVD analysis in DR patients can also assess capillary level ischemia in DR patients [39]. BVD is also sensitive enough to evaluate central/branch VOs [32, 46, 78–80]. Recent studies also showed that vascular distortions could manifest in different localized regions due to different diseases. For example, localized VCI measurements have been demonstrated to be successful at identifying DR progression to PDR [50]; BVD measurements also revealed that the perifoveal region is the most sensitive region for classifying NPDR stages [39, 81]. On the other hand, for SCR, the parafoveal temporal retina is the most sensitive local region for SCR staging [40]. This creates the possibility for a multitude of interesting investigations of OCTA features to customize and tailor features based on disease pathology.

Differential A–V analysis is also one of the new developments in quantitative OCTA analysis that has been demonstrated to improve OCTA feature sensitivity for retinopathy staging [41, 42]. Pathological deformations in the artery–vein morphology can be affected in distinct trends [82]. DR may induce increased venous beading, venous dilation, arterial narrowing and arterial tortuosity [83]. The sensitivity of the OCTA features may be compromised if an average global value of A–V distortions is compared between healthy and diseased eye. Two recent studies revealed two differential A–V features, i.e. A–V ratio of BVC (AVR–BVC) and BVT (AVR–BVT) to improve the performance for DR and SCR staging [41, 42].

For AMD assessment, predominantly quantitative analyses of choroidal vasculature have been explored [60, 84]. Early stage or dry AMD is normally characterized by decreased choroidal blood flow and drusen. More advanced dry AMD is characterized by geographic atrophy. In the case of late stage or wet AMD, CNVs are the most common biomarkers [60, 85]. Jia *et al* first presented an OCTA study to detect and classify CNV types (I and II) [17], reporting decreased choroidal flow near the CNV in all cases. In following studies [57], for quantification of CNV area and connectivity, researchers analyzed artifact removal algorithms, and projection resolved OCTA (PR-OCTA). Miller *et al* compared

Table 7.1. Summary of quantitative OCTA features.

Features	Applications	Clinical OCTA with built-in software	Clinical OCTA with custom software	Custom OCTA
BVD	To quantify vessel density and identify ischemic regions	[27, 33, 48, 52, 87–98]	[23, 28–31, 34, 36–40, 42, 43, 46, 70, 81, 93, 95, 99–133]	[134, 135]
BVC	To quantify blood vessel width, shrinkage or dilation	[136]	[36–42, 44, 70, 106, 108, 109, 111, 116, 117, 120, 124, 131, 137, 138]	[134, 135, 139]
BVT	To quantify blood vessel tortuosity because of change in morphology	[140]	[24, 31, 37, 39–42, 119, 123, 138, 141–144]	[135]
VPI	To quantify changes in the blood vessel perimeter	—	[36, 37, 39, 40, 43, 116, 117, 124, 131, 133]	—
FAZ-A	To quantify area changes in foveal	[24–27, 32, 33, 45, 48, 52, 78, 87–94, 96, 98, 136, 140, 145–157]	[24, 31, 34, 37, 39, 40, 44–47, 70, 81, 93, 95, 101, 110–115, 119, 121–123, 125, 127, 129, 131, 140, 152, 158–169]	—
FAZ-CI	To quantify complexity changes in foveal contour	[48, 87, 88, 91–93, 156, 164]	[47, 81, 110, 111, 113, 122, 131, 152, 159, 162, 163, 165]	—
VCI	To quantify complexity changes in retinal vasculature	—	[28, 30, 38, 51, 81, 99–109, 111, 124, 126, 170–175]	—
BPA	To quantify changes in vascular branching geometry bifurcation	—	[54]	—
A-V analysis	To achieve differential artery and vein analysis in DR, and SCR, etc	—	[41, 42, 137]	[65]
Flow analysis	To quantify blood flow changes	[33, 145]	[34, 55–57, 128, 130, 158, 176, 177]	[17, 60, 178]
CNV analysis	To quantify neovascularization morphology change	—	[28, 57, 62, 63, 84, 86, 102, 103, 138, 170, 172, 177, 179]	[17, 60, 86]

BVD: blood vessel density; BVC: blood vessel caliber; BVT: blood vessel tortuosity; VPI: vessel perimeter index; FAZ: foveal avascular zone; CI: contour irregularity; FD: fractal dimension; VCI: vessel complexity index; LAC: lacunarity; BPA: branchpoint analysis; A–V: artery–vein; AVR: artery–vein ratio; CNV: choroidal neovasculature..

CNV using spectral domain (SD) and swept source (SS)-OCTA [86], and reported statistically significant differences between lesion areas measured by SS and SD OCTA. SS-OCTA showed larger lesion area in both 3×3 mm^2 and 6×6 mm^2, and larger differences were reported in the 6×6 mm^2 OCTA scans. Another parameter studied for AMD is FD; Al-Sheikh *et al* studied CNV lesions in pre- and post-treatment AMD and reported a lower FD value in the inner part of the lesion post treatment [28].

Quantitative OCTA has opened a unique opportunity for computer-aided detection and AI classification in different eye diseases, as described in section 7.3 in detail. Machine learning techniques have been also explored to segment microvasculature [73], non-prefusion areas [74], optical nerve head (ONH) [76]. Although, the limited size of currently available OCTA database remains a major challenging for deep learning-based studies with OCTA, with the increasing interest in AI applications in ophthalmology, there will be a widespread incentive to expand the available OCTA datasets and multi-center collaborations. Therefore, we anticipate that deep learning-based classification (already well established for fundus image application) will play an important role to enable future investigations of automated OCTA detection and classification of eye conditions. Concurrently, transfer learning based technology [180] might also find valuable applications in the near future to foster deep learning based OCTA studies.

References

[1] Wang W and Lo A C 2018 Diabetic retinopathy: pathophysiology and treatments *Int. J. Mol. Sci.* **19** 1816

[2] Karia N 2010 Retinal vein occlusion: pathophysiology and treatment options *Clin. Ophthalmol.* **4** 809

[3] Sharp P F and Manivannan A 1997 The scanning laser ophthalmoscope *Phys. Med. Biol.* **42** 951–66

[4] Kobayashi K and Asakura T 1995 Imaging techniques and applications of the scanning laser ophthalmoscope *Opt. Eng.* **34** 717–26

[5] Liang J Z, Williams D R and Miller D T 1997 High resolution imaging of the living human retina with adaptive optics *Investig. Ophthalmol. Vis. Sci.* **38** 55

[6] Burns S A, Elsner A E, Sapoznik K A, Warner R L and Gast T J 2019 Adaptive optics imaging of the human retina *Prog. Retin. Eye Res.* **68** 1–30

[7] Paques M, Meimon S, Rossant F, Rosenbaum D, Mrejen S, Sennlaub F and Grieve K 2018 Adaptive optics ophthalmoscopy: application to age-related macular degeneration and vascular diseases *Prog. Retin. Eye Res.* **66** 1–16

[8] Banda H K, Shah G K and Blinder K J 2019 Applications of fundus autofluorescence and widefield angiography in clinical practice *Can. J. Ophthalmol.* **54** 11–9

[9] Mehta S 2006 Fundus fluorescein angiography of choroidal tubercles: case reports and review of literature *Indian J. Ophthalmol.* **54** 273–5

[10] Huang D *et al* 1991 Optical coherence tomography *Science* **254** 1178–81

[11] Mailankody P, Lenka A and Pal P K 2019 The role of optical coherence tomography in parkinsonism: a critical review *J. Neurol. Sci.* **403** 67–74

[12] Ruia S, Saxena S, Gemmy Cheung C M, Gilhotra J S and Lai T Y 2016 Spectral domain optical coherence tomography features and classification systems for diabetic macular edema: a review *Asia Pac. J. Ophthalmol. (Phila.)* **5** 360–7
[13] Sohrab M A and Fawzi A A 2013 Review of en-face choroidal imaging using spectral-domain optical coherence tomography *Med. Hypothesis Discov. Innov. Ophthalmol.* **2** 69–73
[14] Gao S S, Jia Y, Zhang M, Su J P, Liu G, Hwang T S, Bailey S T and Huang D 2016 Optical coherence tomography angiography *Invest. Ophthalmol. Vis. Sci.* **57** OCT27–36
[15] Zhang A, Zhang Q, Chen C L and Wang R K 2015 Methods and algorithms for optical coherence tomography-based angiography: a review and comparison *J. Biomed. Opt.* **20** 100901
[16] Bandello F, Corbelli E, Carnevali A, Pierro L and Querques G 2016 Optical coherence tomography angiography of diabetic retinopathy *Dev. Ophthalmol.* **56** 107–12
[17] Jia Y, Bailey S T, Wilson D J, Tan O, Klein M L, Flaxel C J, Potsaid B, Liu J J, Lu C D and Kraus M F 2014 Quantitative optical coherence tomography angiography of choroidal neovascularization in age-related macular degeneration *Ophthalmology* **121** 1435–44
[18] Dastiridou A and Chopra V 2018 Potential applications of optical coherence tomography angiography in glaucoma *Curr. Opin. Ophthalmol.* **29** 226–33
[19] Bojikian K D, Chen P P and Wen J C 2019 Optical coherence tomography angiography in glaucoma *Curr. Opin. Ophthalmol.* **30** 110–6
[20] Leitao Guerra R L, Leitao Guerra C L, Bastos M G, de Oliveira A H P and Salles C 2019 Sickle cell retinopathy: what we now understand using optical coherence tomography angiography. A systematic review *Blood Rev.* **35** 32–42
[21] Faatz H, Farecki M L, Rothaus K, Gunnemann F, Gutfleisch M, Lommatzsch A and Pauleikhoff D 2019 Optical coherence tomography angiography of types 1 and 2 choroidal neovascularization in age-related macular degeneration during anti-VEGF therapy: evaluation of a new quantitative method *Eye* **33** 1466–71
[22] Yao X, Alam M N, Le D and Toslak D 2020 Quantitative optical coherence tomography angiography: a review *Exp. Biol. Med. (Maywood)* **245** 301–12
[23] Pedinielli A, Bonnin S, Sanharawi M E, Mane V, Erginay A, Couturier A and Tadayoni R 2017 Three different optical coherence tomography angiography measurement methods for assessing capillary density changes in diabetic retinopathy *Ophthalmic Surg. Lasers Imaging Retina* **48** 378–84
[24] Mastropasqua R, Toto L, Mastropasqua A, Aloia R, De Nicola C, Mattei P A, Di Marzio G, Di Nicola M and Di Antonio L 2017 Foveal avascular zone area and parafoveal vessel density measurements in different stages of diabetic retinopathy by optical coherence tomography angiography *Int. J. Ophthalmol.* **10** 1545
[25] Falavarjani K G, Scott A W, Wang K, Han I C, Chen X, Klufas M, Hubschman J-P, Schwartz S D, Sadda S R and Sarraf D 2016 Correlation of multimodal imaging in sickle cell retinopathy *Retina* **36** S111–7
[26] Minvielle W, Caillaux V, Cohen S Y, Chasset F, Zambrowski O, Miere A and Souied E H 2016 Macular microangiopathy in sickle cell disease using optical coherence tomography angiography *Am. J. Ophthalmol.* **164** 137–44 e131
[27] Jhaj G, Glazman S, Shrier E M and Bodis-Wollner I 2017 Non-exudative age-related macular degeneration foveal avascular zone area, foveal vessel density, and ganglion cell complex thickness *Invest. Ophth. Vis. Sci.* **58** 36

[28] Al-Sheikh M, Iafe N A, Phasukkijwatana N, Sadda S R and Sarraf D 2018 Biomarkers of neovascular activity in age-related macular degeneration using optical coherence tomography angiography *Retina* **38** 220–30

[29] Liu L, Jia Y, Takusagawa H L, Pechauer A D, Edmunds B, Lombardi L, Davis E, Morrison J C and Huang D 2015 Optical coherence tomography angiography of the peripapillary retina in glaucoma *JAMA Ophthalmol.* **133** 1045–52

[30] Richter G M, Chang R, Situ B, Chu Z, Burkemper B, Reznik A, Bedrood S, Kashani A H, Varma R and Wang R K 2018 Diagnostic performance of macular versus peripapillary vessel parameters by optical coherence tomography angiography for glaucoma *Transl. Vis. Sci. Technol.* **7** 21

[31] Adhi M *et al* 2016 Retinal capillary network and foveal avascular zone in eyes with vein occlusion and fellow eyes analyzed with optical coherence tomography angiography *Invest. Ophthalmol. Vis. Sci.* **57** OCT486–494

[32] Winegarner A *et al* 2018 Changes in retinal microvasculature and visual acuity after antivascular endothelial growth factor therapy in retinal vein occlusion *Invest. Ophthalmol. Vis. Sci.* **59** 2708–16

[33] Chao S-C, Yang S-J, Chen H-C, Sun C-C, Liu C-H and Lee C-Y 2019 Early macular angiography among patients with glaucoma, ocular hypertension, and normal subjects *J. Ophthalmol.* **2019** 7419470

[34] Nesper P L, Roberts P K, Onishi A C, Chai H, Liu L, Jampol L M and Fawzi A A 2017 Quantifying microvascular abnormalities with increasing severity of diabetic retinopathy using optical coherence tomography angiography *Invest. Ophth. Vis. Sci.* **58** BIO307–15

[35] Camino A, Zhang M, Liu L, Wang J, Jia Y and Huang D 2018 Enhanced quantification of retinal perfusion by improved discrimination of blood flow from bulk motion signal in OCTA *Transl. Vis. Sci. Technol.* **7** 20

[36] Lin T-C, Gogte P, Palejwala N, Shahidzadeh A, Itty S, Humayun M S, Moshfeghi A A, Ameri H, Chu Z and Wang R K 2017 Quantitative spectral-domain optical coherence tomography angiography (OCTA) of diabetic retinopathy (DR) severity *Invest. Ophth. Vis. Sci.* **58** 1653

[37] Alam M, Thapa D, Lim J I, Cao D and Yao X 2017 Computer-aided classification of sickle cell retinopathy using quantitative features in optical coherence tomography angiography *Biomed. Opt. Express* **8** 4206–16

[38] Akagi T, Uji A, Huang A S, Weinreb R N, Yamada T, Miyata M, Kameda T, Ikeda H O and Tsujikawa A 2018 Conjunctival and intrascleral vasculatures assessed using anterior segment optical coherence tomography angiography in normal eyes *Am. J. Ophthalmol.* **196** 1–9

[39] Alam M, Zhang Y, Lim J I, Chan R V, Yang M and Yao X 2019 Quantitative optical coherence tomography angiography features for objective classification and staging of diabetic retinopathy *Retina* **40** 322–32

[40] Alam M, Thapa D, Lim J I, Cao D and Yao X 2017 Quantitative characteristics of sickle cell retinopathy in optical coherence tomography angiography *Biomed. Opt. Express* **8** 1741–53

[41] Alam M, Lim J I, Toslak D and Yao X 2019 Differential artery–vein analysis improves the performance of OCTA staging of sickle cell retinopathy *Transl. Vis. Sci. Technol.* **8** 3

[42] Alam M, Toslak D, Lim J I and Yao X 2018 Color fundus image guided artery-vein differentiation in optical coherence tomography angiography *Investig. Ophthalmol. Vis. Sci.* **59** 4953–62

[43] Ye H, Zheng C, Lan X, Zhao L, Qiao T, Li X and Zhang Y 2019 Evaluation of retinal vasculature before and after treatment of children with obstructive sleep apnea-hypopnea syndrome by optical coherence tomography angiography *Graefe's Arch. Clin. Exp. Ophthalmol.* **257** 543–8

[44] Chu Z, Lin J, Gao C, Xin C, Zhang Q, Chen C-L, Roisman L, Gregori G, Rosenfeld P J and Wang R K 2016 *Quantitative Assessment of the Retinal Microvasculature Using Optical Coherence Tomography Angiography* (Bellingham, WA: SPIE)vol 21 1–13 13

[45] Durbin M K, An L, Shemonski N D, Soares M, Santos T, Lopes M, Neves C and Cunha-Vaz J 2017 Quantification of retinal microvascular density in optical coherence tomographic angiography images in diabetic retinopathy *JAMA Ophthalmol.* **135** 370–6

[46] Falavarjani K G, Iafe N A, Hubschman J-P, Tsui I, Sadda S R and Sarraf D 2017 Optical coherence tomography angiography analysis of the foveal avascular zone and macular vessel density after anti-VEGF therapy in eyes with diabetic macular edema and retinal vein occlusion *Invest. Ophth. Vis. Sci.* **58** 30–4

[47] Krawitz B D, Mo S, Geyman L S, Agemy S A, Scripsema N K, Garcia P M, Chui T Y and Rosen R B 2017 Acircularity index and axis ratio of the foveal avascular zone in diabetic eyes and healthy controls measured by optical coherence tomography angiography *Vis. Res.* **139** 177–86

[48] Li Z, Alzogool M, Xiao J, Zhang S, Zeng P and Lan Y 2018 Optical coherence tomography angiography findings of neurovascular changes in type 2 diabetes mellitus patients without clinical diabetic retinopathy *Acta diabetol.* **55** 1075–82

[49] Philip S, Najafi A, Tantraworasin A, Chui T Y P, Rosen R B and Ritch R 2019 Macula vessel density and foveal avascular zone parameters in exfoliation glaucoma compared to primary open-angle glaucoma *Invest. Ophthalmol. Vis. Sci.* **60** 1244–53

[50] Alam M, Le D, Lim J I and Yao X Vascular complexity analysis in optical coherence tomography angiography of diabetic retinopathy *Retina* **41** 538–45

[51] Zahid S *et al* 2016 Fractal dimensional analysis of optical coherence tomography angiography in eyes with diabetic retinopathy *Invest. Ophthalmol. Vis. Sci.* **57** 4940–7

[52] Cabral D, Coscas F, Glacet-Bernard A, Pereira T, Geraldes C, Cachado F, Papoila A, Coscas G and Souied E 2019 Biomarkers of peripheral nonperfusion in retinal venous occlusions using optical coherence tomography angiography *Transl. Vis. Sci. Technol.* **8** 7

[53] Murray C D 1926 The physiological principle of minimum work: I. The vascular system and the cost of blood volume *Proc. Natl. Acad. Sci. USA* **12** 207–14

[54] Le D, Alam M, Miao B A, Lim J I and Yao X 2019 Fully automated geometric feature analysis in optical coherence tomography angiography for objective classification of diabetic retinopathy *Biomed. Opt. Express* **10** 2493–503

[55] Wang X, Jia Y, Spain R, Potsaid B, Liu J J, Baumann B, Hornegger J, Fujimoto J G, Wu Q and Huang D 2014 Optical coherence tomography angiography of optic nerve head and parafovea in multiple sclerosis *Br. J. Ophthalmol.* **98** 1368–73

[56] Zhang Q *et al* 2018 A novel strategy for quantifying choriocapillaris flow voids using swept-source OCT angiography quantifying choriocapillaris flow voids using SS-OCTA *Invest. Ophth. Vis. Sci.* **59** 203–11

[57] Patel R, Wang J, Campbell J P, Kiang L, Lauer A, Flaxel C, Hwang T, Lujan B, Huang D and Bailey S T 2018 Classification of choroidal neovascularization using projection-resolved optical coherence tomographic angiography *Invest. Ophth. Vis. Sci.* **59** 4285–91

[58] Pechauer A D, Jia Y, Liu L, Gao S S, Jiang C and Huang D 2015 Optical coherence tomography angiography of peripapillary retinal blood flow response to hyperoxia OCT angiography of retinal blood flow during hyperoxia *Invest. Ophth. Vis. Sci.* **56** 3287–91
[59] Baumann B, Potsaid B, Kraus M F, Liu J J, Huang D, Hornegger J, Cable A E, Duker J S and Fujimoto J G 2011 Total retinal blood flow measurement with ultrahigh speed swept source/Fourier domain OCT *Biomed. Opt. Express* **2** 1539–52
[60] Moult E *et al* 2014 Ultrahigh-speed swept-source OCT angiography in exudative AMD *Ophthalmic Surg. Lasers Imaging Retina* **45** 496–505
[61] Ploner S B *et al* 2016 Toward quantitative optical coherence tomography angiography: visualizing blood flow speeds in ocular pathology using variable interscan time analysis *Retina* **36** S118–26
[62] Uchida A, Hu M, Babiuch A, Srivastava S K, Singh R P, Kaiser P K, Talcott K, Rachitskaya A and Ehlers J P 2019 Optical coherence tomography angiography characteristics of choroidal neovascularization requiring varied dosing frequencies in treat-and-extend management: an analysis of the AVATAR study *PLoS One* **14** e0218889
[63] Carnevali A, Cicinelli M V, Capuano V, Corvi F, Mazzaferro A, Querques L, Scorcia V, Souied E H, Bandello F and Querques G 2016 Optical coherence tomography angiography: a useful tool for diagnosis of treatment-naïve quiescent choroidal neovascularization *Am. J. Ophthalmol.* **169** 189–98
[64] Alam M, Toslak D, Lim J I and Yao X 2019 OCT feature analysis guided artery-vein differentiation in OCTA *Biomed. Opt. Express* **10** 2055–66
[65] Son T, Alam M, Kim T-H, Liu C, Toslak D and Yao X 2019 Near infrared oximetry-guided artery–vein classification in optical coherence tomography angiography *Exp. Biol. Med.* **244** 813–8
[66] Son T, Alam M, Liu C, Toslak D and Yao X 2019 Optical coherence tomography guided artery-vein classification in retinal OCT angiography of macular region *Ophthalmic Technologies XXIX* (Bellingham, WA: International Society for Optics and Photonics) 108581L
[67] Camino A, Wang Z, Wang J, Pennesi M E, Yang P, Huang D, Li D and Jia Y 2018 Deep learning for the segmentation of preserved photoreceptors on en face optical coherence tomography in two inherited retinal diseases *Biomed. Opt. Express* **9** 3092–105
[68] Fang L, Cunefare D, Wang C, Guymer R H, Li S and Farsiu S 2017 Automatic segmentation of nine retinal layer boundaries in OCT images of non-exudative AMD patients using deep learning and graph search *Biomed. Opt. Express* **8** 2732–44
[69] Roy A G, Conjeti S, Karri S P K, Sheet D, Katouzian A, Wachinger C and Navab N 2017 ReLayNet: retinal layer and fluid segmentation of macular optical coherence tomography using fully convolutional networks *Biomed. Opt. Express* **8** 3627–42
[70] Sandhu H S, Eladawi N, Elmogy M, Keynton R, Helmy O, Schaal S and El-Baz A 2018 Automated diabetic retinopathy detection using optical coherence tomography angiography: a pilot study *Br. J. Ophthalmol.* **102** 1564–9
[71] Heisler M, Karst S, Lo J, Mammo Z, Yu T, Warner S, Maberley D, Beg M F, Navajas E V and Sarunic M V 2020 Ensemble deep learning for diabetic retinopathy detection using optical coherence tomography angiography *Transl. Vis. Sci. Technol.* **9** 20
[72] Le D, Alam M, Yao C K, Lim J I, Hsieh Y-T, Chan R V, Toslak D and Yao X 2020 Transfer learning for automated OCTA detection of diabetic retinopathy *Transl. Vis. Sci. Technol.* **9** 35

[73] Prentašić P, Heisler M, Mammo Z, Lee S, Merkur A, Navajas E, Beg M F, Šarunic M and Lončarić S 2016 Segmentation of the foveal microvasculature using deep learning networks *J. Biomed. Opt.* **21** 075008
[74] Guo Y, Camino A, Wang J, Huang D, Hwang T S and Jia Y 2018 MEDnet, a neural network for automated detection of avascular area in OCT angiography *Biomed. Opt. Express* **9** 5147–58
[75] Guo Y, Hormel T T, Xiong H, Wang B, Camino A, Wang J, Huang D, Hwang T S and Jia Y 2019 Development and validation of a deep learning algorithm for distinguishing the nonperfusion area from signal reduction artifacts on oct angiography *Biomed. Opt. Express* **10** 3257–68
[76] Kasaragod D, Makita S, Hong Y-J and Yasuno Y 2018 Machine-learning based segmentation of the optic nerve head using multi-contrast Jones matrix optical coherence tomography with semi-automatic training dataset generation *Biomed. Opt. Express* **9** 3220–43
[77] Alam M, Le D, Lim J I, Chan R V P and Yao X C 2019 Supervised machine learning based multi-task artificial intelligence classification of retinopathies *J. Clin. Med.* **8** 872
[78] Samara W A, Shahlaee A, Sridhar J, Khan M A, Ho A C and Hsu J 2016 Quantitative optical coherence tomography angiography features and visual function in eyes with branch retinal vein occlusion *Am. J. Ophthalmol.* **166** 76–83
[79] Chen J L, Zahid S, Alam M N, Yao X and Lim J I 2018 Assessment of quantitative optical coherence tomography angiography parameters in branch retinal vein occlusion and monitoring response to treatment *Invest. Ophth. Vis. Sci.* **59** 5458
[80] Zahid S, Alam M N, Yao X and Lim J I 2018 Quantitative optical coherence tomography angiography parameters in central retinal vein occlusion *Invest. Ophth. Vis. Sci.* **59** 5427
[81] Ashraf M, Nesper P L, Jampol L M, Yu F and Fawzi A A 2018 Statistical model of optical coherence tomography angiography parameters that correlate with severity of diabetic retinopathy *Invest. Ophth. Vis. Sci.* **59** 4292–8
[82] Ouyang Y, Shao Q, Scharf D, Joussen A M and Heussen F M 2014 An easy method to differentiate retinal arteries from veins by spectral domain optical coherence tomography: retrospective, observational case series *BMC Ophthalmol.* **14** 66
[83] Joshi V S, Reinhardt J M, Garvin M K and Abramoff M D 2014 Automated method for identification and artery-venous classification of vessel trees in retinal vessel networks *PLoS One* **9** e88061
[84] Zheng F *et al* 2017 Comparison of neovascular lesion area measurements from different swept-source OCT angiographic scan patterns in age-related macular degeneration *Invest. Ophthalmol. Vis. Sci.* **58** 5098–104
[85] de Carlo T E, Romano A, Waheed N K and Duker J S 2015 A review of optical coherence tomography angiography (OCTA) *Int. J. Retin. Vitr.* **1** 5
[86] Miller A R, Roisman L, Zhang Q, Zheng F, de Oliveira Dias J R, Yehoshua Z, Schaal K B, Feuer W, Gregori G and Chu Z 2017 Comparison between spectral-domain and swept-source optical coherence tomography angiographic imaging of choroidal neovascularization *Invest. Ophth. Vis. Sci.* **58** 1499–505
[87] Lim H B, Kim Y W, Kim J M, Jo Y J and Kim J Y 2018 The importance of signal strength in quantitative assessment of retinal vessel density using optical coherence tomography angiography *Sci. Rep.* **8** 12897

[88] Liu L, Gao J, Bao W, Hu C, Xu Y, Zhao B, Zheng J, Fan L and Sun Y 2018 Analysis of foveal microvascular abnormalities in diabetic retinopathy using optical coherence tomography angiography with projection artifact removal *J. Ophthalmol.* **2018** 3926745

[89] Yu S, Frueh B E, Steinmair D, Ebneter A, Wolf S, Zinkernagel M S and Munk M R 2018 Cataract significantly influences quantitative measurements on swept-source optical coherence tomography angiography imaging *PLoS One* **13** e0204501

[90] Corvi F, Pellegrini M, Erba S, Cozzi M, Staurenghi G and Giani A 2018 Reproducibility of vessel density, fractal dimension, and foveal avascular zone using 7 different optical coherence tomography angiography devices *Am. J. Ophthalmol.* **186** 25–31

[91] Tian M, Tappeiner C, Zinkernagel M S, Wolf S and Munk M R 2019 Swept-source optical coherence tomography angiography reveals vascular changes in intermediate uveitis *Acta Ophthalmol* **97** e785–91

[92] Lavia C, Bonnin S, Maule M, Erginay A, Tadayoni R and Gaudric A 2019 Vessel density of superficial, intermediate, and deep capillary plexuses using optical coherence tomography angiography *Retina* **39** 247

[93] Li T, Jia Y, Wang S, Wang A, Gao L, Yang C and Zou H 2019 Retinal microvascular abnormalities in children with type 1 diabetes mellitus without visual impairment or diabetic retinopathy *Invest. Ophth. Vis. Sci.* **60** 990–8

[94] Cao D, Yang D, Huang Z, Zeng Y, Wang J, Hu Y and Zhang L 2018 Optical coherence tomography angiography discerns preclinical diabetic retinopathy in eyes of patients with type 2 diabetes without clinical diabetic retinopathy *Acta Diabetol.* **55** 469–77

[95] Garrity S T, Iafe N A, Phasukkijwatana N, Chen X and Sarraf D 2017 Quantitative analysis of three distinct retinal capillary plexuses in healthy eyes using optical coherence tomography angiography *Invest. Ophthalmol. Vis. Sci.* **58** 5548–55

[96] Fujiwara A, Mengxuan L, Morizane Y, Hosogi M, Kimura S, Hosokawa M, Shiode Y, Masayuki H, Doi S and Toshima S 2018 Interocular symmetry of foveal avascular zone area in healthy eyes: an examination using swept-source optical coherence tomography angiography *Invest. Ophth. Vis. Sci.* **59** 2882

[97] Eastline M, Munk M R, Wolf S, Schaal K B, Ebneter A, Tian M, Giannakaki-Zimmermann H and Zinkernagel M S 2019 Repeatability of wide-field optical coherence tomography angiography in normal retina *Transl. Vis. Sci. Technol.* **8** 6

[98] Lutsenko N, Rudycheva O and Isakova O 2019 Assessing OCTA changes in morphology and structure of retinal microvascular bed in patients with exudative AMD *J. Ophthalmol. (Ukraine)* **2** 7–13

[99] Agarwal A, Aggarwal K, Akella M, Agrawal R, Khandelwal N, Bansal R, Singh R and Gupta V 2018 Fractal dimension and optical coherence tomography angiography features of the central macula after repair of rhegmatogenous retinal detachments *Retina* **39** 2167–77

[100] Richter G M *et al* 2018 Structural and functional associations of macular microcirculation in the ganglion cell-inner plexiform layer in glaucoma using optical coherence tomography angiography *J. Glaucoma* **27** 281–90

[101] Kim S V, Semoun O, Pedinielli A, Jung C, Miere A and Souied E H 2019 Optical coherence tomography angiography quantitative assessment of exercise-induced variations in retinal vascular plexa of healthy subjects *Invest. Ophth. Vis. Sci.* **60** 1412–19

[102] Coscas F, Cabral D, Pereira T, Geraldes C, Narotamo H, Miere A, Lupidi M, Sellam A, Papoila A and Coscas G 2018 Quantitative optical coherence tomography angiography biomarkers for neovascular age-related macular degeneration in remission *PLoS One* **13** e0205513

[103] Murakawa S, Maruko I, Kawano T, Hasegawa T and Iida T 2019 Choroidal neovascularization imaging using multiple en face optical coherence tomography angiography image averaging *Graefe's Arch. Clin. Exp. Ophthalmol.* **257** 1119–25

[104] Dave V P, Pappuru R R, Gindra R, Ananthakrishnan A, Modi S, Trivedi M and Harikumar P 2018 OCT angiography fractal analysis-based quantification of macular vascular density in BRVO eyes *Can. J. Ophthalmol.* **54** 297–300

[105] Hirano T, Kitahara J, Toriyama Y, Kasamatsu H, Murata T and Sadda S 2019 Quantifying vascular density and morphology using different swept-source optical coherence tomography angiographic scan patterns in diabetic retinopathy *Br. J. Ophthalmol.* **103** 216–21

[106] Koulisis N, Kim A Y, Chu Z, Shahidzadeh A, Burkemper B, de Koo L C O, Moshfeghi A A, Ameri H, Puliafito C A and Isozaki V L 2017 Quantitative microvascular analysis of retinal venous occlusions by spectral domain optical coherence tomography angiography *PLoS One* **12** e0176404

[107] Al-Sheikh M, Phasukkijwatana N, Dolz-Marco R, Rahimi M, Iafe N A, Freund K B, Sadda S R and Sarraf D 2017 Quantitative OCT angiography of the retinal microvasculature and the choriocapillaris in myopic eyes *Invest. Ophth. Vis. Sci.* **58** 2063–9

[108] Kim A Y, Rodger D C, Shahidzadeh A, Chu Z, Koulisis N, Burkemper B, Jiang X, Pepple K L, Wang R K and Puliafito C A 2016 Quantifying retinal microvascular changes in uveitis using spectral-domain optical coherence tomography angiography *Am. J. Ophthalmol.* **171** 101–12

[109] Kim A Y, Chu Z, Shahidzadeh A, Wang R K, Puliafito C A and Kashani A H 2016 Quantifying microvascular density and morphology in diabetic retinopathy using spectral-domain optical coherence tomography angiography *Invest. Ophth. Vis. Sci.* **57** OCT362–70

[110] Rosen R B, Krawitz B, Mo S, Geyman L, Phillips E, Carroll J, Weitz R and Chui T Y P 2016 Age-related variations in foveal avascular zone geometry and vessel density-an optical coherence tomography angiography (OCTA) study x *Invest. Ophth. Vis. Sci.* **57** 5501

[111] Fang D, Tang F Y, Huang H, Cheung C Y and Chen H 2019 Repeatability, interocular correlation and agreement of quantitative swept-source optical coherence tomography angiography macular metrics in healthy subjects *Br. J. Ophthalmol.* **103** 415–20

[112] Rabiolo A, Gelormini F, Marchese A, Cicinelli M V, Triolo G, Sacconi R, Querques L, Bandello F and Querques G 2018 Macular perfusion parameters in different angiocube sizes: does the size matter in quantitative optical coherence tomography angiography? *Invest. Ophthalmol. Vis. Sci.* **59** 231–7

[113] Mo S, Krawitz B, Efstathiadis E, Geyman L, Weitz R, Chui T Y, Carroll J, Dubra A and Rosen R B 2016 Imaging foveal microvasculature: optical coherence tomography angiography versus adaptive optics scanning light ophthalmoscope fluorescein angiography *Invest. Ophth. Vis. Sci.* **57** OCT130–40

[114] Iafe N A, Phasukkijwatana N, Chen X and Sarraf D 2016 Retinal capillary density and foveal avascular zone area are age-dependent: quantitative analysis using optical coherence tomography angiography *Invest. Ophth. Vis. Sci.* **57** 5780–7

[115] Kwon J, Choi J, Shin J W, Lee J and Kook M S 2018 An optical coherence tomography angiography study of the relationship between foveal avascular zone size and retinal vessel density *Invest. Ophth. Vis. Sci.* **59** 4143–53

[116] Zhang Y, Do B, Mustafi D, Kogachi K, Chu Z, Wang R K, Rodger D C, Rao N A and Kashani A H 2018 Assessing response of retinal microvasculature density and morphology

to intra-vitreal dexamethasone implant injections for ocular inflammatory disease with spectral domain optical coherence tomography angiography (SD-OCTA) *Invest. Ophth. Vis. Sci.* **59** 2835
[117] Shahidzadeh A, Chen Y C, Chu Z, Wang R K and Kashani A H 2018 Correlation of macular thickness with microvascular density and morphology using spectral domain optical coherence tomography angiography (SD-OCTA) among normal subjects *Invest. Ophth. Vis. Sci.* **59** 2869
[118] Falavarjani K G, Tian J J, Akil H, Garcia G A, Sadda S R and Sadun A A 2016 Swept-source optical coherence tomography angiography of the optic disk in optic neuropathy *Retina* **36** S168–77
[119] Talisa E *et al* 2015 Detection of microvascular changes in eyes of patients with diabetes but not clinical diabetic retinopathy using optical coherence tomography angiography *Retina* **35** 2364–70
[120] Lei J, Yi E, Suo Y, Chen C, Xu X, Ding W, Abdelfattah N S, Fan X and Lu H 2018 Distinctive analysis of macular superficial capillaries and large vessels using optical coherence tomographic angiography in healthy and diabetic eyes *Invest. Ophth. Vis. Sci.* **59** 1937–43
[121] Kaizu Y, Nakao S, Yoshida S, Hayami T, Arima M, Yamaguchi M, Wada I, Hisatomi T, Ikeda Y and Ishibashi T 2017 Optical coherence tomography angiography reveals spatial bias of macular capillary dropout in diabetic retinopathy *Invest. Ophth. Vis. Sci.* **58** 4889
[122] Kwon J, Choi J, Shin J W, Lee J and Kook M S 2017 Alterations of the foveal avascular zone measured by optical coherence tomography angiography in glaucoma patients with central visual field defects *Invest. Ophth. Vis. Sci.* **58** 1637–45
[123] Kim J-A, Kim T-W, Lee E J, Girard M J and Mari J M 2018 Microvascular changes in peripapillary and optic nerve head tissues after trabeculectomy in primary open-angle glaucoma *Invest. Ophth. Vis. Sci.* **59** 4614–21
[124] Muraoka Y, Uji A, Ishikura M, Iida Y, Ooto S and Tsujikawa A 2018 Segmentation of the four-layered retinal vasculature using high-resolution optical coherence tomography angiography reveals the microcirculation unit *Invest. Ophth. Vis. Sci.* **59** 5847–53
[125] Anegondi N, Kshirsagar A, Mochi T B and Roy A S 2018 Quantitative comparison of retinal vascular features in optical coherence tomography angiography images from three different devices *Ophthalmic Sur. Lasers Imaging Retina* **49** 488–96
[126] Gadde S G, Anegondi N, Bhanushali D, Chidambara L, Yadav N K, Khurana A and Roy A S 2016 Quantification of vessel density in retinal optical coherence tomography angiography images using local fractal dimension *Invest. Ophth. Vis. Sci.* **57** 246–52
[127] Wang Q, Chan S, Yang J Y, You B, Wang Y X, Jonas J B and Wei W B 2016 Vascular density in retina and choriocapillaris as measured by optical coherence tomography angiography *Am. J. Ophthalmol.* **168** 95–109
[128] Onishi A C, Nesper P L, Roberts P K, Moharram G A, Chai H, Liu L, Jampol L M and Fawzi A A 2018 Importance of considering the middle capillary plexus on OCT angiography in diabetic retinopathy *Invest. Ophth. Vis. Sci.* **59** 2167–76
[129] Hwang T S, Gao S S, Liu L, Lauer A K, Bailey S T, Flaxel C J, Wilson D J, Huang D and Jia Y 2016 Automated quantification of capillary nonperfusion using optical coherence tomography angiography in diabetic retinopathy *JAMA Ophthalmol.* **134** 367
[130] Jia Y, Bailey S T, Hwang T S, McClintic S M, Gao S S, Pennesi M E, Flaxel C J, Lauer A K, Wilson D J and Hornegger J 2015 Quantitative optical coherence tomography angiography of vascular abnormalities in the living human eye *Proc. Natl. Acad. Sci.* **112** E2395–402

[131] Hsieh Y-T, Alam M N, Le D, Hsiao C-C, Yang C-H, Chao D and Yao X 2019 Optical coherence tomography angiography biomarkers for predicting visual outcomes after ranibizumab treatment for diabetic macular edema *Ophthalmol. Retina* **3** 826–34

[132] Skalet A *et al* 2018 Quantitative optical coherence tomography angiography for evaluation of peripapillary retinal capillary circulation after plaque brachytherapy for uveal melanoma *Ophthalmol. Retina* **2** 244–50

[133] Chu Z, Lin J, Gao C, Xin C, Zhang Q, Chen C-L, Roisman L, Gregori G, Rosenfeld P J and Wang R K 2016 Quantitative assessment of the retinal microvasculature using optical coherence tomography angiography *J. Biomed. Opt.* **21** 066008

[134] Kim T-H, Son T, Lu Y, Alam M and Yao X 2018 Comparative optical coherence tomography angiography of wild-type and rd10 mouse retinas *Transl. Vis. Sci. Technol.* **7** 42

[135] Kim Y, Hong H K, Park J R, Choi W, Woo S J, Park K H and Oh W-Y 2018 Oxygen-induced retinopathy and choroidopathy: *in vivo* longitudinal observation of vascular changes using OCTA *Invest. Ophth. Vis. Sci.* **59** 3932–42

[136] Inooka D, Ueno S, Kominami T, Sayo A, Okado S, Ito Y and Terasaki H 2018 Quantification of macular microvascular changes in patients with retinitis pigmentosa using optical coherence tomography angiography *Invest. Ophthalmol. Vis. Sci.* **59** 433–8

[137] Arthur E, Elsner A E, Sapoznik K A, Papay J A, Muller M S and Burns S A 2019 Distances from capillaries to arterioles or venules measured using OCTA and AOSLO *Invest. Ophth. Vis. Sci.* **60** 1833–44

[138] Brunner M, Romano V, Steger B, Vinciguerra R, Lawman S, Williams B, Hicks N, Czanner G, Zheng Y and Willoughby C E 2018 Imaging of corneal neovascularization: optical coherence tomography angiography and fluorescence angiography *Invest. Ophth. Vis. Sci.* **59** 1263–9

[139] Hosseinaee Z, Tan B, Martinez A and Bizheva K K 2018 Comparative study of optical coherence tomography angiography and phase-resolved doppler optical coherence tomography for measurement of retinal blood vessels caliber *Transl. Vis. Sci. Technol.* **7** 18

[140] Vujosevic S, Muraca A, Alkabes M, Villani E, Cavarzeran F, Rossetti L and De Cilla S 2019 Early microvascular and neural changes in patients with type 1 and type 2 diabetes mellitus without clinical signs of diabetic retinopathy *Retina* **39** 435–45

[141] Chu S, Nesper P L, Soetikno B T, Bakri S J and Fawzi A A 2018 Projection-resolved OCT angiography of microvascular changes in paracentral acute middle maculopathy and acute macular neuroretinopathy *Invest. Ophth. Vis. Sci.* **59** 2913–22

[142] Dansingani K K and Freund K B 2015 Optical coherence tomography angiography reveals mature, tangled vascular networks in eyes with neovascular age-related macular degeneration showing resistance to geographic atrophy *Ophthalmic Sur. Lasers Imaging Retina* **46** 907–12

[143] Saraf S S, Tyring A J, Chen C-L, Le T P, Kalina R E, Wang R and Chao J R 2019 Familial retinal arteriolar tortuosity and quantification of vascular tortuosity using swept-source optical coherence tomography angiography *Am. J. Ophthalmol. Case Rep.* **14** 74–8

[144] Khansari M M, O'Neill W, Lim J and Shahidi M 2017 Method for quantitative assessment of retinal vessel tortuosity in optical coherence tomography angiography applied to sickle cell retinopathy *Biomed. Opt. Express* **8** 3796–806

[145] Bulut M, Kurtuluş F, Gözkaya O, Erol M K, Cengiz A, Akıdan M and Yaman A 2018 Evaluation of optical coherence tomography angiographic findings in Alzheimer's type dementia *Br. J. Ophthalmol.* **102** 233–37

[146] Chang M Y, Phasukkijwatana N, Garrity S, Pineles S L, Rahimi M, Sarraf D, Johnston M, Charles A and Arnold A C 2017 Foveal and peripapillary vascular decrement in migraine with aura demonstrated by optical coherence tomography angiography *Invest. Ophth. Vis. Sci.* **58** 5477–84

[147] Conti F F, Qin V L, Rodrigues E B, Sharma S, Rachitskaya A V, Ehlers J P and Singh R P 2019 Choriocapillaris and retinal vascular plexus density of diabetic eyes using split-spectrum amplitude decorrelation spectral-domain optical coherence tomography angiography *Br. J. Ophthalmol.* **103** 452–6

[148] Dimitrova G, Chihara E, Takahashi H, Amano H and Okazaki K 2017 Quantitative retinal optical coherence tomography angiography in patients with diabetes without diabetic retinopathy *Invest. Ophth. Vis. Sci.* **58** 190–6

[149] Garrity S T, Iafe N A, Phasukkijwatana N, Chen X and Sarraf D 2017 Quantitative analysis of retinal capillary density and foveal avascular zone area using optical coherence tomography angiography of normal eyes *Invest. Ophth. Vis. Sci.* **58** 3646

[150] Goudot M M, Sikorav A, Semoun O, Miere A, Jung C, Courbebaisse B, Srour M, Freiha J G and Souied E H 2017 Parafoveal OCT angiography features in diabetic patients without clinical diabetic retinopathy: a qualitative and quantitative analysis *J. Ophthalmol.* **2017** 8676091

[151] Kim K, Kim E S and Yu S-Y 2018 Optical coherence tomography angiography analysis of foveal microvascular changes and inner retinal layer thinning in patients with diabetes *Br. J. Ophthalmol.* **102** 1226–31

[152] Linderman R, Salmon A E, Strampe M, Russillo M, Khan J and Carroll J 2017 Assessing the accuracy of foveal avascular zone measurements using optical coherence tomography angiography: segmentation and scaling *Transl. Vis. Sci. Technol.* **6** 16

[153] Samara W A, Shahlaee A, Adam M K, Khan M A, Chiang A, Maguire J I, Hsu J and Ho A C 2017 Quantification of diabetic macular ischemia using optical coherence tomography angiography and its relationship with visual acuity *Ophthalmology* **124** 235–44

[154] Scarinci F, Picconi F, Giorno P, Boccassini B, De Geronimo D, Varano M, Frontoni S and Parravano M 2018 Deep capillary plexus impairment in patients with type 1 diabetes mellitus with no signs of diabetic retinopathy revealed using optical coherence tomography angiography *Acta Ophthalmol.* **96** e264–5

[155] Simonett J M, Scarinci F, Picconi F, Giorno P, De Geronimo D, Di Renzo A, Varano M, Frontoni S and Parravano M 2017 Early microvascular retinal changes in optical coherence tomography angiography in patients with type 1 diabetes mellitus *Acta Ophthalmol.* **95** e751–5

[156] Yeung L, Wu I W, Sun C C, Liu C F, Chen S Y, Tseng C H, Lee H C and Lee C C 2019 Early retinal microvascular abnormalities in patients with chronic kidney disease *Microcirculation* e12555

[157] Yoon Y S, Woo J M, Woo J E and Min J K 2018 Superficial foveal avascular zone area changes before and after idiopathic epiretinal membrane surgery *Int. J. Ophthalmol.* **11** 1711

[158] Alibhai A Y, Moult E M, Shahzad R, Rebhun C B, Moreira-Neto C, McGowan M, Lee D, Lee B, Baumal C R and Witkin A J 2018 Quantifying microvascular changes using OCT angiography in diabetic eyes without clinical evidence of retinopathy *Ophthalmol. Retina* **2** 418–27

[159] Andrade Romo J S, Linderman R E, Pinhas A, Carroll J, Rosen R B and Chui T Y P 2019 Novel development of parafoveal capillary density deviation mapping using an age-group and eccentricity matched normative OCT angiography database *Transl. Vis. Sci. Technol.* **8** 1

[160] Hwang T S, Jia Y, Gao S S, Bailey S T, Lauer A K, Flaxel C J, Wilson D J and Huang D 2015 Optical coherence tomography angiography features of diabetic retinopathy *Retina* **35** 2371

[161] Kaizu Y, Nakao S, Wada I, Yamaguchi M, Fujiwara K, Yoshida S, Hisatomi T, Ikeda Y, Hayami T and Ishibashi T 2017 Imaging of retinal vascular layers: adaptive optics scanning laser ophthalmoscopy versus optical coherence tomography angiography *Transl. Vis. Sci. Technol.* **6** 2

[162] Krawitz B D, Phillips E, Bavier R D, Mo S, Carroll J, Rosen R B and Chui T Y 2018 Parafoveal nonperfusion analysis in diabetic retinopathy using optical coherence tomography angiography *Transl. Vis. Sci. Technol.* **7** 4

[163] Lu Y, Simonett J M, Wang J, Zhang M, Hwang T, Hagag A M, Huang D, Li D and Jia Y 2018 Evaluation of automatically quantified foveal avascular zone metrics for diagnosis of diabetic retinopathy using optical coherence tomography angiography *Invest. Ophth. Vis. Sci.* **59** 2212–21

[164] Philip S, Najafi A, Tantraworasin A, Chui T Y, Rosen R B and Ritch R 2019 Macula vessel density and foveal avascular zone parameters in exfoliation glaucoma compared to primary open-angle glaucoma *Invest. Ophth. Vis. Sci.* **60** 1244–53

[165] Rosen R B, Andrade Romo J S, Krawitz B D, Mo S, Fawzi A A, Linderman R E, Carroll J, Pinhas A and Chui T Y P 2019 Earliest evidence of preclinical diabetic retinopathy revealed using optical coherence tomography angiography perfused capillary density *Am. J. Ophthalmol.* **203** P103–15

[166] Samara W A, Say E A, Khoo C T, Higgins T P, Magrath G, Ferenczy S and Shields C L 2015 Correlation of foveal avascular zone size with foveal morphology in normal eyes using optical coherence tomography angiography *Retina* **35** 2188–95

[167] Schmidt T G, Linderman R E, Strampe M R, Chui T Y P, Rosen R B and Carroll J 2019 The utility of frame averaging for automated algorithms in analyzing retinal vascular biomarkers in angioVue OCTA *Transl. Vis. Sci. Technol.* **8** 10

[168] Takase N, Nozaki M, Kato A, Ozeki H, Yoshida M and Ogura Y 2015 Enlargement of foveal avascular zone in diabetic eyes evaluated by en face optical coherence tomography angiography *Retina* **35** 2377–83

[169] Tan C S, Lim L W, Chow V S, Chay I W, Tan S, Cheong K X, Tan G T and Sadda S R 2016 Optical coherence tomography angiography evaluation of the parafoveal vasculature and its relationship with ocular factors *Invest. Ophthalmol. Vis. Sci.* **57** OCT224–234

[170] Anegondi N, Chidambara L, Bhanushali D, Gadde S G, Yadav N K and Sinha Roy A 2018 An automated framework to quantify areas of regional ischemia in retinal vascular diseases with OCT angiography *J. Biophotonics* **11** e201600312

[171] Sioufi K, Shahlaee A, Say E A T, Ferenczy S, Hsu J and Shields C L 2017 Fractal dimension analysis of parafoveal microvascular anatomy using optical coherence tomography angiography in two machines *Invest. Ophth. Vis. Sci.* **58** 1674

[172] Faatz H, Farecki M-L, Rothaus K, Gunnemann F, Gutfleisch M, Lommatzsch A and Pauleikhoff D 2019 Optical coherence tomography angiography of types 1 and 2 choroidal neovascularization in age-related macular degeneration during anti-VEGF therapy: evaluation of a new quantitative method *Eye* **33** 1466–71

[173] Ting D S W, Tan G S W, Agrawal R, Yanagi Y, Sie N M, Wong C W, San Yeo I Y, Lee S Y, Cheung C M G and Wong T Y 2017 Optical coherence tomographic angiography in type 2 diabetes and diabetic retinopathy *JAMA Ophthalmol.* **135** 306–12

[174] Chen Q, Ma Q, Wu C, Tan F, Chen F, Wu Q, Zhou R, Zhuang X, Lu F and Qu J 2017 Macular vascular fractal dimension in the deep capillary layer as an early indicator of microvascular loss for retinopathy in type 2 diabetic patients *Invest. Ophth. Vis. Sci.* **58** 3785–94

[175] Bhardwaj S, Tsui E, Zahid S, Young E, Mehta N, Agemy S, Garcia P, Rosen R B and Young J A 2018 Value of fractal analysis of optical coherence tomography angiography in various stages of diabetic retinopathy *Retina* **38** 1816–23

[176] Spain R I, Liu L, Zhang X, Jia Y, Tan O, Bourdette D and Huang D 2018 Optical coherence tomography angiography enhances the detection of optic nerve damage in multiple sclerosis *Br. J. Ophthalmol.* **102** 520–4

[177] Ishibazawa A, Nagaoka T, Takahashi A, Omae T, Tani T, Sogawa K, Yokota H and Yoshida A 2015 Optical coherence tomography angiography in diabetic retinopathy: a prospective pilot study *Am. J. Ophthalmol.* **160** 35–44 e31

[178] Rebhun C B *et al* 2017 Polypoidal choroidal vasculopathy on swept-source optical coherence tomography angiography with variable interscan time analysis *Transl. Vis. Sci. Technol.* **6** 4

[179] Talisa E, Bonini Filho M A, Chin A T, Adhi M, Ferrara D, Baumal C R, Witkin A J, Reichel E, Duker J S and Waheed N K 2015 Spectral-domain optical coherence tomography angiography of choroidal neovascularization *Ophthalmology* **122** 1228–38

[180] Kermany D S, Goldbaum M, Cai W, Valentim C C, Liang H, Baxter S L, McKeown A, Yang G, Wu X and Yan F 2018 Identifying medical diagnoses and treatable diseases by image-based deep learning *Cell* **172** 1122–31 e1129

IOP Publishing

Photo Acoustic and Optical Coherence Tomography Imaging, Volume 3
Angiography: an application in vessel imaging
Ayman El-Baz and Jasjit S Suri

Chapter 8

Clinical utility of OCT-angio in age-related macular degeneration

Hashim Ali Khan, Smaha Jahangir and Julie Friedman Rodman

Age-related macular denegation (AMD) is a leading cause of blindness among the elderly in developed countries. Beginning as an outer retinal and choriocapillaris (CC) metabolical imbalance, it progresses into one of the two advanced forms, i.e., geographic atrophy or choroidal neovascularization. Drusen are an early sign of AMD. Optical coherence tomography (OCT) has been used for evaluation and documentation of changes associated with drusen, pigment epithelial detachments, choroidal neovascularization as well as being helpful in decision making. A significant number of eyes develop choroidal neovascularization and require frequent intraocular injections of vascular endothelial growth factor inhibitors. Label based angiographic techniques are typically used in diagnosing and management of such cases while most practices supplement the angiographic studies with OCT. OCT-angio (OCTA) is evolving as an indispensable tool in providing high resolution, three-dimensional vascular flow maps as well as structural information of choroidal retinal tissue; thus providing important biomarkers in diagnosing and guiding the treatment decision. The diagnostic accuracy of OCTA is comparable to or better (in some instances) than the current, invasive vascular imaging techniques.

Detailed visualization of the CC was not achievable with other vascular imaging modalities. OCTA, offers the details of the CC in high resolution. Besides its role in clinical management of AMD, it has expanded our cognizance about the role of the CC in pathogenesis and progression of the disease.

In this chapter, we describe clinical as well as OCTA features of different lesions associated with AMD. In addition, recent classification and lexicons for describing these entities are discussed.

AMD is the leading cause of irreversible visual loss in the elderly and its prevalence is estimated to double by 2050. It is derived from several genetic and

doi:10.1088/978-0-7503-2060-3ch8

modifiable environmental factors. AMD can be categorized into early and late stages. Early and intermediate AMD is characterized by good visual functions, focal sub-retinal pigment epithelium (RPE) deposits and a variable degree of pigmentary changes. Late-stage AMD assumes one of the two prevalent forms as choroidal neovascularization (CNV) and geographic atrophy (GA).

Photoreceptors, RPE, Bruch's membrane (BM) and the CC are involved subclinically earlier in the disease course. While one of the entities associated with AMD (end-stage disease) is geographic atrophy, most eyes with CNV, if untreated, rapidly progress to disciform scarring of the macula. Early diagnosis and treatment of the disease may help avoid many complications and slow down the progression significantly.

OCT is the standard imaging modality for investigating the structural changes of retinal tissue while fundus fluorescein angiography (FFA) and indocyanine green angiography (ICGA) are used for evaluation of retinal and choroidal vasculature. Often, both OCT and FFA are needed to diagnose AMD and differentiate between the types. However, ICGA is frequently needed because the details of chroidal vascular changes are not well exhibited by FFA, specifically in occult membranes that reside below the RPE; because fluorescein dye does not penetrate past the blood retinal barrier into the choroid.

OCTA, as mentioned elsewhere in this book, can render depth-resolved, high resolution retinal and choroidal vascular flow maps along with structural information. Side-by-side study of retinal morphology and flow characteristics increases our diagnostic capabilities and accuracy several fold and is very helpful in the early diagnosis and differentiation from other masquerading clinical entities.

8.1 Clinical features and OCTA correlates of early AMD

Drusen are the hallmark clinicopathological feature of dry AMD. The term dry emphasizes the absence of fluid around the neurosensory retina that is typical in the wet form of the disease.

Multiple variants of drusen are described in dry AMD. Nonetheless, drusen-like lesions are seen in other retinal conditions also. Clinicians must be able to differentiate between the types of drusen and comparable retinal lesions for appropriate management of these cases. Structural OCT is used for depth localization and measurement, classification and to evaluate the associated changes in morphology within the photorecptors and adjacent layers.

Hard drusen, also known as drupelets, are well defined, small, yellow-white deposits between the RPE and BM. Their size is less than 63 μm. Hard drusen, on OCT, are seen as variably hyperreflective, small sub-RPE deposits while soft drusen assume a mound-shaped, hyperreflective elevation of RPE of variable size.

Soft drusen are larger in size (> 63 μm), yellowish-white deposits with indistinct margins. Often the adjacent soft drusen coalesce to form confluent soft drusen also known as drusenoid pigment epithelial detachment (PEDs) (figure 8.1). Drusenoid PEDs are difficult to distinguish from large soft drusens clinically as well as on OCT.

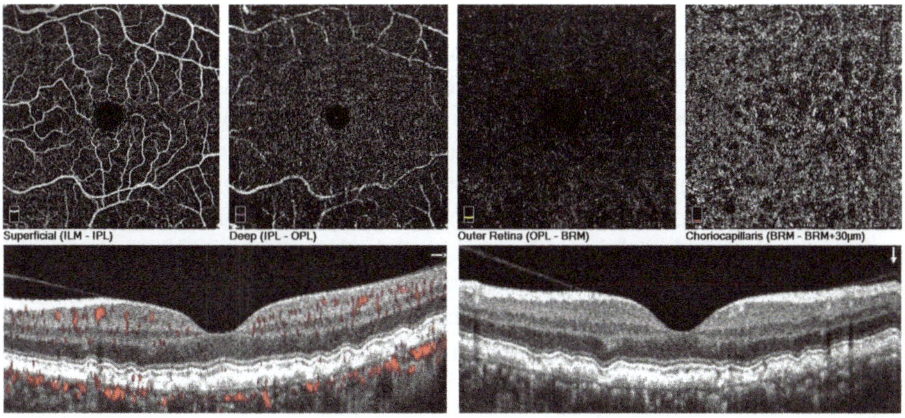

Figure 8.1. Shows OCT and OCTA of confluent soft drusen. Normal vascular density in superficial and deep slabs and no abnormal flow signal in the outer retina. Multiple hyperreflective RPE deposits on OCT cross-sections (bottom panels) and corresponding dark spots (false low flow versus low flow) on CC flow map (top right).

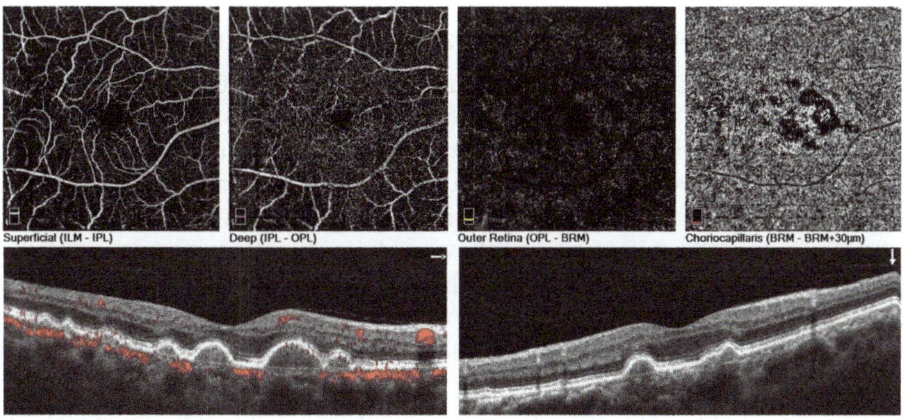

Figure 8.2. Multiple drusenoid PEDs with moderate internal reflectivity on cross sections (bottom panels) with normal retinal flow maps and a variable degree of underlying CC shadowing or flow deficits (top right panel).

They are associated with an increased risk of CNV. There may be an associated shallow serous retinal detachment in between the large drusens (figure 8.2).

Cuticular or basal laminar drusen are seen as multiple, uniform, yellowish-white, round punctate densely packed sub RPE deposits representing the nodular excrescences of BM. They are better characterized by fluorescein angiography (FA) as numerous hyperfluorescent foci taking on the appearance of the starry sky. Cuticular drusens on OCT are prolate shaped, moderately reflective deposits that usually erode the RPE monolayer assuming a sawtooth appearance. These are associated with an increased risk of CNV and pseudo-vitelliform macular detachment. The regression of drusen leads to reversal of RPE changes, however, it is an indicator of progression to advanced disease (i.e. geographic atrophy).

Reticular pseudodrusen (RPD) are characterized by an interlacing pattern of subretinal yellowish deposits. Unlike other types of drusen, they are not hyperfluorescent on FA. Suzuki *et al* [1] described three distinct morphological types of RPD on OCT. Dot pseudodrusen are distributed in the perifovea as discrete whitish accumulations. The second type termed as ribbon pseudodrusen are characterized by interconnected bands of yellowish-white material in a reticular pattern.

The third uncommon type, termed as peripheral pseudodrusen, is defined by the presence of yellow-white globules located outside the major vascular arcades [2–4] Querques *et al* [5] described OCT-based staging for the RPD. Granular hyperreflective material between the RPE and the ellipsoid zone (EZ) is defined as stage I. Stage II RPD are mounds of hyperreflective material with alterations in EZ band contour; that progresses to a thicker hyperreflective material with a conical morphology eroding through EZ in stage III. The material fades and resorbs/migrates into the inner retina in stage IV. Light stripes in choroid are observed under a subset of RPD corresponding to areas of structural damage in the deeper choroid in these eyes. However, this phenomenon is independent of the size of these deposits (figure 8.3).

8.1.1 Role of chorioretinal flow characteristic in pathogenesis of drusen

With the advent of OCTA, much attention has been paid to the role of the CC in the pathogenesis of AMD. While OCTA is becoming an indespensible tool for diagnosis and management of choroidal neovascularization; the relationship between drusen and chorioretinal flow characteristics on OCTA has been studied by numerous authors. CC flow deficit underneath drusen is universal. However, it is debated if drusen are associated with true hypoflow or false hypoflow signals induced by shadowing or signal attenuation caused by drusen [6]. Vascular flow information acquired with spectral-domain (SD) OCTA devices is more susceptible to shadowing and signal attenuation, giving rise to false hypoflow signals. Swept-source (SS) OCTA has enhanced tissue penetration and is least prone to shadowing and masking, yet a possible pseudo flow impairment can not be ruled out. Nonetheless, the true flow impairment was demonstrated in the CC under drusen using variable interscan time analysis (VISTA) on SS-OCTA devices [7–9]. CC flow impairment may be a result of reduced flow speeds (slower than the decorrelation speed), capillary dropout, or reduced vascular caliber. All these findings are consistent with past histological evidence about CC hypoxia being a major driver

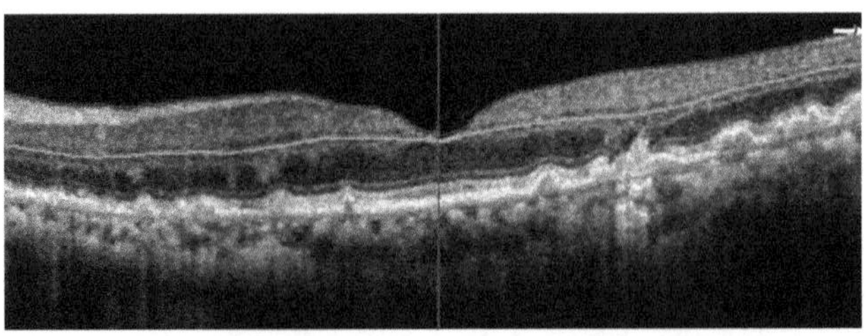

Figure 8.3. Structural OCT of soft and reticular pseudo-drusen.

behind the metabolic compromise of RPE, BM and outer retina with consequent drusen formation.

Clear evidence exists supporting the CC flow characteristic as surrogate to RPE and outer retinal metabolism. Several studies have confirmed drusen formation at the site of previous CC compromise and there exists a close relationship between CC insufficiency related to GA development and progression of GA.

Besides drusenoid PEDs, RPD and cuticular drusen are associated with increased risk of CNV and GA. OCTA may help in diagnosing CNV earlier in the course of the disease and result in better visual outcomes.

8.2 Geographic atrophy

Geographic atrophy is one form of advanced dry AMD characterized by a well-demarcated, round or oval area of outer retinal and RPE atrophy seen as a hypopigmented/depigmented lesion with enhanced visibility of the underlying choroidal vasculature, that gradually progresses. The fovea may be spared initially and involved later.

A number of imaging strategies including color fundus images, fundus auto-fluorescence (AF), multicolor confocal scanning laser ophthalmoscopy (CSLO) and structural OCT are used to classify the geographic atrophy and document progression.

Two prevalent forms of GA are drusen associated geographic atrophy (DAGA) and nascent geographic atrophy (nGA). DAGA is characterized by overt atrophy of RPE and outer retina while nGA is subsidence of retinal tissue from outer plexiform layer (OPL) to inner nuclear layer (INL) on OCT. However, the classification of atrophy meeting (CAM) group has devised a consensus statement on OCT-based definition and grading of AMD-associated macular atrophy. The group proposed four categories as complete RPE and outer retinal atrophy (cRORA), incomplete RPE and outer retinal atrophy (iRORA), complete outer retinal atrophy (cORA), and incomplete outer retinal atrophy (iORA) [10] (table 8.1).

OCTA has furthered our insights into the pathogenesis of dry AMD. Many studies have verified the ability of OCTA to detect CC flow impairment under and beyond the GA margins and have postulated CC flow compromise as a key factor in the progression and development of GA. GA develops in the outer retina corresponding to the area of previous CC flow impairment. Considering the CC flow impairment beyond GA margins, investigators now advocate the use of OCTA for delineating the extent of atrophy and to monitor progression [11].

Nonetheless, vascular rarefactions reportedly appear initially in the deep capillary plexus (DCP) and then in the superficial vascular plexus (SVP) even at intermediate stages of AMD.

It is worth mentioning that OCTA images in drusen, as well as GA, are prone to artifacts inherent to OCT as well as OCTA. The details of these artifacts have been mentioned elsewhere in this book. The clinician should take these artifacts into account while interpreting OCTA images.

Table 8.1. Description and schematic of normal and all categories of macular atrophy on OCT.

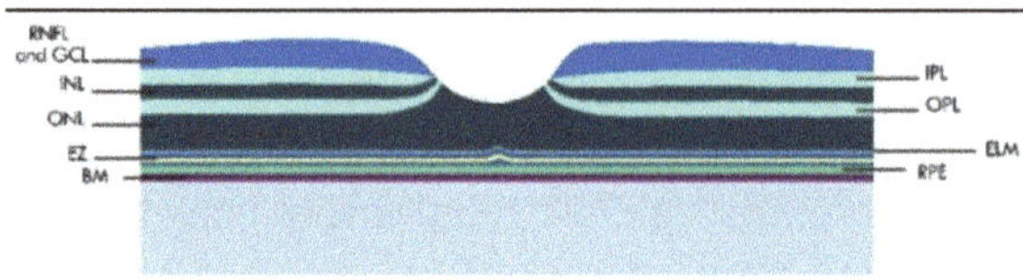	Normal macular morphology. All retinal layers are well defined and visible.	Normal
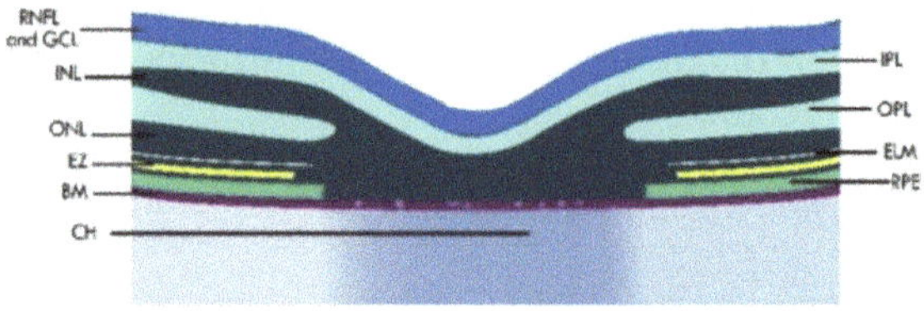	Complete homogenous choroidal hyper-transmission. Complete loss of RPE, photoreceptors, EZ band, ELM and absence of ONL.	cRORA
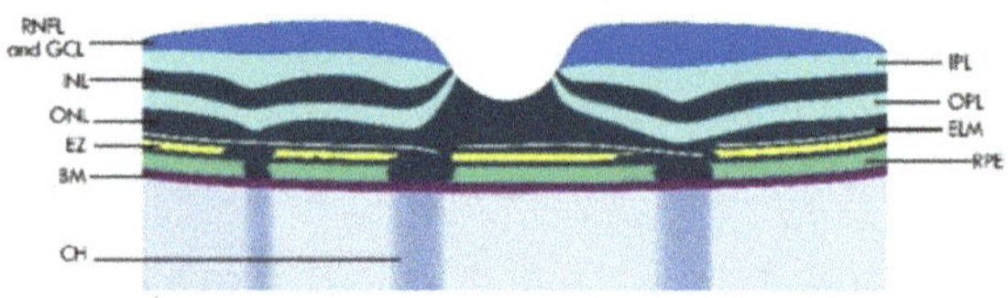	Discontinuous choroidal hyper-transmission. Discontinuous RPE band. Interrupted photoreceptor layer, EZ band and ELM. Subsidence of INL & ONL.	iRORA
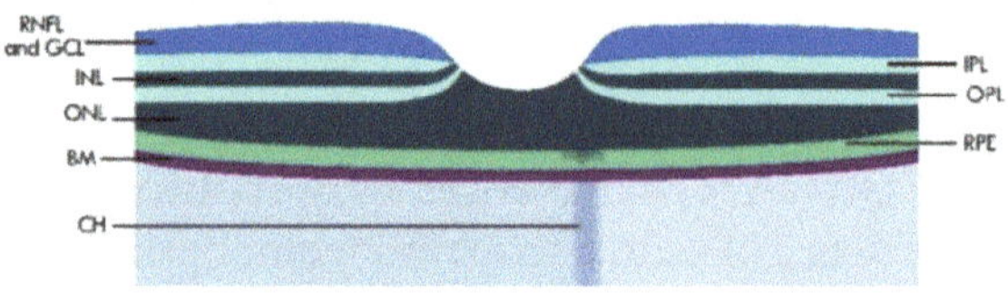	None to intermittent hyper-transmission. Intact RPE band, photoreceptors, EZ band and ELM are absent. Sever thinning and interruption of ONL.	cORA
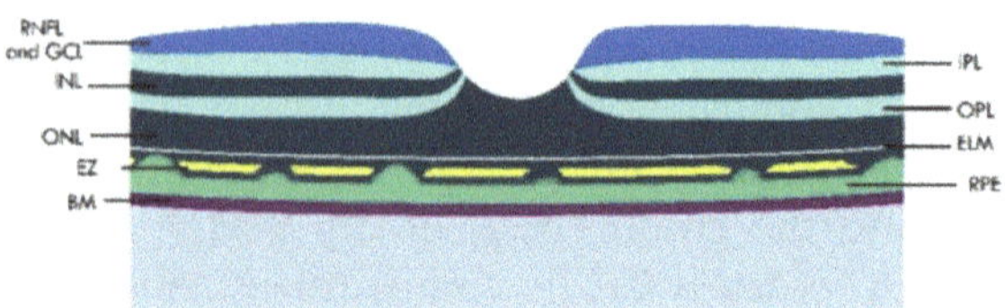	No choroidal hyper-transmission. Intact RPE and ELM bands. EZ may be disrupted in presence of regressing SDD. Detectable thinning of ONL.	iORA

cRORA = complete RPE and outer retinal atrophy, iRORA = incomplete RPE and outer retinal atrophy, cORA = complete outer retinal atrophy, iORA = incomplete outer retinal atrophy, RPE = retinal pigment epithelium, ELM = external limiting membrane, EZ = ellipsoid zone, INL = inner nuclear layer, ONL = outer nuclear layer. RNFL: retinal nerve fiber layer, GCL: ganglion cell layer, OPL: outer plexiform layer, IPL: inner plexiform layer, BM: Bruch's membrane, CH: choroidal hyper-transmission, SDD: subretinal drusenoid deposits..

8.3 Neovascular AMD

Choroidal neovascularization (CNV), now called macular neovascularization (MNV), is characterized by the formation of aberrant vascular proliferation under the RPE or sometimes above the RPE, which is the other end of AMD. Although most of the neovascularization in neovascular AMD (nvAMD) originates in the choroid, a significant number of neovascular membranes originate in intraretinal tissue. Consensus from the neovascular age-related macular degeneration nomenclature (CONAN) group has recently developed a lexicon for standard terminology in nvAMD. We are using the standard terms described by the group and describe the older titles for these entities.

8.3.1 Type 1 MNV

Previously known as occult CNV or type 1 CNV, comprises all kinds of neovascularization that originate in the CC and remain confined to the sub-RPE space and associated with varying degrees of PED. Lesion growth occurs concurrently with remodeling and broadening of feeder vessels both within the lesion and choroid [12].

Based on ill-defined areas of multipunctate hyperfluorescence/leakage on FA, the term occult CNV was coined to describe this subtype. ICGA offers relatively better visualization of these lesions as early hyperfluorescent spots, plaque-like hyperfluorescence, and later hyperfluorescence with indistinct margins. Yet the fine details of neovascular complex cannot be visualized using ICGA.

On structural OCT, this MNV subtype is associated with a PED in 98% of cases either as incomplete RPE elevation (in up to 35% cases) or complete RPE detachment (in 63%) with accompanying subretinal fluid and intraretinal fluid, in retinal cystic changes. A double-layer sign, i.e., two hyperreflective bands indicating an inner RPE and outer BM band separated by a hyporeflective space, is often seen. The double-layer sign denotes a shallow sub-RPE lesion associated with non-exudative neovascular membranes [12, 13].

Unlike ICGA and FA, OCTA clearly displays a distinct pattern of a well-organized vascular network in CC slabs, with evident feeder trunks in most cases. Type 1 MNV takes one of the three appearances on CC segmentation:

- Medusa pattern: a central trunk vessel with thin branches radiating centrifugally.
- Sea-fan pattern: an eccentric trunk vessel with most of the neovessels radiating to one side
- Indistinct pattern: there is no distinguishable pattern and a feeder vessel is not visible.

Most type 1 MNVs assume a medusa pattern of clear branching vascular network or larger caliber, more mature vessels confined to sub-RPE space [6, 14–16] (figure 8.4).

In up to 70% of cases, type 1 neovascular complexes grow significantly under RPE without causing visual symptoms and anatomical alterations. These neovascular membranes were termed as quiescent CNV. However, the nomenclature group has suggested calling it type 1 MNV omitting the term quiescent [6, 14–16].

Polypoidal choroidal vasculopathy (PCV) is another variant of wet AMD, accounting for half of the macular neovascularization in the Asian population. It is characterized by subretinal neovascularization seen as branching vascular network (BVN) with nodular dilatations of vessels called polyps. It is best depicted on ICGA

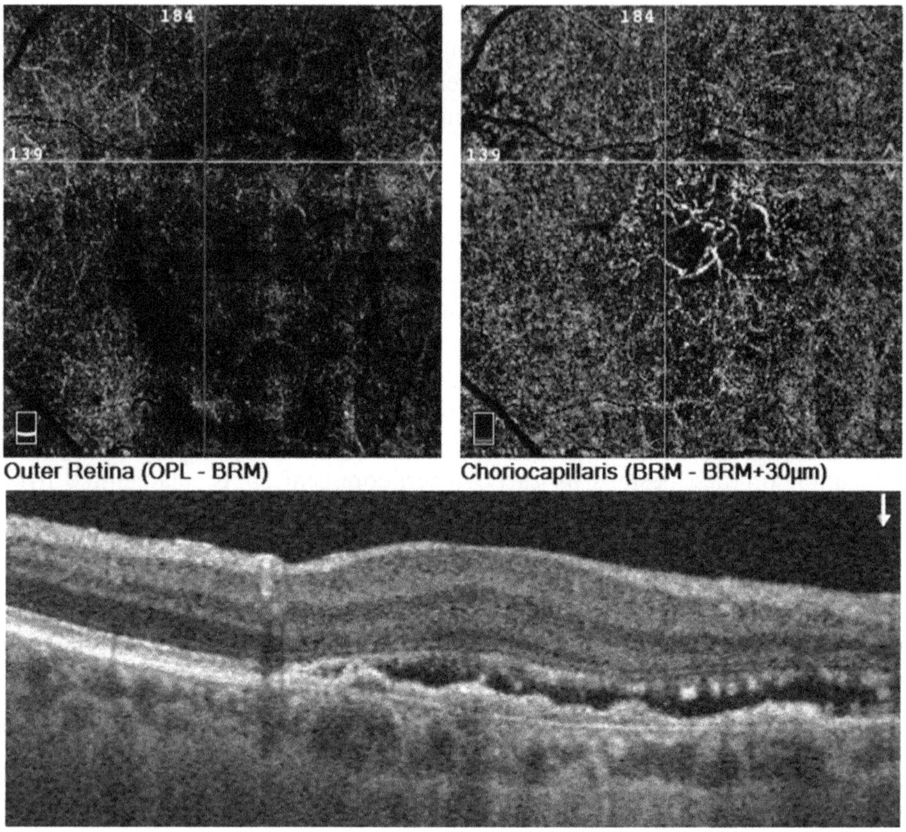

Figure 8.4. Showing hyperreflective lesion under RPE on OCT (bottom panel) and corresponding network of wide caliber vessels in CC slabs (top right panel) and normal outer retinal flow map (top left) on OCTA. Double-layer sign is evident on OCT.

as initial filling of BVN followed by staining of polyps. The neovascular complex gradually progresses underneath the Sub-RPE space to a significant size before causing visual symptoms. Polyps are predisposed to bleed and frequently cause hemorrhagic (serosanguinous PEDs) PED in these eyes. Although BVN can be clearly visualized on CC slabs on OCTA, the slow flow in polyps cannot produce a decorrelation of sufficient strength and are not detected. However, most studies evaluating polypoidal lesions studied the CC slabs while polyps are usually located at the top of PED and not segmented on default CC slabs but are better visualized on avascular outer retinal (AOR) segmentation [17, 18]. Shortly after the initial description of PCV as polyps, histological studies evidenced their origin from outpouching/ outgrowth of cavernous thin-walled vasculature underneath RPE. Considering the histological data, some authors called it aneurysmal type 1 neovascularization. But, the CONAN group classifies polypoidal choroidal neovasculopathy also as type 1 MNV because the neovascular complex does not break RPE and has its origin from choroidal vasculature (figure 8.5).

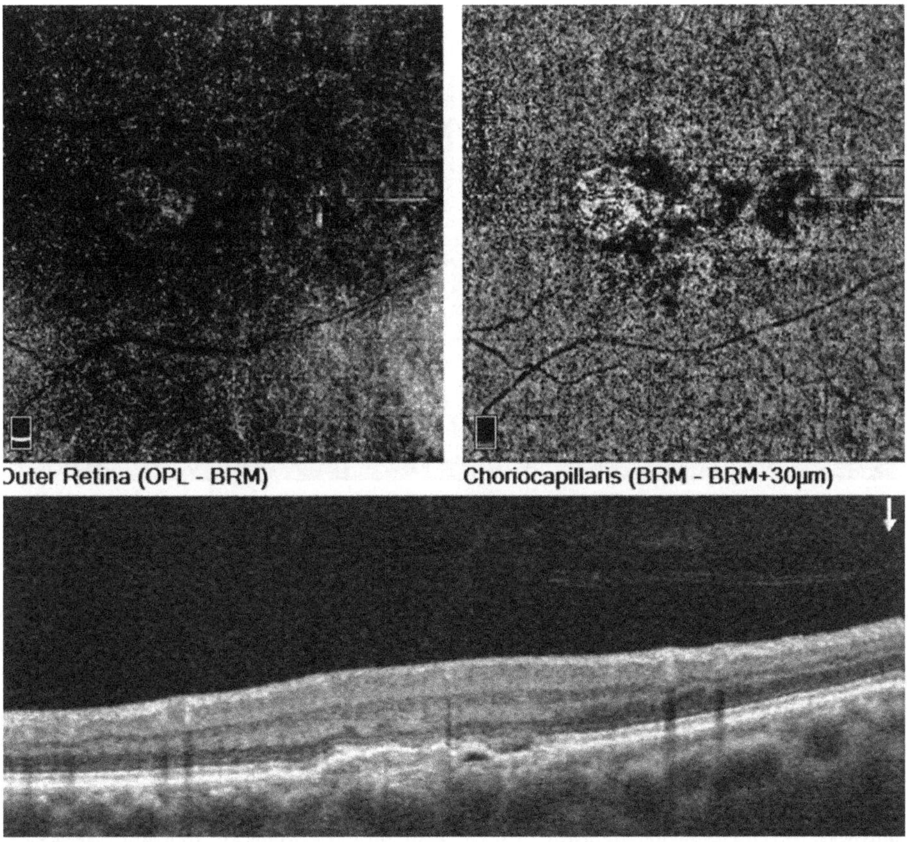

Figure 8.5. Another example of type 1 MNV.

8.3.2 Type 2 MNV

The type 2 neovascularization originates in the CC and traverses the RPE/BM complex to grow into the subretinal space. Despite the neovascular process beginning in CC, the pre-epithelial component leads to exudation and hemorrhaging in the outer retina with resultant visual symptoms and anatomical changes in retinal tissue. A well-defined, hyperfluorescent vascular lesion in the early phase of FFA with subsequent progressive leakage is typical of type 2 MNV [12].

Structural OCT has long been used in diagnosing the disease and guiding treatment decisions in clinical as well as research settings. On OCT, type 2 neovascular membranes are ascertained as the hyperreflective area between the RPE/BM and outer retina with varying degrees of associated posterior shadowing. Disease activity is described by the presence of subretinal and intraretinal fluid, with cystic changes leading to alterations in macular morphology.

OCTA exhibits one of the two morphological patterns (i.e. medusa or glomerular) on AOR slabs and the CC as a well-organized network of vessels giving rise to hyperflow signals. Often one or more, large caliber feeder trunks are seen anastomosing the MNV to deeper choroidal vasculature. Compared to other vessels,

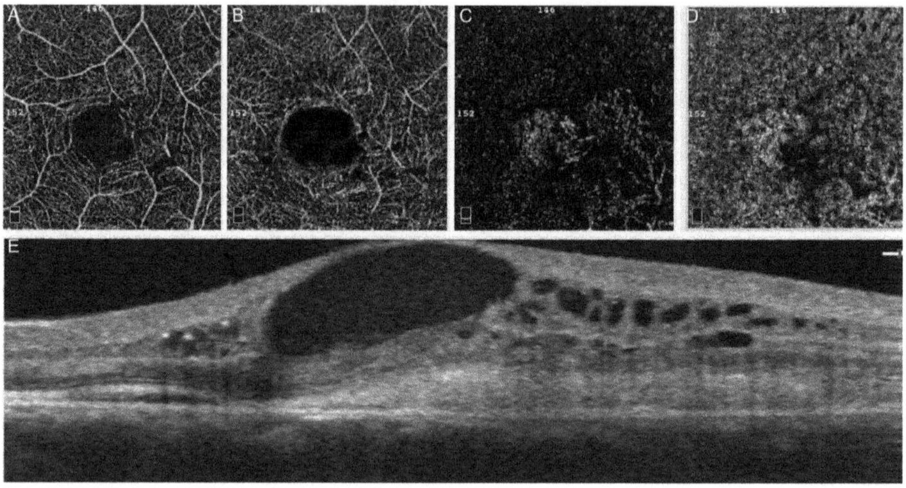

Figure 8.6. OCT of type 2 MNV (E) showing a hyperreflective mass with overlying cystic changes. Hyperflow signal corresponding to neovascular complex can be seen in the outer retina (C) and CC (D).

trunk vessels are larger in diameter. They sprout and give rise to smaller vessels of neovascular complex but have minimal branching and choroidal prominent anastomosis with larger vessels [15]. A perilesional dark halo is often seen which corresponds to CC flow deficit around the lesion (figure 8.6).

The hypoxic environment resulting from CC compromise triggers the release of vascular endothelial growth factor (VEGF) and other angiogenic growth factors from overlying RPE cells giving rise to MNV [6, 9, 14–16, 19–21]. It can be assumed that MNV develops at the site of previous/longstanding flow deficit and grows in areas corresponding to CC flow loss. Longitudinal studies may be needed to study this assumption.

8.3.3 Type 3 MNV

Type 3 neovascularization, previously known as retinal angiomatous proliferation, occult chorioretinal anastomosis, retinal vascular anomalous complex and retinal choroidal anastomosis, originates in the retinal DCP and extends toward the subretinal space forming retinal-retinal anastomosis and subsequently giving rise to a connected subretinal neovascular complex followed by proliferation past RPE. The neovascular complex is prone to leak and bleed causing intraretinal hemorrhages, edema and cystic changes. Over time, retinal vasculature supplying the neovascular component undergoes telangiectatic remodeling to compensate flow requirement. Although there may be a stimulating initial or concurrent choroidal component, the main feature is proliferation within the retina [22].

During the intraretinal stage, focal area of intraretinal staining with indistinct margin is seen on FFA corresponding to the focal intense hyperfluorescent hot spot on ICGA. A vascularized PED is often seen. Due to progressive leakage and retinal edema, visualization of the anastomosis is obscured on FA and ICGA [12].

OCT is used to study the morphological changes associated with MNV. A hyperreflective intraretinal mass is the OCT correlate of the intraretinal component. It is also helpful in documenting the monitoring of retinal edema and differentiating vascularized PEDs from serous [22].

Experience with OCTA in type 3 MNV is limited. It takes a distinctive appearance on OCTA as a small tuft of high flow small-caliber vessels in AOR segmentation originating from DCP. The retinal-retinal anastomosis can be identified on OCTA. Abnormal hyperflow signals can also be appreciated in eyes with PED and sub-RPE proliferation [6, 16, 23, 24]. Tan *et al* [24] using OCTA and co-registered OCT Bscans, described two distinct flow patterns in the eye with type 3 MNV. Abnormal hyperflow signal was limited to the neurosensory retina in up to 11% of eyes while in the rest the flow signal extended through the RPE. In a study by Miere *et al* [25], OCTA was able to demonstrate type 3 MNV in all 18 eyes as a distinct tuft of hyperflow-vascular network in AOR abutting posteriorly in sub-RPE space and anteriorly anastomosing to DCP. Also, a small clew-like hyperflow lesion was seen in 15 eyes anastomosing to choroid in two eyes [25] (figure 8.7).

OCTA has furthered our understanding of the pathogenic mechanisms associated with different types of MNVs. As discussed above, both type 1 and type 2 MNVs originate in the CC and later proliferate past the RPE into the subretinal space; both demonstrate different features on CC segmentations. Contrary to type 2 lesions, the demarcation between type 1 neovascularization and adjacent CC is not well-defined. Compared to type 2, Type 1 MNVs, as evidenced by OCTA, have greater linear dimensions and wider vascular caliber. The trunk vessels supplying type 1 MNVs have larger caliber than those of type 2 MNVs. These data establish that type 1 MNVs bear more mature vessels which are resistant to treatment with antiVEGF drugs. Type 1 MNVs grow up significantly under RPE before leading to symptoms and remain undetected for long. Formation of these membranes is in in-part driven by VEGF during angiogenesis and later by other growth factors through arteriogenesis and vascular normalization [15].

8.4 Diagnostic accuracy of OCTA

Multiple studies have compared OCTA with FA and ICGA for detection of MNV demonstrating diagnostic accuracy comparable to FA or ICGA. However, being non-invasive, quick and supplemented by structural information of chorioretinal tissue, OCTA is more advantageous when compared to FA or ICGA alone [6, 26].

Wang *et al* [27] performed a metanalysis of 16 studies reporting the diagnostic accuracy of OCTA compared to either FA or ICGA. They reported a pooled sensitivity and specificity as 0.87 and 0.96. Others have reported sensitivities as low as 50% and as high as 90% [14, 28–30]. Nevertheless, *en face* OCTA flow maps have diagnostic capabilities comparable to either FFA or ICGA but when combined with structural OCT and cross-sectional flow maps OCTA had greater diagnostic accuracy compared to either FFA or ICGA alone.

Most of the MNVs are associated with macular edema; and leakage of dye makes it difficult to ascertain vascular features of neovascular complexes using dye-based vascular imaging modalities. However on OCTA, vascular flowmap extraction is not

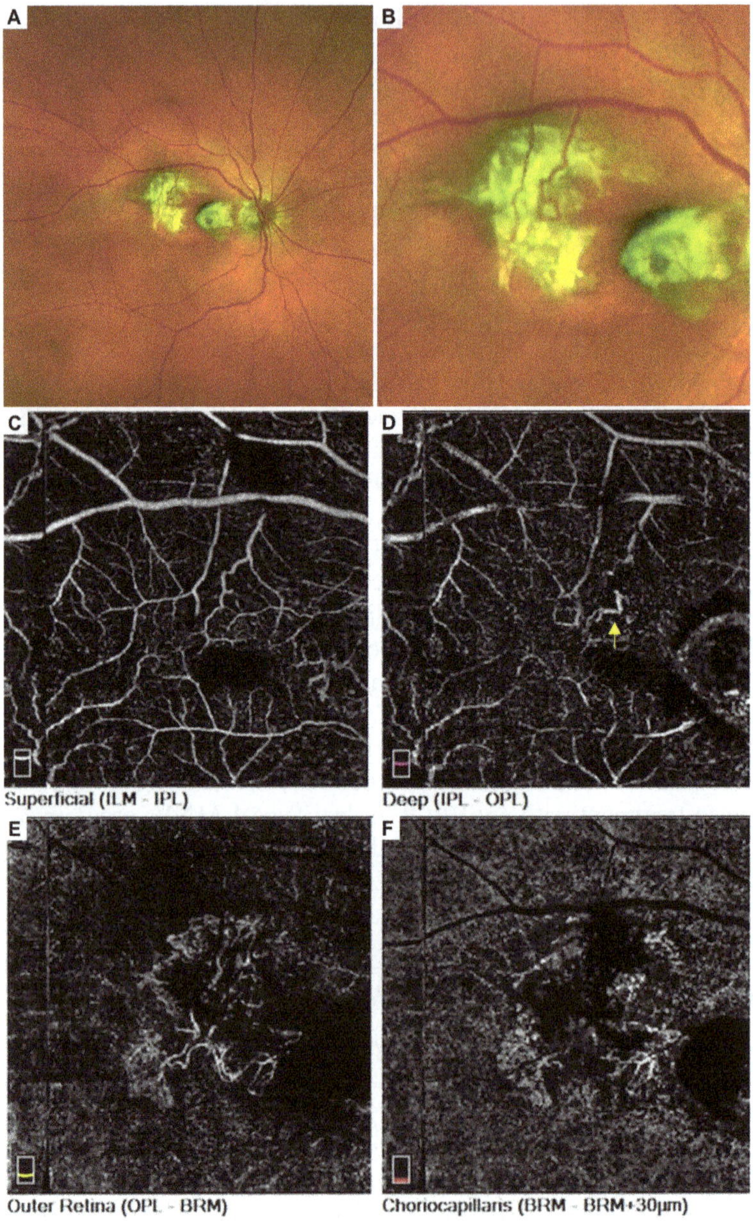

Figure 8.7. Type 3 MNV on color fundus images (a and b). Abnormal hyperflow signal neovascular network in outer retina (e) and CC (f). Arrow points to the anastomosing trunk from deep capillary plexus (d) with outer retina.

affected significantly and detailed visualization of neovascular complexes and associated vascular changes is more readily visible. Furthermore, early MNVs associated with drusenoid or other types of small PEDs can be missed on FA but a careful evaluation of cross-sectional and enface flowmaps can be very helpful in detecting them early.

While interpreting OCTA images from different platforms or comparing them to dye based imaging methods, it must be taken into account that the MNV sizes measured using SD-OCTA are significantly smaller than those acquired using SS-OCT and dye-based methods [14].

Many factors influence the diagnostic accuracy of OCTA in detecting MNV; signal strength, retinal architecture changes, pigmentary abnormalities, and artifacts are most important. It is important to look and correct for segmentation artifacts while grading the OCTA image. Structural information from OCT B scans in addition to *en face* OCTA flow maps improves the detection of MNVs significantly [31]. (see the detailed discussion elsewhere in this book).

8.5 OCTA for guiding treatment of AMD

FA and ICGA are gold-standard imaging modalities for the diagnosis of macular neovascularization. Nonetheless, with the advent of anti-VEGF injections for neovascular AMD, structural OCT has become a standard for monitoring the anatomical changes as a response to treatment. The decision to treat and monitor is partially guided by retinal thickness measured by OCT. As frequent injections are needed to eradicate neovascular activity, clinicians may need to acquire FA or ICGA repeatedly for studying the impact of treatment on the neovascular membranes. Owing to the invasive nature of both of these tests, there is a risk of associated adverse events which increases with every repetition of the tests.

It is known that OCTA cannot detect the neovascular complexes and membranes with low flow or no flow, and FFA or ICGA demonstrate better sensitivity for such neovascular membranes. However, we believe that MNVs with no flow or slow flow are fibrosed complexes which do not respond better to treatment and are usually monitored [6, 32]. OCTA can be used for monitoring these membranes for both structural changes (exudation, retinal thickness changes, morphological variations) and for changes in flow characteristics that also aids in making treatment decisions.

OCTA can provide flow information both as cross-sectional and depth-resolved *en face* images at any desired depth in chorioretinal tissue. Multiple objective metrics are available on commercial devices which can be used to gauge the treatment effects on the neovascular complex as well as retinal morphology. MNV flow area, vascular density and retinal thickness are automatically measured by some of the commercially available OCTA devices (only Optovue has analytics available at this time) with high accuracy and reproducibility. Attributable to their high resolution, these objective metrics can be used to objectively evaluate the treatment response.

References

[1] Suzuki M, Sato T and Spaide R F 2014 Pseudodrusen subtypes as delineated by multimodal imaging of the fundus *Am. J. Ophthalmol.* **157** 1005–12

[2] Spaide R F and Curcio C A 2010 Drusen characterization with multimodal imaging *Retina* **30** 1441–54

[3] Schmitz-Valckenberg S, Steinberg J S, Fleckenstein M, Visvalingam S, Brinkmann C K and Holz F G 2010 Combined confocal scanning laser ophthalmoscopy and spectral-domain

optical coherence tomography imaging of reticular drusen associated with age-related macular degeneration *Ophthalmology* **117** 1169–76

[4] Zweifel S A, Spaide R F, Curcio C A, Malek G and Imamura Y 2010 Reticular pseudodrusen are subretinal drusenoid deposits *Ophthalmology* **117** 303–12

[5] Querques G, Canouï-Poitrine F and Coscas F *et al* 2012 Analysis of progression of reticular pseudodrusen by spectral domain-optical coherence tomography *Investig. Ophthalmol. Vis. Sci.* **53** 1264–70

[6] Khan H A, Mehmood A and Khan Q A *et al* 2017 A major review of optical coherence tomography angiography *Expert Rev. Ophthalmol.* **12** 373–85

[7] Kashani A H, Chen C-L and Gahm J K *et al* 2017 Optical coherence tomography angiography: a comprehensive review of current methods and clinical applications *Prog. Retin. Eye Res.* **60** 66–100

[8] Moult E M, Waheed N K and Novais E A *et al* 2016 Swept-source optical coherence tomography angiography reveals choriocapillaris alterations in eyes with nascent geographic atrophy and drusen-associated geographic atrophy *Retina* **36** S2–11

[9] Lauermann J L, Eter N and Alten F 2018 Optical coherence tomography angiography offers new insights into choriocapillaris perfusion *Ophthalmologica* **239** 74–84

[10] Sadda S R, Guymer R and Holz F G *et al* 2018 Consensus definition for atrophy associated with age-related macular degeneration on OCT: classification of atrophy report 3 *Ophthalmology* **125** 537–48

[11] Sacconi R, Borrelli E and Carnevali A *et al* 2020 Optical coherence tomography angiography in type 3 neovascularization *Diabetes and Fundus OCT* Amsterdam: Elsevier pp 321–41

[12] Spaide R F, Jaffe G J and Sarraf D *et al* 2020 Consensus nomenclature for reporting neovascular age-related macular degeneration data: consensus on neovascular age-related macular degeneration nomenclature study group *Ophthalmology* **127** 616–36

[13] Shi Y, Motulsky E H and Goldhardt R *et al* 2019 Predictive value of the OCT double-layer sign for identifying subclinical neovascularization in age-related macular degeneration *Ophthalmol. Retin.* **3** 211–19

[14] Told R, Sacu S and Hecht A *et al* 2018 Comparison of SD-optical coherence tomography angiography and indocyanine green angiography in type 1 and 2 neovascular age-related macular degeneration *Investig. Ophthalmol. Vis. Sci.* **59** 2393–400

[15] Zhao Z, Yang F and Gong Y *et al* 2019 The comparison of morphologic characteristics of type 1 and type 2 choroidal neovascularization in eyes with neovascular age-related macular degeneration using optical coherence tomography angiography *Ophthalmologica.* **242** 178–86

[16] Nagiel A, Bansal M and Lenis T *et al* 2015 Morphological analysis of type 1, type 2, and type 3 neovascularization in exudative age-related macular degeneration using OCT angiography *Investig. Ophthalmol. Vis. Sci.* **56** 3965–5

[17] Wang M, Zhou Y and Gao S S *et al* 2016 Evaluating polypoidal choroidal vasculopathy with optical coherence tomography angiography *Investig. Ophthalmol. Vis. Sci.* **57** OCT526–32

[18] Tomiyasu T, Nozaki M, Yoshida M and Ogura Y 2016 Characteristics of polypoidal choroidal vasculopathy evaluated by optical coherence tomography angiography *Investig. Ophthalmol. Vis. Sci.* **57** OCT324–30

[19] Moult E, Choi W and Waheed N K *et al* 2014 Ultrahigh-speed swept-source OCT angiography in exudative AMD *Ophthalmic Surg. Lasers Imaging Retin.* **45** 496–505

[20] El Ameen A, Cohen S Y and Semoun O *et al* 2015 Type 2 Neovascularization secondary to age-related macular degeneration imaged by optical coherence tomography angiography *Retina* **35** 2212–18
[21] Souied E H, El Ameen A, Semoun O, Miere A, Querques G and Cohen S Y 2016 Optical coherence tomography angiography of type 2 neovascularization in age-related macular degeneration *Dev. Ophthalmol.* **56** 52–6
[22] Tsai A S H, Cheung N and Gan A T L *et al* 2017 Retinal angiomatous proliferation *Surv. Ophthalmol.* **62** 462–92
[23] Kuehlewein L, Dansingani K K and de Carlo T E *et al* 2015 Optical coherence tomography angiography of type 3 neovascularization secondary to age-related macular degeneration *Retina* **35** 2229–35
[24] Tan A C, Dansingani K K, Yannuzzi L A, Sarraf D and Freund K B 2017 Type 3 Neovascularization imaged with cross-sectional and en face optical coherence tomography angiography *Retina* **37** 234–46
[25] Miere A, Querques G, Semoun O, El Ameen A, Capuano V and Souied E H 2015 Optical coherence tomography angiography in early type 3 neovascularization *Retina* **35** 2236–41
[26] de Carlo T E, Romano A, Waheed N K and Duker J S 2015 A review of optical coherence tomography angiography (OCTA) *Int. J. Retin. Vitr.* **1** 5
[27] Wang R, Liang Z and Liu X 2019 Diagnostic accuracy of optical coherence tomography angiography for choroidal neovascularization: a systematic review and meta-analysis *BMC Ophthalmol.* **19** 162
[28] Perrott-Reynolds R, Cann R and Cronbach N *et al* 2019 The diagnostic accuracy of OCT angiography in naive and treated neovascular age-related macular degeneration: a review *Eye (Lond.)* **33** 274–82
[29] Schneider E W and Fowler S C 2018 Optical coherence tomography angiography in the management of age-related macular degeneration *Curr. Opin. Ophthalmol.* **29** 217–25
[30] Inoue M, Jung J J and Balaratnasingam C *et al* 2016 A comparison between optical coherence tomography angiography and fluorescein angiography for the imaging of type 1 neovascularization *Investig. Ophthalmol. Vis. Sci.* **57** OCT314–23
[31] Faridi A, Jia Y and Gao S S *et al* 2017 Sensitivity and specificity of OCT angiography to detect choroidal neovascularization *Ophthalmol. Retin.* **1** 294–303
[32] Al-Sheikh M, Iafe N A, Phasukkijwatana N, Sadda S R and Sarraf D 2018 Biomarkers of neovascular activity in age-related macular degeneration using optical coherence tomography angiography *Retina* **38** 220–30

IOP Publishing

Photo Acoustic and Optical Coherence Tomography Imaging, Volume 3

Angiography: an application in vessel imaging

Ayman El-Baz and Jasjit S Suri

Chapter 9

Optical coherence tomography angiography imaging in age-related macular degeneration

Onur Gokmen and Muhammet Derda Ozer

Optical coherence tomography angiography (OCTA) is becoming a popular method that enables non-invasive evaluation of retinal vascular structures. OCTA has started to be used as an alternative method in age-related macular degeneration (AMD) or in conjunction with classical fundus fluorescein angiography (FFA). Although OCTA is a new method, it has not been able to replace FFA in AMD completely. The limitations of OCTA for AMD can be listed as being highly affected by image artifacts, providing a relatively small field of view, and the failure of segmentation algorithms that can confuse the interpretation of the findings. In this chapter, literature summaries about the usage areas and clinical application of OCTA, especially in neovascular AMD, will be mentioned.

9.1 Introduction

AMD is a disease affecting the macula with progressive central vision loss. It is the most common cause of irreversible vision loss in adults over 50 age in developed countries and ranks third worldwide. Approximately 11 million people were affected by this disease in the USA alone, and the prevalence of AMD in the world is 170 million [1]. Diabetes, hypertension, smoking, aging, ultraviolet exposure, obesity are thought to be the risk factors in AMD development [2].

9.2 The principles Of OCTA technology and image quantification methods

The most recent development in the OCT era is OCTA. We can obtain high quality, 3D angiographic images of the retinal and choroidal vascular layers through this new technology. Despite its insensitivity to determining leakage and the other

doi:10.1088/978-0-7503-2060-3ch9

dynamic features of various retinal lesions seen in FFA, OCTA would likely help to clarify the pathophysiological processes underlying multiple ocular diseases. The repeated high-speed scans at the same location provide blood vessel identification. The blood flow through vessels alters the reflectance signal, leading to variations between the repeated scans at the same location. Quantifying these signal variations makes it possible to identify vascular structures with an objective evaluation of vascular pathologies [3]. 3D flow images can be presented as 2D *en face* images or conventional cross-sectional images. The flow signal in the superficial (between the internal limiting membrane and the outer limit of the outer plexiform layer) and the deep (between the outer plexiform layer and Bruch membrane) retinal layers could be combined and presented as one *en face* image in the originally mentioned way. The colored flow rate could also be overlaid on the conventional OCT image in the latter method. By these imagining techniques, the depth and topographical location of abnormal vessels and blood flow impairment could be determined easily [4]. The most commonly used OCTA parameters are vessel density and flow index [5–10]. The vessel density is defined as the percentage of the occupied area by the vessels, while the flow index is the mean signal intensity of blood flow in a region of interest. The diagnosis and follow-up of retinal vascular pathologies are possible with vessel density analysis [11]. The flow index is more sensitive in determining metabolic alterations in the retina [12].

There are various other biomarkers for quantification of vascular pathologies such as avascular area, perfusion density mapping, neovascularization area that were recently defined and utilized in monitoring the course of multiple diseases [13–18]. The avascular area measurement is a valuable tool for objectively assessing the degree of capillary dropout in certain conditions, for instance, glaucoma and diabetic retinopathy [13–15]. Perfusion mapping of vascular density is another biomarker to evaluate flow impairment in a particular scanned area [17].

The measurement of neovascularization area, which is calculated as the total neovascular area pixel, was defined recently to evaluate and monitor the retinal or choroidal neovascular membranes quantitatively [18].

9.3 FFA, ICG, and OCTA

OCTA is a relatively new method that allows us to see retinal blood vessels' flow by creating high-resolution contrasts, non-invasively. In the b-scan scan in OCTA, the images are obtained by observing the contrast created by erythrocytes' flow in the retinal vessels. For this reason, to create an image in OCTA, unlike many conventional OCTs, it is necessary to obtain high-resolution images much faster [3].

FFA and ICG are conventional and invasive retinal vascular imaging methods, and unlike OCTA, contrast dyes must be injected into the individual's vessels in these methods. Besides, FFA and ICG applications are more costly than OCTA. The contrast agent should be dispersed in the vessels and retina for 10–30 min to take images. This is a time-consuming practice with trained staff, so it is not ideal for routine busy clinical practice. Another handicap of these methods is that even though the dyes used in FFA and ICG are relatively safe, allergic reactions to these contrast agents can still develop, nausea-vomiting and anaphylaxis can occur. For these reasons, these methods should be applied in environments where healthcare personnel can be treated urgently. Besides, since these drugs are excreted from the body through the kidneys, patients should be questioned about renal failure, and applications should be made by obtaining the relevant branch physician's opinion for patients with renal failure. In short, it would be appropriate to perform FFA and ICG applications in clinics where emergency intervention opportunities are available and where an internal medicine or nephrologist's opinion can be obtained [19–21].

Although OCTA is a new imaging method, FFA is still considered as the gold standard in wet AMD or retinal neovascularization. Regarding ICG, it is accepted as the gold standard in imaging the choroidal vessels, especially since the contrast specific to the choroidal vessels is used [22]. In these methods, images are formed in two dimensions, but at the same time, unlike OCTA, images are obtained in a large part of the retina. Unlike OCTA, in FFA and ICG, in diseases in which retinal vascular integrity is impaired, the leaks created by the extravascular blood in the vascular leakage areas and the regions where it is ponding can be analyzed. However, since the images in FFA and ICG are two-dimensional, the desired information about the lesion may not be obtained in any hemorrhage that will develop in the retinal vessels, artifacts in the imaging of the vessels, opacities in the cornea or lens, which are other parts of the eye, and defects that may affect the image contrast below or above the retinal vessels. Also, changes in the retinal architectural structure due to atrophy and fibrosis may make it challenging to interpret OCTA and vascularity. As a result, FFA and ICG show the lesion's presence, but they cannot give us enough information about the lesion's axial nature [19, 21].

9.4 AMD types

Dry-type macular degeneration develops from the macula's degeneration caused by the accumulation of lipid deposits called drusen in the pigment epithelium with age. In dry type AMD, the retinal pigment epithelium (RPE) pigmentary changes, changes in the photoreceptor layers, and geographic atrophy are characterized by widespread and progressive and incurable damage RPE, which is called geographic atrophy, in the later stages of the disease can be seen [23, 24].

Wet AMD types are divided into 3. While neovascular vessels are seen in the sub-retinal area in the first type (occult) CNV as type 1, neovascularization is observed in the sub-RPE area in type 2 CNV (classical), and intra-retinal angiomatous proliferation (rap) is observed in the third type. Thanks to OCTA's 3D imaging technology, it can be useful in differentiation in typing according to axial locations in CNVs [25].

9.5 OCTA features in AMD

As mentioned earlier, the AMD is divided into two types during its natural course. The dry type or non-exudative AMD typically affects the choriocapillaris and retinal pigment epithelium layers leading to shrinkage of choriocapillaris and accumulation of drusenoid materials along with RPE alterations/atrophy. The wet type or exudative AMD is related to the neovascularization and is classified into three categories according to the neovascularization location as the occult (under the RPE), the classical (under the retinal layers), and the retinal angiomatous proliferation type. The OCTA imaging has some limitations and superiorities to classical imaging methods (FFA, ICG) in detecting certain features of both dry and wet AMD.

OCTA is a fast, non-invasive imaging technique. It provides high resolution, 3D static volumetric angiographic information from both superficial, intermediate, and deep retinal plexus as well as choriocapillaris. On the other hand, the field of view is limited in OCTA. It could not provide dynamic angiographic information and is very sensitive to the patient's fixation ability and ocular movements compared to the conventional imaging methods. OCTA can provide important information with regard to the diagnosis and treatment of AMD. Its 3D and high-quality image acquisition features are unique for locating the neovascular membrane's axial localization and identifying its distinctive patterns. The static image acquisition feature also provides a follow-up to the alterations that emerged after treatment [26–29]. On the other hand, due to the lack of dynamic imaging features of OCTA compared to FFA, it is challenging to evaluate the extent and severity of the neovascular lesion's leakage status [30]. Besides, the low flow rate cannot be distinguished from flow interruption due to undetectable slow flow rate with interscan timing [31]. The scanned area is relatively small compared to conventional imaging methods, and when one tries to enlarge the scanning area, the image resolution would be significantly decreased due to usage of insufficient A-scans to image a larger cross-sectional area [32–34].

9.6 Dry-type age-related macular degeneration

Dry-type AMD is characterized by drusen, pigmentary alterations, and RPE loss. The role of vascular changes before the development of the alterations, as mentioned earlier, is still controversial, but several histologic studies showed that choriocapillaris loss would have an initiator role in the development of dry-type AMD [35, 36]. Choriocapillaris layer visualization is limited by conventional angiography due to poor visualization. Spectral-domain OCT devices have an inadequate role in obtaining choriocapillaris layer images until a significant degree of RPE loss would occur due to low short-wavelength beam penetration through the intact RPE [30]. After the invention of OCTA, the role of choriocapillaris perfusion in the development of dry-type AMD was understood better. Lane *et al* showed a significant decrease in choriocapillaris flow rate at the drusen site [37]. The generalized decrease in choroidal vessel density that was mainly affecting the choriocapillaris layer has previously been shown, and impairment of blood flow in choriocapillaris is suspected as a crucial step in the pathogenesis of dry-type AMD [38].

The investigators preferred to investigate choroidal vascular alterations in geographic atrophy to avoid visualization difficulties that emerged by signal attenuation due to intact RPE. Choi *et al* compared 12 eyes of seven patients with geographic atrophy with control subjects using OCTA and found a significant flow impairment

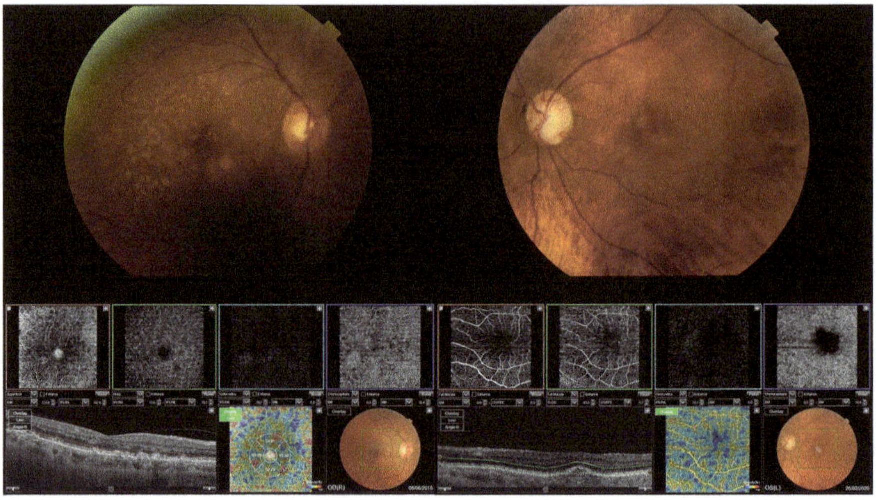

Figure 9.1. Fundus photographs, OCT images and OCTA images of both eyes of a patient with dry-type AMD.

in the choriocapillaris layer under the geographic atrophy area [39]. Besides, recently published data shows a subtle flow impairment and diminished vessel density in the choriocapillaris layer located in the vicinity of geographic atrophy. This implies that the vascular changes would likely emerge before the development of RPE alterations in dry AMD [40, 41]. OCTA also has superior efficacy in determining geographic atrophy progression and excluding the presence of neovascularization by providing additional blood flow parameters [42] (figure 9.1).

9.7 Wet-type age-related macular degeneration

The characteristic feature of wet-type or exudative AMD is neovascularization. It is also further divided into three categories according to the axial location of the choroidal neovascular membrane (CNVM) as occult, classical, and retinal angiomatous proliferation.

As mentioned previously, FFA as a gold standard imaging modality for detecting wet-type AMD has not been replaced by OCTA yet. A series of studies evaluating the sensitivity and specificity of OCTA in determining choroidal neovascularization showed significantly high false-negative rates [43–47]. In other words, a relatively large amount of choroidal neovascularisation that can be identified by FFA could not be determined by OCTA. Based on the previous reports, CNVMs accompanied by extensive subretinal hemorrhage and pigment epithelial detachment are more prone to be missed by OCTA [43, 45, 46, 48]. In summary, type 1 (occult) wet-type AMD has a lower detection rate on OCTA [47, 48]. The low detection rate of occult CNVM by spectral-domain OCTA devices was related to the short wavelength laser beam's decreased penetration rate. Swept-source OCTA devices in which a longer wavelength laser beam source is used may provide better visualization in such conditions due to improved resolution [49]. In comparative studies, the same CNVM was measured by both spectral-domain, and swept-source OCTA and the actual CNVM dimensions were found to be larger with the latter one [50, 51]. A previous study also showed that the measured CNVM dimensions were underestimated by spectral-domain OCTA relative to ICG [52].

The reliability and effectivity of OCTA in determining CNV and measuring actual lesion size are well studied in the literature. The studies subsequently focused on the morphologic analysis of CNVM related to AMD in order to predict activity and prognosis. The CNVMs related to the AMD were classified into two distinct microvascular appearance patterns in further investigations as type 1 and 2 CNVM [27, 53–55]. Type 1 CNVM composed of densely packed vascular nets showing a medusa pattern with extensive anastomosis. Type 2 CNVM has larger and lower branching vascular networks that radiate from the main trunk with a

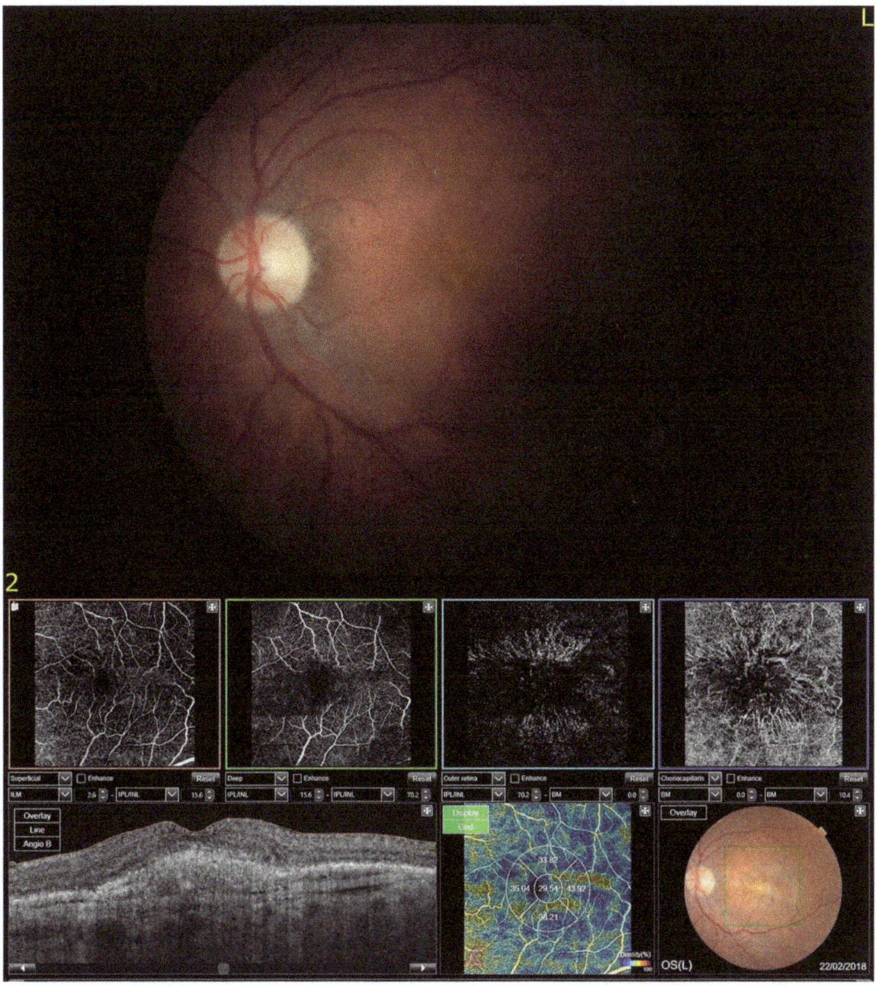

Figure 9.2. Fundus photograph of a patient with inactivate wet-type amd, subretinal druzenoid accumulation in OCT and increased vascularity in deep capillary plexus in OCTA are observed.

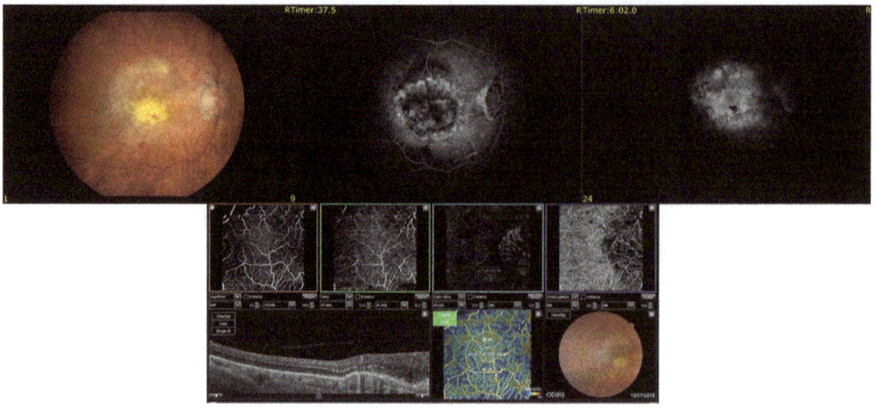

Figure 9.3. Fundus photograph, FFA image, OCT image and neovascularization in deep vascular plexus in OCTA of a patient with active wet AMD.

sharp demarcation and lower anastomosis rate [56–59]. Type 1 CNVM was found to be associated with the activity [57–59]. Another activity sign was reported as the presence of a perilesional, hypointense halo surrounding CNVM [57, 59]. Besides, OCTA can also provide a quantitative assessment of CNVM membrane area and its flow index [60]. These defining features of OCTA facilitate the follow-up of CNVMs related to AMD and decision making on treatment initiation.

The alterations in vascular morphology of AMD related CNVM after anti-VEGF treatment was reported in longitudinal studies [29, 61]. The short term studies showed that anti-VEGF treatment leads to decreased CNVM flow area with closure of anastomosis starting the first day after the treatment [29, 61, 62]. Besides, it is shown that the CNVM begins to get activated after the second week of treatment, which gives us an insight into the need for frequent treatment regimens [61]. A recently published long term longitudinal study suggested that the anti-VEGF treatment predominantly serves as anti-leakage rather than leading to complete resolution of neovascular membranes [63] (figures 9.2 and 9.3).

9.8 Conclusion

As promising imaging technology, OCTA has made it possible to promptly diagnose AMD related CNVM development and distinguish precisely its cross-sectional location throughout the entire retina. Additionally, the quantification of the vessel flow parameters and the definition of distinct features of active and inactive CNVM membrane types by OCTA has facilitated the clinician's capability to notice the progression rates and earlier decision making on reinjection before any further visual deterioration. With the lack of dye usage and the absence of FFA dynamic features (leakage, pooling, etc), the imaging of obscured CNVMs is now possible. On the other hand, there are multiple limitations in the visualization of type 1 CNVM due to low penetration of scanning light under the RPE or the relatively small size of the scanned area compared to conventional imaging methods. These limitations would likely be overcome with the help of technological developments and the invention of more suitable imaging methods like swept-source OCT.

The evolution of our understanding of AMD related CNVM, and geographic atrophy progression has been enhanced after the clinical application of OCTA but is still to be continued. The reliability and consistency of this recent imaging technology on determining the progression of dry-type AMD with its unique quantification features have been proved. Additionally, the types of CNVMs and their response to anti-VEGF treatments could be visualized, and the reactivation signs became easily noticeable imminently. In the future, up to date knowledge would likely be influenced by further improvements in OCTA technology, and proactive treatment options will be discussed in the future.

References

[1] Pennington K L and DeAngelis M M 2016 Epidemiology of age-related macular degeneration (AMD): associations with cardiovascular disease phenotypes and lipid factors *Eye Vis.* **3** 34

[2] Lipecz A, Miller L and Kovacs I *et al* 2019 Microvascular contributions to age-related macular degeneration (AMD): from mechanisms of choriocapillaris aging to novel interventions *GeroScience* **41** 813–45

[3] Jia Y, Tan O and Tokayer J *et al* 2012 Split-spectrum amplitude-decorrelation angiography with optical coherence tomography *Opt. Express* **20** 4710

[4] Zhang M, Wang J and Pechauer A D *et al* 2015 Advanced image processing for optical coherence tomographic angiography of macular diseases *Biomed. Opt. Express* **6** 4661

[5] Jia Y, Bailey S T and Hwang T S *et al* 2015 Quantitative optical coherence tomography angiography of vascular abnormalities in the living human eye *Proc. Natl. Acad. Sci.* **112** E2395–402

[6] Jia Y, Morrison J C and Tokayer J *et al* 2012 Quantitative OCT angiography of optic nerve head blood flow *Biomed. Opt. Express* **3** 3127

[7] Jia Y, Bailey S T and Wilson D J *et al* 2014 Quantitative optical coherence tomography angiography of choroidal neovascularization in age-related macular degeneration *Ophthalmology* **121** 1435–44

[8] Wang X, Jia Y and Spain R *et al* 2014 Optical coherence tomography angiography of optic nerve head and parafovea in multiple sclerosis *Br. J. Ophthalmol* **98** 1368–73

[9] Jia Y, Wei E and Wang X *et al* 2014 Optical coherence tomography angiography of optic disc perfusion in glaucoma *Ophthalmology* **121** 1322–32

[10] Wei E, Jia Y and Tan O *et al* 2013 Parafoveal retinal vascular response to pattern visual stimulation assessed with OCT angiography; S G Solomon *PLoS One* **8** e81343

[11] Al-Sheikh M, Tepelus T C, Nazikyan T and Sadda S R 2017 Repeatability of automated vessel density measurements using optical coherence tomography angiography *Br. J. Ophthalmol* **101** 449–52

[12] Pechauer A D, Jia Y, Liu L, Gao S S, Jiang C and Huang D 2015 Optical coherence tomography angiography of peripapillary retinal blood flow response to hyperoxia *Investig. Opthalmol. Vis. Sci.* **56** 3287

[13] Hwang T S, Gao S S and Liu L *et al* 2016 Automated quantification of capillary nonperfusion using optical coherence tomography angiography in diabetic retinopathy *JAMA Ophthalmol* **134** 367

[14] Liu L, Jia Y and Takusagawa H L *et al* 2015 Optical coherence tomography angiography of the peripapillary retina in glaucoma *JAMA Ophthalmol* **133** 1045
[15] Zhang M, Hwang T S, Dongye C, Wilson D J, Huang D and Jia Y 2016 Automated quantification of nonperfusion in three retinal plexuses using projection-resolved optical coherence tomography angiography in diabetic retinopathy *Investig. Opthalmol. Vis. Sci.* **57** 5101
[16] Jain N, Jia Y and Gao S S *et al* 2016 Optical coherence tomography angiography in choroideremia *JAMA Ophthalmol* **134** 697
[17] Agemy S A, Scripsema N K and Shah C M *et al* 2015 Retinal vascular perfusion density mapping using optical coherence tomography angiography in normals and diabetic retinopathy patients *Retina* **35** 2353–63
[18] Liu L, Gao S S, Bailey S T, Huang D, Li D and Jia Y 2015 Automated choroidal neovascularization detection algorithm for optical coherence tomography angiography *Biomed. Opt. Express* **6** 3564
[19] Bennett T J, Quillen D A and Coronica R Fundamentals of fluorescein angiography *Insight* **41** 5–11 http://ncbi.nlm.nih.gov/pubmed/30230734
[20] Kornblau I S and El-Annan J F 2019 Adverse reactions to fluorescein angiography: A comprehensive review of the literature *Surv. Ophthalmol* **64** 679–93
[21] Slakter J S, Yannuzzi L A, Guyer D R, Sorenson J A and Orlock D A 1995 Indocyanine-green angiography *Curr. Opin. Ophthalmol* **6** 25–32
[22] Desmettre T, Devoisselle J and Mordon S 2000 Fluorescence properties and metabolic features of indocyanine green (ICG) as related to angiography *Surv. Ophthalmol* **45** 15–27
[23] Kuppermann B and Narayanan R 2017 Hot topics in dry AMD *Curr. Pharm. Des.* **23** 542–6
[24] Johnen S and Koutsonas A 2019 Trockene AMD—Zelluläre und gentherapeutische Behandlungsansätze *Klin. Monbl. Augenheilkd.* **236** 1096–102
[25] de Carlo T E, Romano A, Waheed N K and Duker J S 2015 A review of optical coherence tomography angiography (OCTA) *Int. J. Retin. Vitr.* **1** 5
[26] Spaide R F, Klancnik J M and Cooney M J 2015 Retinal vascular layers imaged by fluorescein angiography and optical coherence tomography angiography *JAMA Ophthalmol* **133** 45
[27] Farecki M-L, Gutfleisch M and Faatz H *et al* 2017 Characteristics of type 1 and 2 CNV in exudative AMD in OCT-angiography *Graefe's Arch. Clin. Exp. Ophthalmol.* **255** 913–21
[28] Karacorlu M, Sayman Muslubas I, Arf S, Hocaoglu M and Ersoz M G 2019 Membrane patterns in eyes with choroidal neovascularization on optical coherence tomography angiography *Eye* **33** 1280–9
[29] Lumbroso B, Rispoli M and Savastano M C 2015 Longitudinal optical coherence tomography–angiography study of type 2 naive choroidal neovascularization early response after treatment *Retina* **35** 2242–51
[30] Schneider E W and Fowler S C 2018 Optical coherence tomography angiography in the management of age-related macular degeneration *Curr. Opin. Ophthalmol* **29** 217–25
[31] Couturier A, Mané V and Bonnin S *et al* 2015 Capillary plexus anomalies in diabetic retinopathy on optical coherence tomography angiography *Retina* **35** 2384–91
[32] Huang D, Jia Y, Gao S S, Lumbroso B and Rispoli M 2016 Optical coherence tomography angiography using the optovue device *Dev. Ophthalmol.* **56** 6–12
[33] Rosenfeld P J, Durbin M K and Roisman L *et al* 2016 ZEISS Angioplex^{TM} spectral domain optical coherence tomography angiography: technical aspects *Dev. Ophthalmol.* **56** 18–29

[34] Mase T, Ishibazawa A, Nagaoka T, Yokota H and Yoshida A 2016 Radial peripapillary capillary network visualized using wide-field montage optical coherence tomography angiography *Investig. Opthalmol. Vis. Sci.* **57** OCT504
[35] Biesemeier A, Taubitz T, Julien S, Yoeruek E and Schraermeyer U 2014 Choriocapillaris breakdown precedes retinal degeneration in age-related macular degeneration *Neurobiol. Aging* **35** 2562–73
[36] Seddon J M, McLeod D S and Bhutto I A *et al* 2016 Histopathological insights into choroidal vascular loss in clinically documented cases of age-related macular degeneration *JAMA Ophthalmol* **134** 1272
[37] Lane M, Moult E M and Novais E A *et al* 2016 Visualizing the choriocapillaris under drusen: comparing 1050-nm swept-source versus 840-nm spectral-domain optical coherence tomography angiography *Investig. Opthalmol. Vis. Sci.* **57** OCT585
[38] Cicinelli M V, Rabiolo A and Marchese A *et al* 2017 Choroid morphometric analysis in non-neovascular age-related macular degeneration by means of optical coherence tomography angiography *Br. J. Ophthalmol* **101** 1193–200
[39] Choi W, Moult E M and Waheed N K *et al* 2015 Ultrahigh-speed, swept-source optical coherence tomography angiography in nonexudative age-related macular degeneration with Geographic Atrophy *Ophthalmology* **122** 2532–44
[40] Sacconi R, Corbelli E, Carnevali A, Querques L, Bandello F and Querques G 2018 Optical coherence tomography angiography in geographic atrophy *Retina* **38** 2350–5
[41] Nassisi M, Shi Y and Fan W *et al* 2019 Choriocapillaris impairment around the atrophic lesions in patients with geographic atrophy: a swept-source optical coherence tomography angiography study *Br. J. Ophthalmol* **103** 911–7
[42] Corbelli E, Sacconi R and Rabiolo A *et al* 2017 Optical coherence tomography angiography in the evaluation of geographic atrophy area extension *Investig. Opthalmol. Vis. Sci.* **58** 5201
[43] de Carlo T E, Bonini Filho M A and Chin A T *et al* 2015 Spectral-domain optical coherence tomography angiography of choroidal neovascularization *Ophthalmology* **122** 1228–38
[44] Gong J, Yu S, Gong Y, Wang F and Sun X 2016 The diagnostic accuracy of optical coherence tomography angiography for neovascular age-related macular degeneration: a comparison with fundus fluorescein angiography *J. Ophthalmol.* **2016** 1–8
[45] Inoue M, Jung J J and Balaratnasingam C *et al* 2016 A comparison between optical coherence tomography angiography and fluorescein angiography for the imaging of type 1 neovascularization *Investig. Opthalmol. Vis. Sci.* **57** OCT314
[46] Faridi A, Jia Y and Gao S S *et al* 2017 Sensitivity and specificity of OCT angiography to detect choroidal neovascularization *Ophthalmol. Retin.* **1** 294–303
[47] Malamos P, Tsolkas G and Kanakis M *et al* 2017 OCT-angiography for monitoring and managing neovascular age-related macular degeneration *Curr. Eye Res* **42** 1689–97
[48] Ahmed D, Stattin M and Graf A *et al* 2018 Detection of treatment-naive choroidal neovascularization in age-related macular degeneration by swept source optical coherence tomography angiography *Retina* **38** 2143–9
[49] Považay B, Hermann B and Unterhuber A *et al* 2007 Three-dimensional optical coherence tomography at 1050 nm versus 800 nm in retinal pathologies: enhanced performance and choroidal penetration in cataract patients *J. Biomed. Opt.* **12** 041211
[50] Novais E A, Adhi M and Moult E M *et al* 2016 Choroidal neovascularization analyzed on ultrahigh-speed swept-source optical coherence tomography angiography compared to spectral-domain optical coherence tomography angiography *Am. J. Ophthalmol.* **164** 80–8

[51] Miller A R, Roisman L and Zhang Q *et al* 2017 Comparison between spectral-domain and swept-source optical coherence tomography angiographic imaging of choroidal neovascularization *Investig. Opthalmol. Vis. Sci.* **58** 1499
[52] Eandi C M, Ciardella A and Parravano M *et al* 2017 Indocyanine green angiography and optical coherence tomography angiography of choroidal neovascularization in age-related macular degeneration *Investig. Opthalmol. Vis. Sci.* **58** 3690
[53] Dansingani K K and Freund K B 2015 Optical coherence tomography angiography reveals mature, tangled vascular networks in eyes with neovascular age-related macular degeneration showing resistance to geographic atrophy *Ophthalmic Sur. Lasers Imaging Retin* **46** 907–12
[54] Kuehlewein L, Dansingani K K and de Carlo T E *et al* 2015 Optical coherence tomography angiography of type 3 neovascularization secondary to age-related macular degeneration *Retina* **35** 2229–35
[55] Kuehlewein L, Sadda S R and Sarraf D 2015 OCT angiography and sequential quantitative analysis of type 2 neovascularization after ranibizumab therapy *Eye* **29** 932–5
[56] Sulzbacher F, Pollreisz A, Kaider A, Kickinger S, Sacu S and Schmidt-Erfurth U 2017 Identification and clinical role of choroidal neovascularization characteristics based on optical coherence tomography angiography *Acta Ophthalmol* **95** 414–20
[57] Coscas G J, Lupidi M, Coscas F, Cagini C and Souied E H 2015 Optical coherence tomography angiography versus traditional multimodal imaging in assessing the activity of exudative age-related macular degeneration *Retina* **35** 2219–28
[58] Al-Sheikh M, Iafe N A, Phasukkijwatana N, Sadda S R and Sarraf D 2018 Biomarkers of neovascular activity in age-related macular degeneration using optical coherence tomography angiography *Retina* **38** 220–30
[59] Miere A, Butori P and Cohen S Y *et al* 2019 Vascular remodeling of choroidal neovascularization after anti-vascular endothelial growth factor therapy visualized on optical coherence tomography angiography *Retina* **39** 548–57
[60] Hagag A, Gao S, Jia Y and Huang D 2017 Optical coherence tomography angiography: Technical principles and clinical applications in ophthalmology *Taiwan J. Ophthalmol* **7** 115
[61] Huang D, Jia Y, Rispoli M, Tan O and Lumbroso B 2015 Optical coherence tomography angiography of time course of choroidal neovascularization in response to anti-angiogenic treatment *Retina* **35** 2260–4
[62] Pilotto E, Frizziero L and Daniele A R *et al* 2019 Early OCT angiography changes of type 1 CNV in exudative AMD treated with anti-VEGF *Br. J. Ophthalmol* **103** 67–71
[63] Levine E S, Custo Greig E and Mendonça L S M *et al* 2020 The long-term effects of anti-vascular endothelial growth factor therapy on the optical coherence tomography angiographic appearance of neovascularization in age-related macular degeneration *Int. J. Retin. Vitr* **6** 39

Chapter 10

Optical coherence tomography angiography in ophthalmology: an application in vessel imaging

Anadi Khatri, Araniko Pandey, Gunjan Prasai, Kinsuk Singh, Muna Kharel, Eli Pradhan and Rupesh Agrawal

The posterior segment of the eye is an interface for a highly specialized neural network that sends optical information to the brain. It is also a window to a wide range of local and systemic pathophysiological processes. Employing clinical fundus examination, various imaging technologies and biopsy samples, the retina can be studied in diverse settings.

Because accurate diagnosis is a must for successful timely treatment, the identification can be notably improved by employing disease image modalities such as: optical coherence tomography (OCT), fundus imaging and optical coherence tomography angiography (OCTA). This chapter will focus on OCTA imaging for the diagnosis of retinal diseases.

OCTA is an add-on to OCT that depicts information on retinal and choroidal circulations without the need for invasive procedures such as fluorescein or indocyanine dyes. With the development of OCT devices with high scan rates and further improvements being made to adjust algorithms to make it optimal, OCTA has increasingly become a powerful everyday non-invasive diagnostic tool in ophthalmology.

10.1 Introduction

The posterior segment of the eye is an interface for a highly specialized neural network that sends optical information to the brain [1, 2]. It is also a window to a wide range of local and systemic pathophysiological processes. The retina is a directly visible structure, which is a unique characteristic that allows extensive diagnostic opportunities [3]. Employing clinical fundus examination, various imaging technologies and biopsy samples, the retina can be studied in diverse settings [4, 5].

doi:10.1088/978-0-7503-2060-3ch10

Pathologies affecting the posterior segment of the eye are one of the major causes of blindness in developed countries and are becoming more prevalent due to an increasing population with greater longevity of life. Commonly encountered diseases in the posterior segment worldwide are; uveitis, diabetic retinopathy, macular edema, proliferative retinopathy, age related macular degeneration and glaucoma, among others [6–11]. In recent times, studies from low- and middle-income countries have also shown an increasing trend of retinal pathologies [9, 12].

Because accurate diagnosis is a must for successful timely treatment, the identification can be notably improved by employing disease-specific computer-aided diagnostic (CAD) systems based on different image modalities, such as OCT, fundus imaging and OCTA. This chapter will focus on OCTA imaging for the diagnosis of retinal diseases [13–17].

OCTA is an add-on to OCT that depicts information on retinal and choroidal circulations without the need for invasive procedures such as the fluorescein or the indocyanine dyes [17] (figure 10.1). Development of OCT devices with high scan rates and further improvements being made to adjust algorithms to make it optimal, OCTA has increasingly become a powerful everyday non-invasive diagnostic tool in ophthalmology [18–20].

Since its invention in the early 1990s, OCT has become one of the most important imaging modalities in ophthalmology [21]. OCT is a noniinvasive imaging technology based on low-coherence interferometry. It generates high-resolution cross-sectional images from backscattered light, enabling clinicians to assess structural changes in different retinal diseases. However, structural OCT cannot be used to monitor vascular changes because of the low contrast between capillaries and retinal tissue [22, 23].

Given that many ocular diseases are associated with vascular abnormalities, the ability to visualize and quantify blood flow in the eye is of high importance. Conventionally, fluorescein angiography (FA) and indocyanine green angiography (ICGA) are used for qualitative clinical assessment of retinal and choroidal circulations, respectively [24]. They require intravenous injection of contrast agents, which is time-consuming and can have potentially serious side effects [25]. Moreover, FA and ICGA provide two-dimensional (2D) images of ocular circulations, limiting

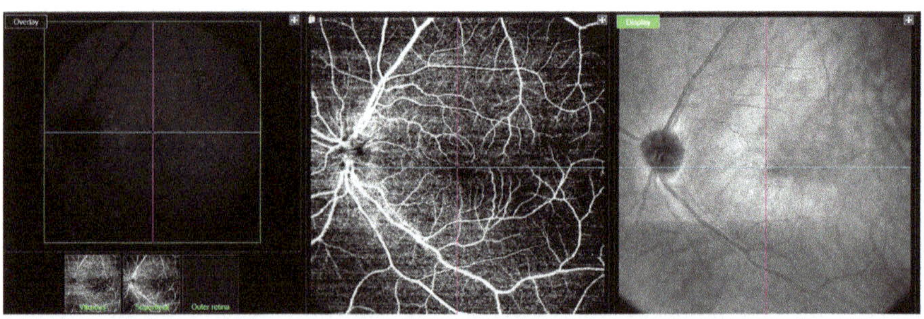

Figure 10.1. OCTA (middle) showing all layers over the 9 mm × 9 mm scan over the posterior pole (inclusive of both the optic nerve head and the macular area). Red free filter (left), infrared imaging (right).

depth perception and detailed investigation of the retinal and choroidal vasculatures. The only limitation of OCTA over dye based angiography is the inability to identify leakage [19, 22, 26, 27].

Multiple non-invasive imaging technologies have been employed in the last few decades to visualize and quantify ocular circulations. Ultrasound color Doppler and functional magnetic resonance imaging have been previously used for research purposes [28]. While they were able to provide good tissue penetration, they did not receive much clinical attention in ophthalmology due to poor resolution and measurement reproducibility. Other imaging modalities such as laser Doppler flowmetry and velocimetry, blue field entoptic technique and laser speckle assessment have been used extensively to study vascular retinal physiology [29]. However, their clinical use has been limited by their complexity, poor reproducibility and wide population variation [30].

Functional extensions of OCT have also been explored for vascular imaging of the eye. Doppler OCT (DOCT) uses the Doppler frequency shift resulting from the movement of red blood cells to quantify volumetric blood flow in large vessels, as well as for total retinal blood flow measurement [31, 32]. However, DOCT is less suitable for investigating the retinal microvasculature. DOCT is sensitive to blood flow parallel to the OCT beam, whereas flow in the retinal microvasculature is mainly perpendicular to the OCT beam. OCT angiography (OCTA) is a more recent development [10] and there are multiple terminologies used in interpretation which are listed in table 10.1. It has the capability of producing high-resolution, 3D angiograms of the retinal and choroidal vascular networks which provides more detailed images of the vasculature.

10.2 Principles of OCTA

OCTA, a functional extension of OCT, generates high resolution volumetric *en face* images from backscattered light. This allows clinicians and researchers to evaluate retinochoroidal vasculature non-invasively. Recent advancements in the techniques of image processing have proven the quality of scans comparable to or even better than dye based procedures [19, 33].

OCTA images are generated by repeated scans of the retina at the same location which detect the motion of RBCs within the vasculature against the static background retinal tissues. Image processing was possible after the introduction of Fourier-domain based OCT, which has a higher scanning speed [17, 34, 35] (figure 10.2).

Various techniques were studied as a clinical tool for non-invasive angiography viz. ultrasound color doppler, functional magnetic resonance imaging and doppler OCT [32]. Poor reproducibility and complexity in the procedure made these techniques less useful as a clinical alternative. Notably, the use of Doppler OCT was deemed limited as it was less suitable for retinal microvasculature [36].

The variations in reflective signals of RBCs detected by the OCT can be quantified on the basis of varying intensity, phase or complex (intensity and phase) signals [27].

Table 10.1. Commonly used terminologies in OCTA.

Term	Definition
A-scan rate	The maximal rate at which A-scans are acquired during a B-scan.
Decorrelation	The change in an area from one sequentially repeated image to another.
Gap defect	An area where the image information is lost image which could be due to various reason—such as eye movement, software rendering errors, etc.
Image artifact	Abnormality in the visual representation of information derived from an object (figure 10.9).
Phase	The position of a wave cycle which describes or illustrates the difference between periodic repetitions present in the wave.
Projection artifact	Light which passes through the vessel fluctuate over time causing anything posterior to the vessel to be illuminated by this fluctuating light. This can be represented as artifact images of vessels and cause it to be seen at deeper locations than they actually inhabit.
Segmentation	Division of larger set into smaller sets according to a set of algorithms/ parameters.
Shadowing	Attenuation of signal behind an obstruction which can be due to various cause.
Stretch artifact	A defect produced when the software attempts to correct eye motion, causing a part of the image to be stretched.
Vessel doubling	A defect produced when the software attempts to correct eye motion, causing each blood vessel to be split and appear in parts or in the whole mapped area (figure 10.6).
White line artifact	White line seen in OCTA images associated with an eye movement.

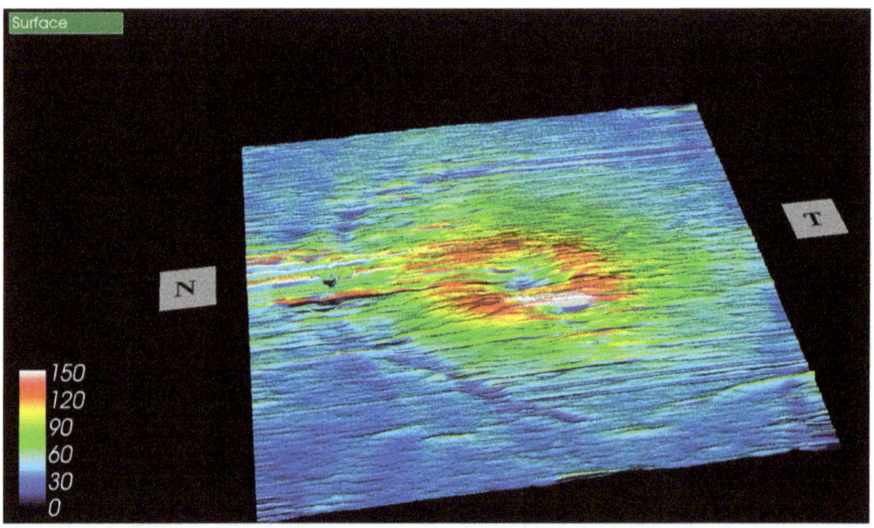

Figure 10.2. Multiple A-scan lines which are combined to form the overall 'surface' image of the retina.

10.2.1 Variants

10.2.1.1 Intensity based OCTA

Uses amplitude or speckle information to generate an image. An intensity based approach has been trialed in time-domain OCT to swept-source OCT systems. A higher image quality and decreased signal-to-noise ratio (SNR) has been demonstrated when any of the system operates with a faster scanning rate. Additional computer algorithms have also been studied for better speckled variance information between successive B-scans. These are correlation mapping and decorrelation mapping—split spectrum amplitude decorrelation angiography (SSADA) [37].

10.2.1.2 Phase based OCTA

Uses the phase signals received from the tissues to generate an image. This was basically used with Doppler OCT, where the Doppler shift of RBCs in motion was recorded [38].

10.2.1.3 Complex signal based OCTA

Uses combined information of phase and amplitude-based variance and provides superior results. The inability of amplitude-based variance to detect slow flowing vessels and phase-based variance to detect microvasculature can be mitigated by a combined approach. Recently, split-spectrum amplitude and phase gradient analysis (SSAPGA) has been introduced, which combines phase variance with SSADA [37]. Complex signal based OCTA algorithms provide the best vascular visualization in terms of connectivity, image contrast-to-noise ratio and SNR [39].

10.2.1.4 Metrics used in OCTA

Various metrics have been used for an objective quantification of vascular pathologies. Vessel density and flow density of *en face* angiograms have been used to clinically evaluate the retinal and choroidal vascular pathophysiological changes. Quantifying avascular zones and neovascularization are other metrics that can be used to monitor the progression of diseases like diabetic retinopathy [22, 27, 32].

10.2.2 Fundamentals of OCTA in ophthalmology

OCTA technology uses laser light reflectance of a moving column of red blood cells (RBCs) to illustrate the vessels through different layers of the eye [22]. These layers can be further divided into sub-layers depending on the tissue which is undergoing the scanning. With current applications, this can be the retina, the choroid or even the neural tissue such as the optic nerve head. The major advantage of this technique is that it is non-invasive—meaning it eliminates the need for intravascular injection of dyes [33].

The OCT scan of a retina consists of multiple amplitude (A)-scans. This information when translated into brightness (B)-scans via algorithms scan provide cross-sectional structural information. The basic principal of OCTA is that it scans an area of the same tissue area multiple times repeatedly and the differences are compared between the scans, thus allowing the device to calculate and produce results [34]. These areas or zones are often divided into flow or no flow zones depending upon the detectable thresholds of the device. Usually, the scan rates, at the time of writing of this chapter, are limited to four scans per second on any segment of the retina [40, 41]. We can highly expect there to be developments of machines with higher scan rates or adjustable scan rates to detect such flows in future.

10.2.3 Scanning technique

Light, usually of wavelength near to 800 nm, or closer to 1050 nm, is emitted depending upon the type of OCT. The former is usually used in spectral domain OCT (SD-OCT) while the latter with a longer wavelength is generally used in swept-source OCT (SS-OCT) [42]. The benefits of having longer wavelengths is that they have deeper tissue penetrance and hence can provide information on deeper tissues. Both the retinal and the choroidal microvasculature can be visualized with help of OCTA. However, this same property leads to a disadvantage of generating a slightly lower axial resolution of the scanned areas [43].

In contrast, if we opt for dye-based angiography, fluorescein angiography FA is generally only applicable for evaluating the status of the retinal vessels and indocyanine green is more ideal for imaging the choroid [44].

OCTA works on the principle of diffractive particle movement detection and mainly uses two methods to detect motion: amplitude decorrelation and phase variance. The former detects amplitude differences of two different B-scans. Phase variance is more related to the properties of the light wave which is emitted. It calculates the variation of phase when moving objects are detected [22, 26, 27].

By using these techniques and incorporating various motion algorithms, OCTA is able to provide flow information. The only major limitation of OCTA is that it cannot appreciate leakage and that is where dye based imaging still outshines it. However, due to its ability to scan various layers and detect flow individually in each of them, OCTA does have an advantage that it can provide approximate delineation and size of retinal or choroidal pathology such as the choroidal neovascularization/ membrane [45] (figure 10.3). This is especially superior in detection and diagnosis of type 1 CNV where localization is inferential and most of the time is very inaccurate

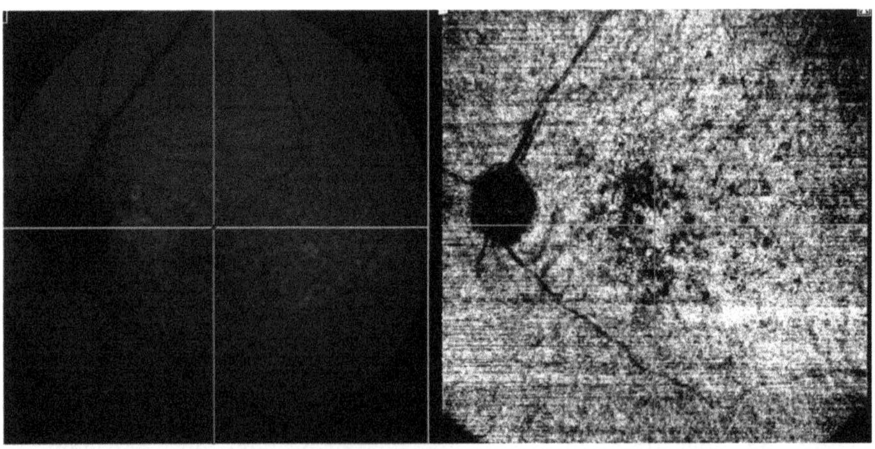

Figure 10.3. OCTA eliciting choroidal neovascularization in the external retinal layer extending from the choriocapillaris. The image color is inversed for easier depiction.

with imaging that uses dyes. However, one must remember that OCTA scans can be obscured by hemorrhage in any layer of the retina or the choroid as they block the scanning light and can give an impression of flow voids [46] (figure 10.4).

In the present scenario, the main problem associated with OCTA imaging is that it is more prone to artifacts than conventional dye-based angiography. The larger retinal vessels produce a 'ghost image'—often called the shadow artifact. This is usually a problem when the outer retinal slabs are being evaluated as these artifacts make it difficult to appreciate the presence of abnormal vessels in the deeper slabs. At the same time, the light incident over the retinal vessels may be refracted or even pass through, causing an increased illumination in the deeper layers. These are termed projection artifacts [32, 40, 46] (figure 10.5).

As discussed above, OCTA uses the principle that movement represents blood flow. This makes it prone to motion artifact also. White bands or lines (which denote decorrelation signal over the entire B-scan) appear if the patient's eye moves due to fixation loss (figure 10.6 top left). Similarly, if the OCTA registers a signal as no movement, there will be the presence of black bands or links. Usually when the patient blinks, the OCT signal is blocked and is unable to detect any movement (figures 10.6 top left and 10.7).

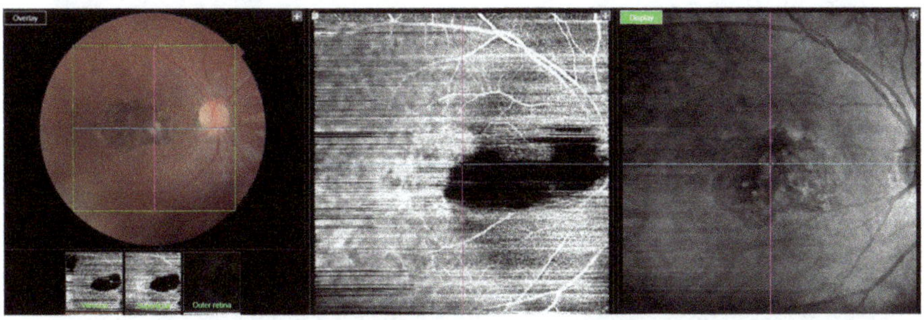

Figure 10.4. OCTA scans of the external retina and choriocapillaris obscured due to scarring which were present in the inner retinal layers. Note the black artifact in the middle image giving a false impression of flow void.

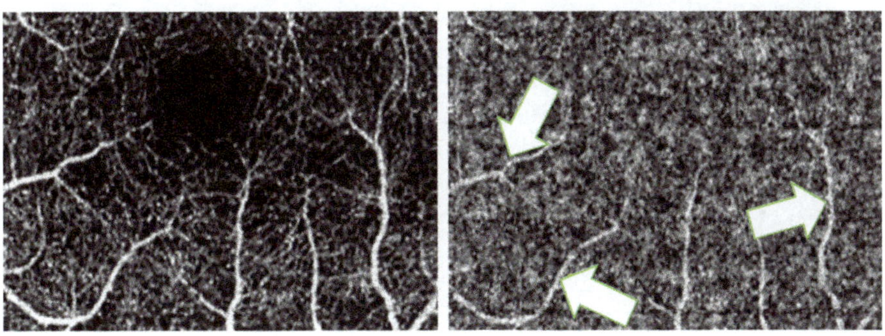

Figure 10.5. Example of projection artifacts where the vessels from the superficial retinal plexuses are being represented in the deeper retinal plexuses due to over illumination due to high flow signal detection which is overexpressed.

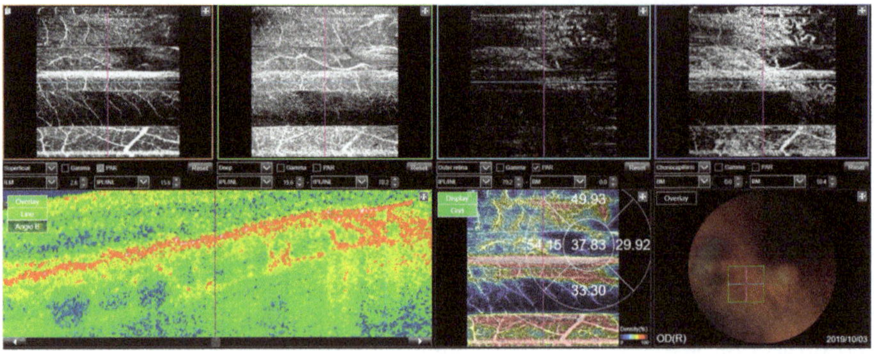

Figure 10.6. Black and white line artifacts which are present in the top left and top right images. Black lines are produced during the blinking phase where the signals are blocked and white lines are produced due to saccades produced by eye movements, which OCTA expresses as a 'particle movement' and registers as a gross flow. Also note the vessel doubling and stretch effects in the vessels due to the attempt by the algorithm to correct and correlate the signals.

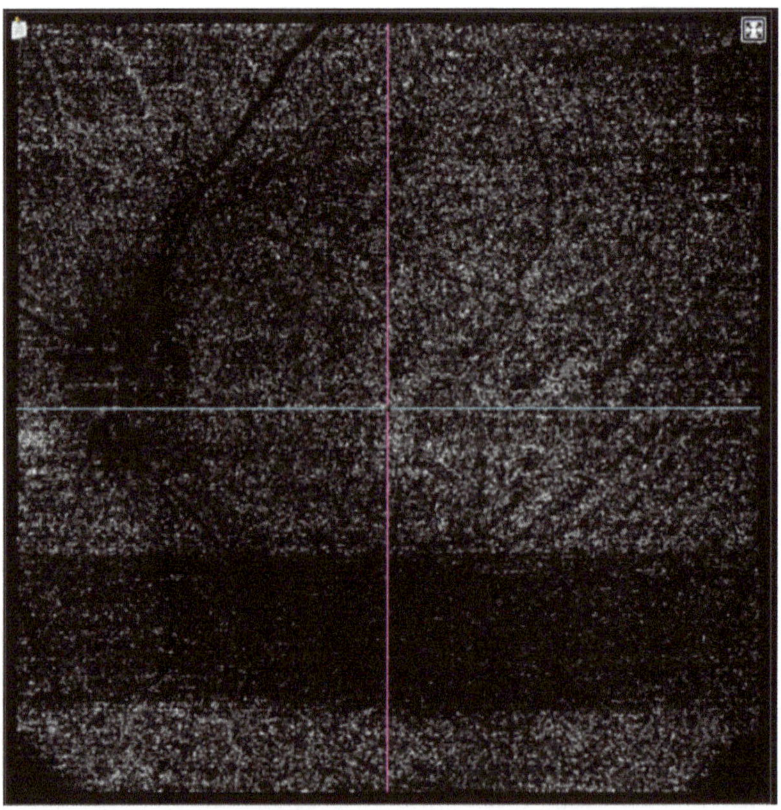

Figure 10.7. Example of black band artifact.

10.3 OCTA of normal eyes

Current OCTA devices can obtain volumetric scans at up to 100 000 A-scans per second in approximately 3–4 s. The scan areas are often offered from an option of 2 × 2 mm, 3 × 3 mm, 6 × 6 mm and 9 × 9 mm OCT angiograms [22, 41, 46] (figures 10.8 and 10.9).

The software automatically segments the scanned areas of the retinal thickness into 'superficial' and 'deep' inner retinal vascular plexuses, outer retina or avascular layer and choriocapillaris. While the OCT angiogram segmentation of the superficial inner retina contains a projection of the vasculature in the retinal nerve fiber layer (RNFL) and ganglion cell layer (GCL), the deeper inner retina slabs are represented as a composite of the vascular plexuses at the border of the inner plexiform layer (IPL) and inner nuclear layer (INL) and the border of the INL and outer plexiform layer (OPL) [47].

The OCTA prototype with the fastest acquisition rate can obtain scans of 500 × 500 A-scans at 400 000 A-scans per second in less than 4 s. This ultra-high speed allows for imaging of wider fields of view and uses 1060 nm wavelength which allows increased light penetration into pigmented tissues and improved choroidal blood flow visualization [40, 48].

10.4 Current applications of OCTA in ophthalmology

10.4.1 OCTA in dry (non-neovascular) AMD

Dry age-related macular degeneration (AMD) is characterized by the presence of drusen, pigmentary changes in retina and loss of photoreceptors and RPE. It is also

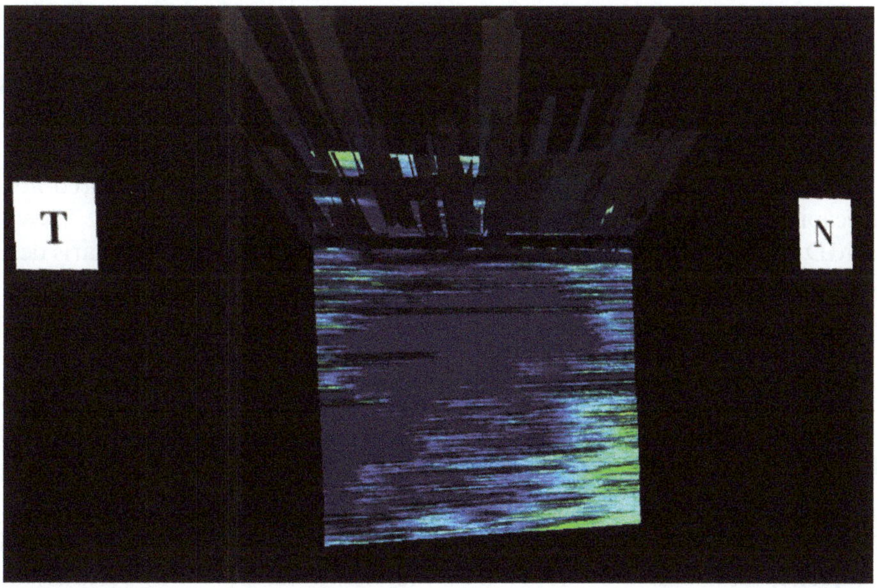

Figure 10.8. OCT A-scan (represented as horizontal lines) and the scan is being represented from nasal to temporal side. On the top, some spikes are present. This is usually due to eye movements or blinking that might have occurred during the scanning of that region.

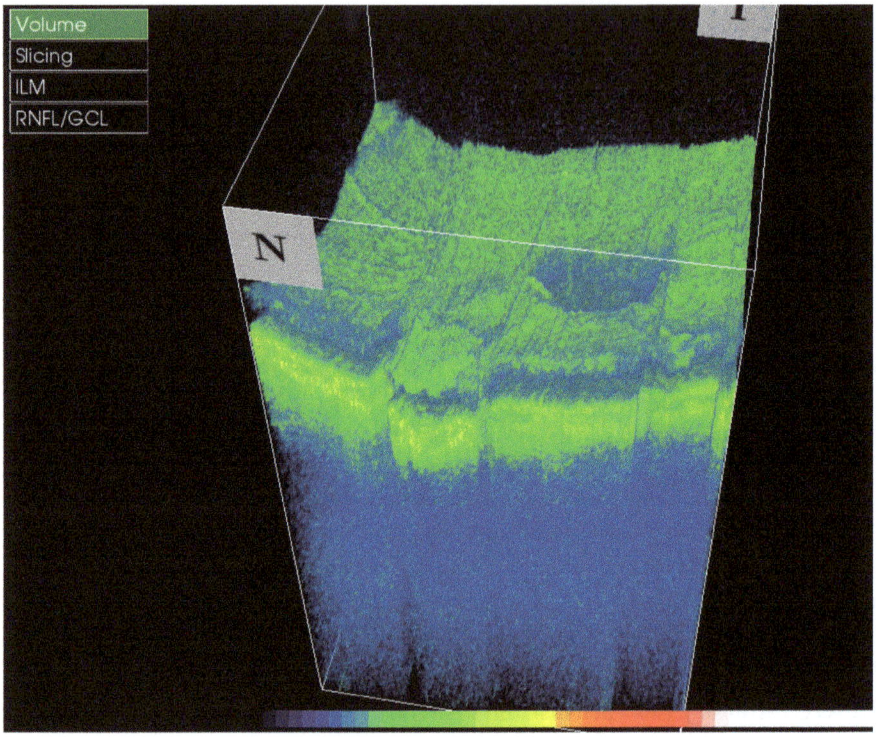

Figure 10.9. An image artifact causing the retinal slabs to be seen as separated in full thickness scan (similar to tectonic plates). These are usually caused when some signal loss occurs during the scan in that area but the OCT then tries to 'stitch' together the scans to create a complete image.

known as geographic atrophy (GA) of the macula. It is a well-established fact that there is decreased foveolar choroidal blood flow in AMD and it has been hypothesized that the evaluation of the choroidal vascularity may help to predict the disease progression [49].

Various literatures have already pointed that there are areas of impaired flow in choriocapillaris which typically extend beyond the delineation of the GA. Eyes with dry AMD were also found to have a generalized decrease in choriocapillaris density and often associated with presence of drusen. Other studies have further added that these alterations were not due to shadowing or signal blockages.

Further studies in future can ascertain these findings to determine if the choriocapillary changes associated with drusen are true areas of flow changes.

10.4.2 OCTA in wet (neovascular) AMD

With the introduction of OCTA, this novel technology has offered further insight into our understanding of the disease pathology and has enabled us to monitor choroidal neovascularization (CNV) in the majority of cases [49–51]. Various studies have already shown that sensitivity and specificity for CNV detection with *en face* OCTA combined with cross-sectional OCTA can be almost as effective as the gold standard fluorescein angiography (FA) with OCT, hence avoiding the need of invasive procedures (figure 10.3).

There have been several publications concerning OCTA of eyes with wet AMD. The ability of OCTA to elicit CNV was first described in 2014 and since then various literatures are present describing the sensitivity and specificity using FFA/ICG as the gold standard for comparison [52]. Much work and effort have also been made to understand and describe characteristics of CNV both quantitatively and qualitatively. It is beyond the scope of this chapter to describe the various types of lesions such as type 2 or type 3 CNV (retinal angiomatous proliferation, RAP), and further reading is recommended from literature available in the research database [53].

The major advantage of OCTA in monitoring the response of treatment or the progression of the pathology is that the patient can undergo eye scan multiple times in every follow-up without the need of dye injection. The eyes can also be analyzed in segments or slabs to fully understand the extent of the disease and damage to the internal retinal elements.

10.4.3 OCTA in diabetic retinopathy

Various researches have demonstrated the advantage and ability of OCTA to detect and even predict the progression of diabetic retinopathy [54, 55]. Depending upon the features such as choriocapillary abnormalities and/or retinal microvascular abnormalities such as microaneurysms, vascular remodeling adjacent to the foveal avascular zone (FAZ), enlarged FAZ, capillary tortuosity and dilation, OCTA has even able to predict the likelihood of progression of diabetic retinopathy. Used in various combinations, it may even predict which of the features can mean the eye is more susceptible to becoming potentially blind [26, 27, 56–58].

Various sizes and shapes of microaneurysms have been analyzed. Microaneurysms have also been classified as high flow and low flow depending upon the signal threshold [59]. The location where they can cause more anatomical and functional damage has been also postulated. Various literatures have already identified that if the microaneurysms are present more on the deep capillary plexus, they cause the retina to be more vulnerable to edema [26, 27, 58] (figure 10.10). Recent studies have shown a

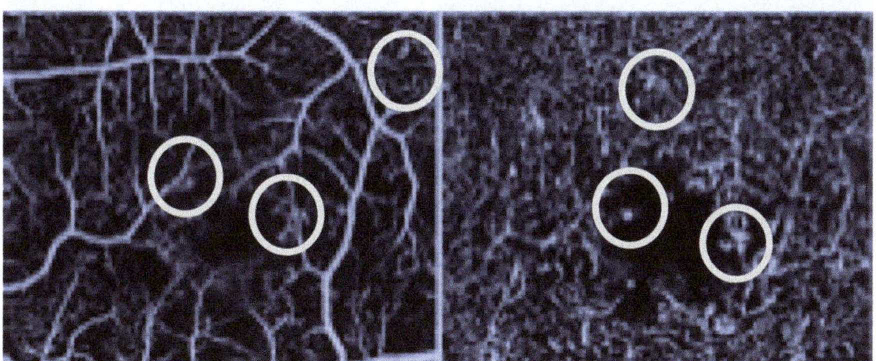

Figure 10.10. Microaneurysm depicted by OCTA at various layers of the retina.

similar distribution of the low flow microaneurysms—meaning those which were able to be caught in the FFA but not in OCTA have been found in areas where the capillary density is decreased [26]. It has been postulated that these may be the areas of decompensated flow and such microaneurysms could mean the areas of impending ischemia and ischemic maculopathy in the long term [27].

Some studies have actually shown that OCTA may be superior in detection of microaneurysms as it was able to appreciate some microaneurysms that were not detected by conventional angiography [60]. OCTA has also been shown to successfully detect other abnormalities which may not be evident on FA such as areas of retinal non-perfusion, reduced capillary density and increased vessel tortuosity.

10.4.4 OCTA in artery and vein occlusion

Retinal vascular occlusions are one of the most common retinal vascular pathologies [61] which are slowly being understood more using OCTA as an imaging modality. Various reports have suggested that OCTA usually shows large areas of capillary non-perfusion (flow voids) in the area of ischemia with clear delineation of the boundary, including vascular abnormalities such as microaneurysms, telangiectasia and anastomoses [62, 63]. Vascular density mapping can demonstrate the decrease in the density of vascular plexuses—which directly is also a representation of the ischemic load. The same area can be measured using an inbuilt caliper or using the disc diameters to estimate the ischemic stress of the retina [64].

Branched retinal artery occlusion demonstrates wedge-shaped areas of capillary non-perfusion that correlates to areas of abnormalities on the retinal thickness map. This illustrates the potential use of OCTA in pinpointing areas of ischemia and edema (figure 10.11).

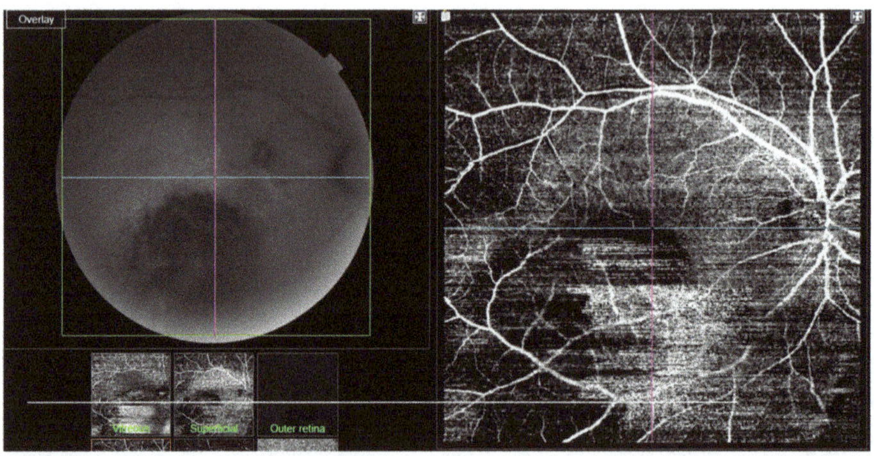

Figure 10.11. An OCTA image showing flow voids in the inferotemporal arcade following inferotemporal tributary vein occlusion.

Central retinal artery occlusions show diffuse capillary flow voids areas supplied by the central retinal artery in both the superficial capillary plexus and the deep capillary plexus [65]. Flow is still seen in the major retinal vessels. There is an absence of blood flow in the superficial disc vasculature supplied by the central retinal artery around the optic disc but the flow around the lamina cribosa remains intact.

One major disadvantage is that OCTA provides a snapshot in time and registers all the movements that are above its detection threshold. Hence, it is not able demonstrate delayed arteriovenous transit time. However, the other major advantage is that OCTA also enables one to view the choriocapillaris and choroidal vascular flow—which are usually found to be minimally affected in arterial or venous occlusions [66, 67].

10.4.5 OCTA in glaucoma

OCTA has been a prolific tool for evaluating optic disc perfusion in eyes suspected with glaucoma or already with signs of the disease. OCTA usually demonstrates the dense peripapillary microvascular network in normal eyes and the current evidence is that this vascular network is attenuated in both the superficial disc vasculature and the deeper lamina cribosa (figure 10.12). With processing after averaging the decorrelation signal in OCT angiograms, it is able to further approximate the area of microvasculature and provide physicians with parameters allowing them to calculate the flow index. This flow index has been depicted to have very high sensitivity and specificity in differentiating glaucomatous eyes from normal eyes and there is evidence that it may even help in segregating eyes at risk of glaucoma in future or eyes with pre-perimetric glaucoma [14, 68].

The concept of assessing glaucomatous damage by evaluating blood flow is not new. Over decades, physicians and researchers working for drug delivery and understanding the pathogenesis of glaucoma have been correlating between blood flow and glaucomatous damage to the optic nerve head. Fluorescein angiography features such as fluorescence and delayed vascular filling have been noted. However,

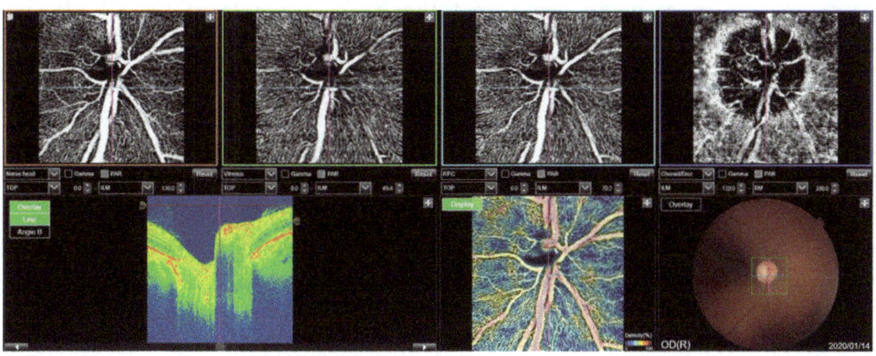

Figure 10.12. OCTA scan showing vascular arrangement at different layers of the optic nerve head (top row). Vascular density elicited by the OCTA using heat mapping colors (lower row middle).

this method is invasive and is not suitable for immediately repeating if the initial results are inconclusive, making it impractical for routine evaluation to understand the trend of the damage [69, 70].

OCTA has the ability to compare both dysfunctional and dead ganglion cells and various hypotheses correlating metabolism demands of such neuronal structures with vascular density and parameters like papillary, para or peripapillary vascular density have established themselves as strong indicators of assessing glaucomatous damage. There are even clinical findings suggesting that reduced capillary plexus could be representing areas of RFL or GCC which are under stress and may undergo apoptosis causing irreversible damage if there is no intervention [71]. Prospective cohorts of these eyes have shown that such areas undergo subsequent thinning of NFL and GCC—further strengthening the association. One must remember that although this thinning can be detected by structural OCT and one can monitor the progression and modify treatment accordingly, irreversible damage will have already occurred [72]. By adding OCTA as a diagnostic modality, it has an advantage of detecting glaucoma earlier, perhaps even before the damage becomes evident, enabling physicians to adopt treatment at primary prevention level.

10.4.6 OCTA in uveitis

10.4.6.1 Using OCTA for retinal imaging in uveitis

Currently, there are various researches which have pointed out the usefulness of OCTA in evaluation of the disease progression or response of treatment of uveitis. This has been limited not just for posterior uveitis but also for anterior uveitis [73].

Recent studies using OCTA have showed that uveitic macular edema is associated with changes in the capillary density or morphology at deep capillary plexus (DCP). Eyes with uveitis and macular edema were found to have a significantly lower vessel density in DCP [15]. Parafoveal capillary loss in the superficial capillary plexus was also observed irrespective of whether the macular edema was present or not. These findings have been further reinforced with a similar finding where remodeling of capillaries and the irregularity and/or enlargement of foveal avascular zone (FAZ) were found to be consistent in patients with uveitis [74].

However, one must again not forget the basic principle of OCTA. It detects movement of blood flow and there is a threshold for which the signal can be reseated [22]. The areas of non-perfusion may only be representing sluggish or a slow flow not definitely a complete capillary dropout/or inadequate capillary perfusion [75]. Irrespective of the pathology, it is also important to remember that most of the vasculopathies related with retina and choroid have shown that the insult in the DCP is of more serious nature [15, 26, 73]. One of the reasons is that the deep capillaries may be more susceptible to ischemia as they are not directly connected to arterioles. However, production of projection artifacts on OCTA which maybe produced by the atrophic or dead retinal tissues must also be considered, which may confuse by indicating nonperfusion/hypoperfusion in normal areas. If such confusions occur, a rescan and careful *en face* revaluation by the physician or technician must be done to assess each of the slabs in the machine and not to rely only on the printed report.

10.4.6.2 Using OCTA for choroidal imaging in uveitis

Evaluation of the choriocapillaris by OCTA has become an important indicator to monitor disease progression. These have been demonstrated in various case reports and literature [13, 76–78]. Choriocapillaris is noted to have loss in the 'fine granular pattern' which it normally elicits and instead becomes irregular with loss of the architecture [77] (figure 10.13). This may further be evaluated by measuring the choroidal thickness. Involvement of the choroid in uveitis and evaluation of thickness have already been established with development of a novel quantitative imaging biomarker, called the choroidal vascularity index (CVI) [78].

The index is defined as the ratio of vascular area under scan to the total choroidal area in reference. CVI has been indicated as a potential tool in establishing early diagnoses, monitoring disease progression and prognosticating patients and has the potential to be a robust marker in numerous retinal and choroidal diseases.

Uveitis is also associated with development of CNV in many of the uveitis disorders such as punctuate inner choroiditis or in ocular histoplasmosis syndrome.

The challenge is, however, differentiating purely inflammatory lesions from inflammatory CNV in uveitis patients [79].

Figure 10.13. Normal OCTA of the choriocapillaris showing granular angioarchitecture. Also note the projection artifacts from the retinal vessels.

OCTA is known to aid early detection of CNV, especially when findings from other imaging modalities become inconclusive. Similar to its application in wet-ARMD, the same property allows it to delineate the neovascular network inside the area of a lesion without being obscured by dye leakage [80]. OCTA is also effective in both detecting and revealing CNV despite existing subretinal fluid or hemorrhage around the area of the lesion which sometimes is obscured, prevailing in such cases over FFA. Its effectiveness in monitoring the progress of CNV lesions that were under treatment and their response to it is also remarkable. Lesions that show no blood flow signal on OCTA can be distinguished from potential CNV lesions and further management can be planned accordingly [81].

A short summary on application of OCTA and findings in each of the scanned slabs in uveitis are listed below [15]:

- **Superficial retinal capillary plexus**—found to be disturbed in inflammatory vasculitis. OCTA reveals flow voids in superficial retinal vessels. May also show capillary remodeling and a lower vessel density. The vascular angioarchitecture can be interpreted, which may be useful in management decisions [74].
- **Deep retinal capillary plexus**—OCTA is able to detect patterns when such eyes are associated with cystoid macular edema. The disease or pathology may be more vision threatening [82].
- **Choriocapillaris**—there is decrease in the choroidal blood flow or ischemia. These areas of flow voids are found to be very consistent with hypo-perfused areas when the same are compared with the findings from the indocyanine-green angiography findings [15].

10.5 The future of OCTA and ophthalmology

10.5.1 Non-invasive imaging and neurovascular coupling

Recently, studies of blood flow dynamics are being concentrated in disruption of autoregulatory mechanisms of the retinal blood flow. Various human as well as animal studies are being conducted to identify quantifiable parameters suggesting neurovascular decoupling, in advance of pathologies like glaucoma and diabetic retinopathy. Neurovascular coupling is the hemodynamic response to light stimulation, which has been shown by high correlation of increased blood flow in the retina and optic nerve head (ONH) to increased neural activity [83, 84].

With the advancement of newer technologies to identify the pattern of blood flow and vascular pathologies in retina or ONH, it has become possible to explore the implications of this physiologic phenomenon. Laser Doppler flowmetry is based on the Doppler effect which uses a fundus camera and a computer system to calculate velocity, blood volume and blood flow around the area of interest. OCTA can be used to compute the flow index, which is an arbitrary measurement of blood flow. Color Doppler imaging is also based on the principle of Doppler frequency shift and can be used to measure the pattern of blood flow changes in color coded format. Laser speckled flowgraphs use speckled contrast images of the retina and ONH to present a temporal waveform pattern of blood flow and hence, provide values of blood flow and velocity for analysis [83].

There are numerous trials that are studying on retinal neurovascular response to visual stimulus. These experiments have presented multiple protocols, to establish a physiological basis of the dose/stimulus-response relationship. The majority of the studies used flicker stimulus in dark adapted and light adapted eyes within a certain time frame. The neurovascular response to visual stimuli are then recorded using various machines ranging from OCTA, laser speckled contrast imaging to scanning laser ophthalmoscope. These experiments are slowly gaining momentum for translating this theory into the mainstream clinical practice [85–89].

Studies have shown changes in the compliance of retinal vessels with stimulation of the ganglion cells to light. The normal response to visual stimuli that demands high metabolism is vasodilation of the retinal blood vessels [83]. This physiologic autoregulatory response can be disrupted early during the pathologic course of diabetic retinopathy or glaucoma, which is termed as neurovascular decoupling. The purpose of these ongoing experiments is to quantitatively identify and use these biomarkers for early diagnosis and treatment [88]. The morphologic areas of interest that can be quantified for correlation can be the retinal vasculature at various levels *viz.* superficial, middle or deep capillary plexus around the parafoveal region, optic nerve head or elsewhere. Change in blood flow can be quantitatively measured with biomarkers that calculate the velocity, blood flow or vascular cross-sectional area and give us values like blood flow velocity, vessel length or vessel length diameter. These values can then be clinically correlated to predict diabetic retinopathy, glaucoma or even response to treatment [90].

Blood vessels dilate to meet the metabolic demands of the retinal neurons in response to visual stimuli [84, 85]. Preliminary studies have shown decreased retinal vascularity in mild DR, in contrast to an increased vessel density in diabetics without DR. These studies postulate varying patterns of decreased retinal vessel density or increased foveal avascular zone at various stages of DR [88].

Similarly, disrupted autoregulatory neurovascular response around the optic nerve head has been postulated in those with glaucoma. This can be of particular interest for developing diagnostic biomarkers in advance of any visual field defect or neuronal damage in glaucoma [83].

In conclusion, efforts have now been concentrated towards developing an easy and replicable protocol with least variability. Pairing this to a cost-effective, non-invasive and user-friendly imaging device with a robust computational analytic system, will certainly narrow down our search for quantifiable biomarkers in the management of ocular neurovascular pathologies.

References

[1] Hogan M J, Alvarado J A and Weddell J 1971 *Histology of the Human Eye* (Philadelphia, PA: W.B. Saunders)

[2] Jakobiec F A 1982 *Ocular Anatomy, Embryology and Teratology* (Philadelphia, PA: Harper and Row)

[3] Hildebrand G D and Fielder A R 2011 Anatomy and physiology of the retina *Pediatric Retina* (Berlin: Springer) pp 39–65

[4] Schneiderman H 1990 The funduscopic examination ed H K Walker, W D Hall and J W Hurst *Clinical Methods: The History, Physical, and Laboratory Examinations* 3rd edn (Boston, MA: Butterworth) ch 117. Available from https://ncbi.nlm.nih.gov/books/NBK221/

[5] Khochtali S, Khairallah-Ksiaa I and Ben Yahia S 2016 Normal fundus fluorescein angiography *Uveitis Atlas [Internet]* ed V Gupta, Q D Nguyen, P LeHoang and C P Herbort Jr (New Delhi: Springer) pp 1–7

[6] Wong W L, Su X, Li X, Cheung C M, Klein R, Cheng C Y and Wong T Y 2014 Global prevalence of age-related macular degeneration and disease burden projection for 2020 and 2040: a systematic review and meta-analysis *Lancet* **2** E106–16

[7] Silva R, Cachulo M L, Fonseca P, Bernardes R, Nunes S, Vilhena N and Faria de Abreu J 2011 Age-related macular degeneration and risk factors for the development of choroidal neovascularisation in the fellow eye: a 3-year follow-up study *Ophthalmologica* **226** 110–8 Epub 2011 Aug 3

[8] Klein R, Klein B E, Moss S E, Davis M D and DeMets D L 1984 The Wisconsin epidemiologic study of diabetic retinopathy. III. Prevalence and risk of diabetic retinopathy when age at diagnosis is 30 or more years *Arch. Ophthalmol.* **102** 527–32

[9] Friedman D S, Ali F and Kourgialis N 2011 Diabetic retinopathy in the developing world: how to approach identifying and treating underserved populations *Am. J. Ophthalmol.* **151** 192–4 e1

[10] Tham Y-C, Li X, Wong T Y, Quigley H A, Aung T and Cheng C-Y 2014 Global prevalence of glaucoma and projections of glaucoma burden through 2040: a systematic review and meta-analysis *Ophthalmology* **121** 2081–90

[11] Robert N and Scott W 2010 *Uveitis: Fundamentals and Clinical Practice* 4th edn (Maryland Heights, MO: Mosby Elsevier) pp 303–18

[12] Hubley J and Gilbert C 2006 Eye health promotion and the prevention of blindness in developing countries: critical issues *The Br. J. Ophthalmol.* **90** 279–84

[13] Agrawal R, Ding J, Sen P, Rousselot A, Chan A and Nivison-Smith L *et al* 2020 Exploring choroidal angioarchitecture in health and disease using choroidal vascularity index *Prog. Retin. Eye Res* **77** 100829

[14] Jia Y, Morrison J C and Tokayer J *et al* 2012 Quantitative OCT angiography of optic nerve head blood flow *Biomed. Opt. Express* **3** 3127–37

[15] Pichi F, Sarraf D and Arepalli S *et al* 2017 The application of optical coherence tomography angiography in uveitis and inflammatory eye diseases *Prog. Retin. Eye Res* **59** 178–201

[16] Koustenis A, Harris A, Gross J, Januleviciene I, Shah A and Siesky B 2017 Optical coherence tomography angiography: an overview of the technology and an assessment of applications for clinical research *Br. J. Ophthalmol* **101** 16–20

[17] Spaide R F, Klancnik J M and Cooney M J 2015 Retinal vascular layers imaged by fluorescein angiography and optical coherence tomography angiography *JAMA Ophthalmol* **133** 45–50

[18] Choi W, Mohler K J, Potsaid B, Lu C D, Liu J J and Jayaraman V *et al* 2013 Choriocapillaris and choroidal microvasculature imaging with ultrahigh speed OCT angiography *PLoS One* **8** e81499

[19] de Carlo T E, Romano A, Waheed N K and Duker J S 2015 A review of optical coherence tomography angiography (OCTA) *Int. J. Retin. Vitr* **1** 5

[20] Matsunaga D, Puliafito C A and Kashani A H 2014 OCT angiography in healthy human subjects *Ophthalmic Surg. Lasers Imaging Retin* **45** 510–5
[21] Gabriele M L, Wollstein G and Ishikawa H *et al* 2011 Optical coherence tomography: history, current status, and laboratory work *Invest. Ophthalmol. Vis. Sci.* **52** 2425–36
[22] Huang D, Swanson E A and Lin C P *et al* 1991 Optical coherence tomography *Science* **254** 1178–81
[23] Fercher A F, Hitzenberger C K, Drexler W, Kamp G and Sattmann H 1993 *In vivo* optical coherence tomography *Am. J. Ophthalmol.* **116** 113–4
[24] Schachat A P Clinical applications of diagnostic indocyanine green angiography *Ryan's Retina* vol 1 6th edn (London: Elsevier) pp 46–76
[25] Hope-Ross M, Yannuzzi L A, Gragoudas E S, Guyer D R, Slakter J S and Sorenson J A *et al* 1994 Adverse reactions due to indocyanine green *Ophthalmology* **101** 529–33
[26] Khatri A, Pradhan E, Bal Kumar KC, Kharel M, Rijal R K and Timalsena S *et al* 2020 Detection, localization, and characterization of vision-threatening features of microaneurysms using optical coherence tomography angiography in diabetic maculopathy *Eur. J. Ophthalmol* **31** 1120672120924609
[27] Khatri A, Bal Kumar K C, Kharel M, Ashma K C and Pradhan E 2020 Analysis of microaneurysms and capillary density quantified by OCT-angiography and its relation to macular edema and macular ischemia in diabetic maculopathy *Eye [Internet]* **35** 1777–9
[28] Lieb W E, Cohen S M, Merton D A, Shields J A, Mitchell D G and Goldberg B B 1991 Color Doppler imaging of the eye and orbit *Arch. Ophthalmol.* **109** 527–31
[29] Neganova A Y, Postnov D D, Jacobsen J C and Sosnovtseva O 2016 Laser speckle analysis of retinal vascular dynamics *Biomed. Opt. Express* **7** 1375–84 Published 2016 Mar 18
[30] Basak K, Manjunatha M and Dutta P K 2012 Review of laser speckle-based analysis in medical imaging *Med. Biol. Eng. Comput.* **50** 547–58
[31] Tan O, Liu G, Liang L, Gao S S, Pechauer A D and Jia Y *et al* 2015 En face Doppler total retinal blood flow measurement with 70 kHz spectral optical coherence tomography *J. Biomed. Opt.* **20** 066004
[32] Wang Y, Fawzi A, Tan O, Gil-Flamer J and Huang D 2009 Retinal blood flow detection in diabetic patients by Doppler Fourier domain optical coherence tomography *Opt. Express* **17** 4061–73
[33] Kwiterovitch K A, Maguire M G and Murphy R P *et al* 1991 Frequency of adverse systemic reactions occurring after fluorescein angiography—results of a prospective study *Ophthalmology* **98** 1139–42
[34] Wylęgała A 2018 Principles of OCTA and applications in clinical neurology *Curr. Neurol. Neurosci. Rep.* **18** 96 Published 2018 Oct 18
[35] Spaide R F, Fujimoto J G, Waheed N K, Sadda S R and Staurenghi G 2018 Optical coherence tomography angiography [Internet] *Prog. Retin. Eye Res.* [cited 2018 Aug 17] p 1–55
[36] Ido M, Osawa S and Fukukita M *et al* 2007 The use of colour Doppler imaging in the diagnosis of retinal detachment *Eye* **21** 1375–8
[37] Jia Y *et al* 2012 Split-spectrum amplitude-decorrelation angiography with optical coherence tomography *Opt. Express* **20** 4710–25
[38] Nadiarnykh O, Davidoiu V, Gräfe M G O, Bosscha M, Moll A C and de Boer J F 2019 Phase-based OCT angiography in diagnostic imaging of pediatric retinoblastoma patients: abnormal blood vessels in post-treatment regression patterns *Biomed. Opt. Express* **10** 2213–26

[39] Chu Z, Chen C L, Zhang Q, Pepple K, Durbin M, Gregori G and Wang R K 2017 Complex signal-based optical coherence tomography angiography enables *in vivo* visualization of choriocapillaris in human choroid *J. Biomed. Opt.* **22** 1–10
[40] Arya M, Rashad R, Sorour O, Moult E M, Fujimoto J G and Waheed N K 2018 Optical coherence tomography angiography (OCTA) flow speed mapping technology for retinal diseases *Expert Rev. Med. Devices* **15** 875–82
[41] Topcon O C T 2020 Angiography with the DRI Swept Source OCT Triton. Accessed Date: 11 September. Available from: http://topconmedical.com/products/ssoctangiotm.htm
[42] Mastropasqua R, Di Antonio L and Di Staso S *et al* 2015 Optical coherence tomography angiography in retinal vascular diseases and choroidal neovascularization *J. Ophthalmol.* **2015** 343515
[43] Choi J, Kwon J, Shin J W, Lee J, Lee S and Kook M S 2017 Quantitative optical coherence tomography angiography of macular vascular structure and foveal avascular zone in glaucoma *PLoS One* **12** e0184948
[44] Watzke R C, Klein M L and Hiner C J *et al* 2000 A comparison of stereoscopic fluorescein angiography with indocyanine green video angiography in age-related macular degeneration *Ophthalmology* **107** 1601–6
[45] de Carlo T E, Bonini Filho M A, Chin A T, Adhi M, Ferrara D, Baumal C R, Witkin A J, Reichel E, Duker J S and Waheed N K 2015 Spectral-domain optical coherence tomography angiography of choroidal neovascularization *Ophthalmology* **122** 1228–38
[46] Spaide R F, Fujimoto J G and Waheed N K 2015 Image artifacts in optical coherence tomography angiography *Retina* **35** 2163–80
[47] Rocholz R, Corvi F, Weichsel J, Schmidt S and Staurenghi G 2019 OCT angiography (OCTA) in retinal diagnostics *High Resolution Imaging in Microscopy and Ophthalmology* ed J Bille (Cham: Springer)
[48] Wei X, Hormel T T, Pi S, Guo Y, Jian Y and Jia Y 2019 High dynamic range optical coherence tomography angiography (HDR-OCTA) *Biomed. Opt. Express* **10** 3560–71
[49] Bressler N M and Bressler S B 2013 Neovascular (Exudative or 'Wet') age-related macular degeneration *Retina* ed S J Ryan, S R Sadda and D R Hinton (London: Elsevier Saunders) pp 1183–212
[50] Jia Y, Bailey S T, Wilson D J, Tan O, Klein M L and Flaxel C J *et al* 2014 Quantitative optical coherence tomography angiography of choroidal neovascularization in age-related macular degeneration *Ophthalmology* **121** 1435–44
[51] Joshi K, KC A, Rijal R, Rai S, Lammichane G and Yadav R 2020 Compliance and visual outcome of 'Treat-And-Extend' versus 'Pro Re Nata' dosing of intravitreal bevacizumab in wet age-related macular degeneration—a perspective from a developing country *Birat. J. Health Sci.* **5** 921–26
[52] Moult E, Choi W, Waheed N K, Adhi M, Lee B and Lu C D *et al* 2014 Ultrahigh-speed swept-source OCT angiography in exudative AMD *Ophthalmic Surg. Lasers Imaging Retina* **45** 496–505
[53] Uchida A, Hu M and Babiuch A *et al* 2019 Optical coherence tomography angiography characteristics of choroidal neovascularization requiring varied dosing frequencies in treat-and-extend management: an analysis of the AVATAR study *PLoS One* **14** e0218889
[54] Couturier A, Mané V and Bonnin S *et al* 2015 Capillary plexus anomalies in diabetic retinopathy on optical coherence tomography angiography *Retina* **35** 2384–91

[55] Hasegawa N, Nozaki M, Takase N, Yoshida M and Ogura Y 2016 New insights into microaneurysms in the deep capillary plexus detected by optical coherence tomography angiography in diabetic macular edema *Invest. Ophthalmol. Vis. Sci.* **57** OCT348–55
[56] Hwang T S, Jia Y, Gao S S, Bailey S T, Lauer A K and Flaxel C J *et al* 2015 Optical coherence tomography angiography features of diabetic retinopathy *Retina* **35** 2371–6
[57] Moore J, Bagley S, Ireland G, McLeod D and Boulton M E 1999 Three dimensional analysis of microaneurysms in the human diabetic retina *J. Anat* **194** 89–100
[58] Parravano M, De Geronimo D, Scarinci F, Virgili G, Querques L, Varano M, Bandello F and Querques G 2018 Progression of diabetic microaneurysms according to the internal reflectivity on structural OCT and visibility on OCT angiography *Am. J. Ophthalmol.* **198** P8–16
[59] Rodrigues T M, Marques J P, Soares M, Simão S, Melo P and Martins A *et al* 2019 Macular OCT-angiography parameters to predict the clinical stage of nonproliferative diabetic retinopathy: an exploratory analysis *Eye [Internet]* **33** 1240–7
[60] Karti O, Ipek S C and Saatci A O 2020 Multimodal imaging characteristics of a large retinal capillary macroaneurysm in an eye with severe diabetic macular edema: a case presentation and literature review *Med. Hypothesis Discov. Innov. Ophthalmol* **9** 33–7
[61] Khatri A, Karki P, Joshi S, Khatri B K, Kharel M, Banstola A and Shrestha S M 2020 Macular profile of eyes developing macular holes in cases of central retinal vein occlusion treated with Bevacizumab *J. Kathmandu Medical College* **9** 13–9
[62] Tsai G, Banaee T, Conti F F and Singh R P 2018 Optical coherence tomography angiography in eyes with retinal vein occlusion *J. Ophthalmic Vis. Res* **13** 315–32
[63] Khatri A, Timalsena S, Gautam S and Kharel M 2019 Varicella retinal vasculopathy: unilateral cilioretinal artery occlusion despite acyclovir therapy caught using optical coherence tomography-angiography (OCTA) *Case Reports in Ophthalmology Medicine* ed S G Schwartz (London: Hindawi)
[64] Novais E A and Waheed N K 2016 Optical coherence tomography angiography of retinal vein occlusion *Dev. Ophthalmol* **56** 132–8
[65] Çelik T, Bilen F, Yalçındağ F N and Atilla H 2018 Optical coherence tomography angiography in branch retinal artery occlusion *Turk. J. Ophthalmol* **48** 150–4
[66] Bonini Filho M A *et al* 2015 Optical coherence tomography angiography in retinal artery occlusion *Retina* **35** 2339–46
[67] Ferrara D, Waheed N K and Duker J S 2016 Investigating the choriocapillaris and choroidal vasculature with new optical coherence tomography technologies *Prog. Retin. Eye Res* **52** 130–55
[68] Jia Y, Wei E, Wang X, Zhang X, Morrison J C and Parikh M *et al* 2014 Optical coherence tomography angiography of optic disc perfusion in glaucoma *Ophthalmology* **7** 1322–32
[69] Akil H, Huang A S, Francis B A, Sadda S R and Chopra V 2017 Retinal vessel density from optical coherence tomography angiography to differentiate early glaucoma, pre-perimetric glaucoma and normal eyes *PLoS One* **12** e0170476
[70] Yarmohammadi A, Zangwill L M and Manalastas P I C *et al* 2018 Peripapillary and macular vessel density in patients with primary open-angle glaucoma and unilateral visual field loss *Ophthalmology* **125** 578–87
[71] Zhang X, Dastiridou A and Francis B A *et al* 2017 Comparison of glaucoma progression detection by optical coherence tomography and visual field *Am. J. Ophthalmol.* **184** 63–74

[72] Wu Z, Saunders L J, Daga F B, Diniz-Filho A and Medeiros F A 2017 Frequency of testing to detect visual field progression derived using a longitudinal cohort of glaucoma patients *Ophthalmology* **124** 786–92

[73] Tranos P, Karasavvidou E M, Gkorou O and Pavesio C 2019 Optical coherence tomography angiography in uveitis *J. Ophthalmic Inflamm. Infect.* **9** 21

[74] Kim A Y, Rodger D C and Shahidzadeh A *et al* 2016 Quantifying retinal microvascular changes in uveitis using spectral-domain optical coherence tomography angiography *Am. J. Ophthalmol.* **171** 101–12

[75] Khairallah M, Abroug N and Khochtali S *et al* 2017 Optical coherence tomography angiography in patients with behçet uveitis *Retina* **37** 1678–91

[76] Khatri A, Wagle B, Hony K C, Chaurasiya B D, Timalsena S and Singh K *et al* 2021 Post typhoid fever neuroretinitis with serous retinal detachment and choroidal involvement: a case report *Am. J. Ophthalmol. Case Rep.* **21** 101025

[77] Khatri A, Timalsena S and Khatri B K *et al* 2020 A rare entity: sympathetic ophthalmia presumably after blunt trauma to the phthisical eye and optical coherence tomography angiography metrics to monitor response to treatment *Clin. Case Rep.* **8** 149–54

[78] Agrawal R, Salman M, Tan K-A, Karampelas M, Sim D A and Keane P A *et al* 2016 Choroidal vascularity index (CVI)—a novel optical coherence tomography parameter for monitoring patients with panuveitis? *PLoS One* **11** e0146344

[79] Zahid S, Chen K C and Jung J J *et al* 2017 Optical coherence tomography angiography of chorioretinal lesions due to idiopathic multifocal choroiditis *Retina* **37** 1451–63

[80] Cheng L, Chen X and Weng S *et al* 2016 Spectral-domain optical coherence tomography angiography findings in multifocal choroiditis with active lesions *Am. J. Ophthalmol.* **169** 145–61

[81] Klufas M A, Phasukkijwatana N and Iafe N A *et al* 2017 Optical coherence tomography angiography reveals choriocapillaris flow reduction in placoid chorioretinitis *Ophthalmol. Retin* **1** 77–91

[82] Phasukkijwatana N, Iafe N and Sarraf D 2017 Optical coherence tomography angiography of A29 birdshot chorioretinopathy complicated by retinal neovascularization *Retin. Cases Brief Rep.* **11** S68–72

[83] Prada D, Harris A, Guidoboni G, Siesky B, Huang A M and Arciero J 2016 Autoregulation and neurovascular coupling in the optic nerve head *Surv. Ophthalmol* **61** 164–86

[84] Garhöfer G, Chua J, Tan B, Wong D, Schmidl D and Schmetterer L 2020 Retinal neurovascular coupling in diabetes *J. Clin. Med* **9** 2829

[85] Warner R L, de Castro A and Sawides L *et al* 2020 Full-field flicker evoked changes in parafoveal retinal blood flow *Sci. Rep.* **10** 16051

[86] Aschinger G, Schmetterer L, Fondi K, Bata A, dos Santos V A, Garhofer G and Werkmeister R M 2016 Quantification of total retinal blood flow via dual-beam bidirectional Doppler optical coherence tomography for the assessment of neurovascular coupling in the human retina *Invest. Ophthalmol. Vis. Sci.* **57** 4619

[87] Cho K A, Rege A and Jing Y *et al* 2020 Portable, non-invasive video imaging of retinal blood flow dynamics *Sci. Rep.* **10** 20236

[88] Zhang Y S, Mucollari I, Kwan C C, Dingillo G, Amar J, Schwartz G W and Fawzi A A 2020 Reversed neurovascular coupling on optical coherence tomography angiography is the earliest detectable abnormality before clinical diabetic retinopathy *J. Clin. Med* **9** 3523

[89] Nesper P L, Lee H E, Fayed A E, Schwartz G W, Yu F and Fawzi A A 2019 Hemodynamic response of the three macular capillary plexuses in dark adaptation and flicker stimulation using optical coherence tomography angiography *Invest. Ophthalmol. Vis. Sci.* **60** 694–703
[90] Fondi K, Bata A M, Luft N, Witkowska K J, Werkmeister R M and Schmidl D *et al* 2018 Evaluation of flicker induced hyperemia in the retina and optic nerve head measured by Laser Speckle Flowgraphy *PLoS One* **13** e0207525

IOP Publishing

Photo Acoustic and Optical Coherence Tomography Imaging, Volume 3
Angiography: an application in vessel imaging
Ayman El-Baz and Jasjit S Suri

Chapter 11

Optical coherence tomography angiography: principles and clinical application

Ogugua Ndubuisi Okonkwo and Adekunle Olubola Hassan

Optical coherence tomography angiography (OCTA) is an imaging technology that evolved due to the faster scan speed of available optical coherence tomograms. Using this technology dramatically enhances understanding of the retinal vascular network in health and several disease states. Since it is non-invasive, easily reproducible, and provides three-dimensional images of the retinal, choroidal and peripapillary network, several clinicians and researchers have investigated its use in diseases such as diabetic retinopathy, macular degeneration, retinal vascular occlusion, and glaucoma. We provide an easy-to-understand review of the basic principles of OCTA, its clinical application, and unmet needs. An assembly of OCTA images, with corresponding *en face* and cross-sectional OCT images, is provided to illustrate the synergy between information on the retinal and choroidal structure and vascular provided by OCT and OCTA, respectively.

Abbreviations

AMD:	Age-related macular degeneration
BRM:	Bruch's membrane
CNV:	Choroidal neovascularization
DCP:	Deep capillary pexus
DR:	Diabetic retinopathy
ERM:	Epiretinal membrane
FAZ:	Foveal avascular zone
FA:	Fluorescein angiography
FFA:	Fundus fluorescein angiography
ILM:	Internal limiting membrane
INL:	Inner nuclear layer

doi:10.1088/978-0-7503-2060-3ch11

IPL:	Inner plexiform layer
ICP:	Intermediate capillary plexus
ICGA:	Indocyanine green angiography
MH:	Macular hole
NFLVP:	Nerve fiber layer vascular plexus
OCT:	Optical coherence tomography
OCTA:	Optical coherence tomography angiography
OCTARA:	OCT angiography ratio analysis
OPL:	Outer plexiform layer
PED:	Pigment epithelial detachment
PCV:	Polypoidal choroidal vasculopathy
RBC:	Red blood cell
RPE:	Retinal pigment epithelium
RVO:	Retinal vein occlusion
RAO:	Retinal artery occlusion
RNFL:	Retinal nerve fiber layer
SSADA:	Split spectrum amplitude decorrelation algorithm
SS:	Swept-source
SD:	Spectral domain
SCP:	Superficial capillary plexus
VCSEL:	Vertical-cavity surface-emitting lasers
VEGF:	Vascular endothelial growth factor

11.1 Introduction

Advances in OCT imaging and existing age-long drawbacks of conventional dye angiography in imaging the retina and choroid using fundus fluorescein angiography (FFA) and indocyanine green angiography (ICGA) have resulted in the development of OCTA, and its rapid evolution and widespread acceptance as an ophthalmic diagnostic tool [1, 2]. OCTA offers a non-invasive, fast to produce, easily repeatable, reproducible, and non-dye technology for assessing retinal vasculature and blood flow within the retina's microvasculature, optic disc, and other ocular structures in health and several disease states [1, 3]. In summary, OCTA detects motion within the blood vessel lumen by measuring the variation in reflected OCT signal amplitude between consecutive cross-sectional B-scans. It does this by using laser light reflectance of the surface of moving red blood cells to depict vessels accurately [4]. This field has undergone rapid improvements with application in several areas, including ophthalmology and retinal medicine, research, and clinical practice [2, 4]. We present a review of OCTA's principle and its clinical uses. In some instances we include OCTA images, with accompanying *en face* and cross-sectional OCT images.

11.2 Principles of OCTA

11.2.1 Basics

OCTA evolved from Doppler based imaging of flow within the vasculature [4, 5]. With the technological advancements in newer OCT, such as the swept-source (SS) OCT, with Fourier domain features, which produce faster A-scans due to more rapid scanning speeds and enhanced axial resolution, three-dimensional OCT images became possible [1, 2, 4].

OCTA relies on using the OCT signal variation caused by moving particles, such as red blood cells (RBCs), as the contrast mechanism for imaging blood flow [4]. The backscattering between moving RBCs within the blood vessels contrasts from backscattering from a static, more stationary interface. Simultaneously, the fixed interface signal remains stable, while the RBC signal continually changes. This difference results in a motion contrast signal which is detectable [6].

Moving particles can be differentiated from static tissue by repeated scans performed at the same location. Temporal changes of the OCT signal in subsequent scans caused by the moving particles generate the angiographic contrast, enabling visualization of the microvasculature.

In summary, In OCTA, multiple repeated B-scans are performed in the same location, and the structural images are compared on a pixel-by-pixel basis to detect signal changes that occur because of flowing erythrocytes. A motion contrast image is seen as the difference between these rapidly repeated B-scans and is displayed. B-scans consist of multiple A-scans. The rate of A-scans in available OCT scans ranges from 70 000 to 100 000 scans per second [7]. The time interval between two B-scans at the same scan location is known as the interscan time. The interscan time plays a crucial role, reducing motion sensitivity and reducing motion artifacts. OCTA is a rapidly evolving imaging technology and has gained popularity due to dye's non-use and eliminating all the attendant complications of dye usage [1]. Therefore, unlike conventional dye-based angiography, OCTA can be repeated with reasonable safety.

11.2.2 Image segmentation

This is the process of partitioning a digital image into multiple segments. The goal of segmentation is to simplify and change an image's representation into something more meaningful and more comfortable to analyze. In OCTA, as in some other digital or computer acquired images, segmentation is used to locate boundaries (lines and curves) [8]. In general, commercial OCTA software separates the retinal vasculature into a 'superficial' and 'deep' layer. The software predefined the retinal layers included in these retinal slabs, used for the output images, as shown in figure 11.1. The user has little input in this decision, and re-segmenting the retinal layers is a laborious and often impossible task without custom analysis software [8].

The segmented OCTA provides information about the superficial and deep vascular plexus, outer retina, Bruch's membrane, and the choriocapillaris. The segmentation varies between the various OCTA machines. For example, the OptoVue AngioVue TM system designates the vasculature as either superficial or deep plexus depending on the layer boundaries. The superficial slab or 'plexus' is located between the inner limiting membrane (ILM) and the posterior border of the inner plexiform layer (IPL), as shown in figure 11.1. The deep 'plexus' is the capillaries between the posterior boundary of the IPL and the rear boundary of the outer plexiform layer (OPL). Similarly, the choriocapillaris is the layer of capillaries

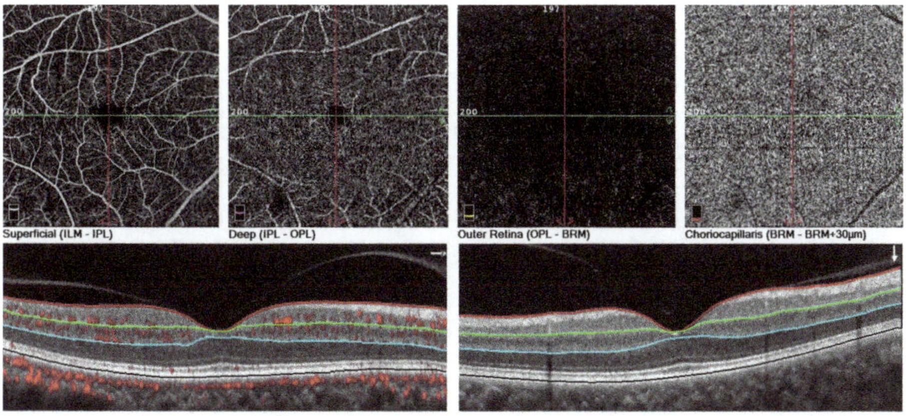

Figure 11.1. Illustrates segmentation of an OCTA image into component layers, superficial and deep plexus, outer retina, and choriocapillaris. The cross-sectional OCT image shows the layers of segmentation of the component slabs, which are superficial plexus (ILM—IPL), deep plexus (IPL—OPL), outer retina (OPL—BRM), and choriocapillaris (BRM—BRM + 30 microns).

in a 30 μm thick section located immediately posterior to the RPE—BRM [9]. One can appreciate the entire segmented layers of a standard OCT in figure 11.1.

The segmentation feature seen in OCTA is an advantage over FFA, which images only in two dimensions and demonstrates pathology in the superficial vascular plexus [10]. In contrast, the segmentation of OCTA images allows for 3D reconstruction and precise localization of pathology. For example, figure 11.2 represents an eye harboring an active choroidal neovascular membrane (CNVM), and one can quickly identify this in the sub RPE location (a type 1 CNVM). The sequential OCT shows a pigment epithelial detachment (PED) in the same area.

11.2.3 *En face* imaging

'*En face*' means 'facing forward.' In this case, the OCT acquires an image cube of the retina using a dense raster scan. Computer software is then used to reconstruct C scan images on the coronal plane. This computer program allows the view of a specific retinal layer as a transverse image. Whereas traditionally, most clinicians are used to cross-sectional OCT scans, the *en face* image views the retina in a coronal plane, with the entire surface visible to the clinician all at once, figure 11.2. This type of OCT permits a more accurate mapping of pathological and structural changes in the examined layer. Therefore, computer software reconstruction of 3D images of OCT images is the basis of *en face* imaging. The reader can identify *en face* OCT images in figure 11.2. This computer technology is incorporated into both OCT technology and OCTA, as seen in figure 11.2. Its usefulness is further illustrated when combined with the segmentation features of the OCTA. OCTA image grading is performed on these *en face* image slabs obtained by segmentation of the different retinal layers into various horizontal depth-resolved slabs, including superficial and deep retinal plexus, the choriocapillaris, and choroid. In this manner, it can be used for clinical assessment and evaluating treatment response to therapeutics.

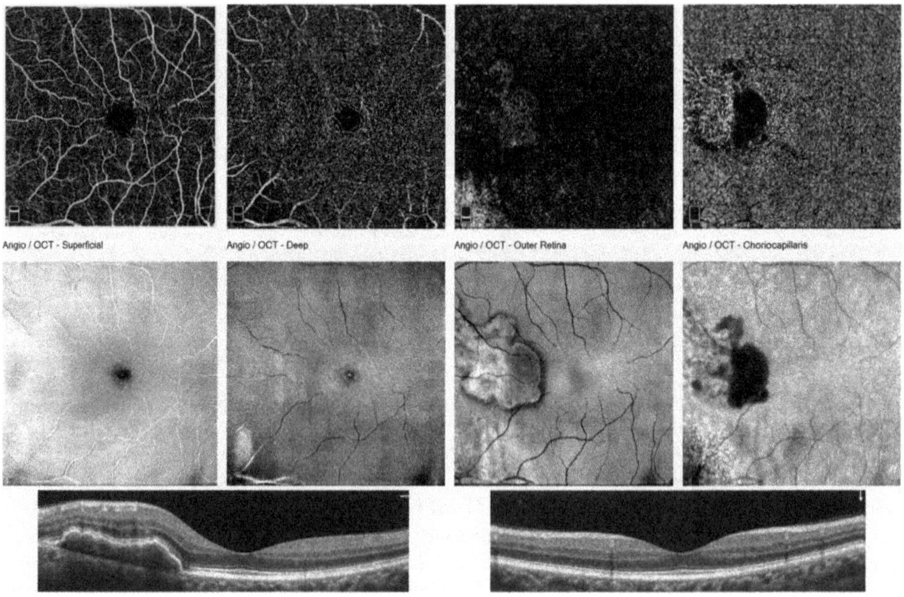

Figure 11.2. An eye with an active choroidal neovascular membrane (CNVM) in the sub RPE location (a type 1 CNVM). The *en face* and cross-sectional OCT image further localizes the CNVM and identifies a pigment epithelial detachment (PED) harboring this CNVM. *En face* and cross-sectional images allow for precise localization of lesions within specific subretinal layers, using their axial location on OCT cross-sections.

High-quality *en face* images offer a broader assessment area at the segmentation level and enable timely diagnosis while reducing to a minimum missed pathology. Some of the advantages of *en face* imaging include the ability to precisely localize lesions within specific subretinal layers, using their axial location on OCT cross-sections. Also, it offers the ability to register projected OCT images to other fundus imaging modalities, using retinal vessels as landmarks.

11.3 OCTA technology: software and hardware

Several algorithms exist for using repeated B-scan OCT images to calculate motion contrast. OCTA algorithms can use the OCT signal amplitude, the OCT signal phase, or both amplitude and phase (known as complex amplitude) to detect motion perpendicular to the OCT beam's direction. The most widely used OCTA algorithms use the OCT signal amplitude.

An algorithm called split spectrum amplitude decorrelation algorithm (SSADA) was demonstrated to improve the signal-to-noise ratio and reduce sensitivity to bulk eye motion [11, 12]. The commercially available Angiovue software on the Avanti RTVue XR device (Optovue, Fremont, CA) uses the SSADA software. Other algorithms in use include the optical micro-angiography (OMAG) algorithm, [13, 14], which uses both phase and intensity information from sequential OCT B-scans (used in the Zeiss Cirrus HD-OCT 5000 (Zeiss Meditec, Dublin, CA) with

Angioplex) [14]. The OCT angiography ratio analysis (OCTARA) algorithm uses intensity information to be more sensitive to low blood flow [15].

SSADA was the first algorithm to be extensively used in a commercially available ophthalmic OCTA instrument. Decorrelation is a mathematical function that quantifies variation without being affected by the average signal strength (as long as the signal is strong enough to predominate over optical and electronic noise). SSADA uses signal processing methods to divide the OCT spectrum into multiple narrow-band spectra, which reduces the axial image resolution to minimize sensitivity to eye motion and match the transverse OCT image resolution. Speckle decorrelations are calculated on a B-scan-to-B-scan basis between the split spectral data and then combined to generate a single data set with an increased signal-to-noise ratio.

There has been rapid commercial development of OCTA in recent years. Optovue introduced the first commercial OCTA product, the AngioVue, based on the SD-OCT model. Topcon introduced the first commercial SS-OCTA as the Atlantis and Triton product line, which image at 1050 nm wavelength and 100 000 A-scans per second. Zeiss introduced the AngioVue OCTA using the SD-OCT platform and the AngioPlex OCTA using SS-OCT. A wide array of OCTA algorithms and display methods exist for various OCTA technologies, varying significantly between different instrument manufacturers. Therefore, it is reasonable to exercise caution when comparing different OCTA instruments' results. The development of commercially available and easy to use software and hardware meant that more clinicians could use OCTA, which resulted in an acceleration in this technology's growth.

The development of vertical-cavity surface-emitting lasers (VCSELs) enabled imaging speeds of 400 000 A-scans per second, suggesting that higher imaging speeds will be commercially available in the future [1, 16].

11.4 FA versus OCTA comparison

Though conventional FA can image only the superficial vascular plexus, it has played a significant role in diagnosis and disease monitoring for several years and has not been replaced by OCTA. The introduction of OCTA offers quick imaging of retinal vasculature method that FA or ICGA methods cannot image. OCTA images the superficial vascular plexus and shows similar patterns seen with FA in this layer (see figures 11.3(a) and (b)). OCTA allows imaging of the intermediate, deeper, and choroidal vascular layers. OCTA can also image the radial peripapillary vascular network of vessels (figure 11.4). The rapid acquisition of images by OCTA means that it can obtain serial photos of the same pathology several times, e.g., daily imaging or several times in a day. This was not possible with FA. Therefore, OCTA offers a wealth of clinical information from following the evolution of pathology or monitoring the outcome of interventions.

OCTA shows superiority over FA and ICG in providing more explicit images of vascular pathology, e.g., neovascular network, since there is no dye leakage and therefore no hyperfluorescence to obscure the neovascularization (figure 11.5). Due to this fact, one can visualize high-resolution, well-defined microvasculature borders.

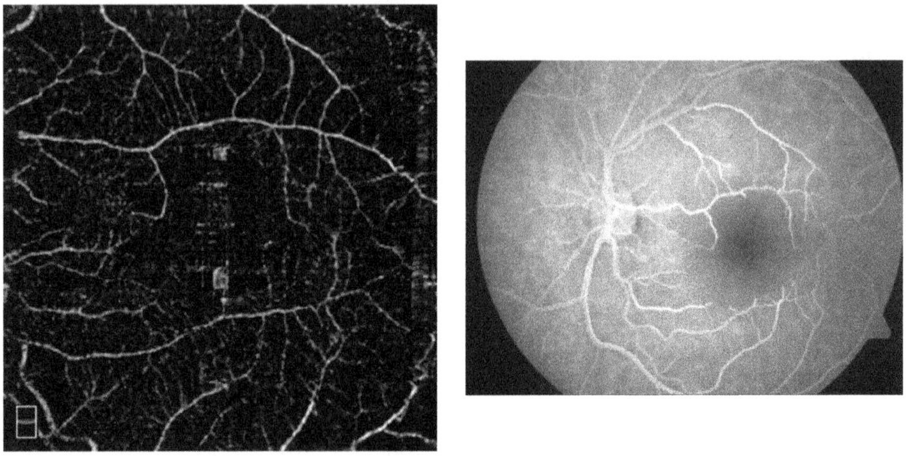

(a) (b)

Figure 11.3. (a) OCTA image of left eye central retinal artery occlusion with significant macular non-perfusion, ischemia, and perifoveal vasculature absence. There is a grossly enlarged FAZ in the deep capillary plexus. The OCTA image shows similarity to that seen in (b), which is a late venous phase of the same eye's conventional fluorescein angiography.

Also, the tissue segmentation seen with OCT is possible with OCTA. Volumetric data can undergo segmentation to visualize the vascular network at different layers of the retina, choriocapillaris, and vascular pathology such as neovascularization. The pathology's actual depth or layer is seen in cross-sectional OCT images, as shown in figure 11.2, where a subretinal pigment epithelium (sub-RPE) choroidal neovascular membrane (CNVM type 1) is apparent.

OCTA is accompanied by the sequential acquisition of structural OCT, which includes cross-sectional images that help co-localize pathology.

A summary of the comparison of OCTA over FFA and ICG includes the following.

OCTA can be acquired in seconds. Since there is no need for intravenous injection of dye, there is no nausea, vomiting or anaphylaxis, and other life-threatening adverse events.

As OCTA is performed with considerable ease, patients have frequent follow-up scans done.

OCTA can detect vascular abnormalities based on depth and vascular patterns present above the RPE (type II) and between the Bruch's membrane and the RPE and beneath the RPE (type I).

OCTA affords image segmentation, which helps the three-dimensional assessment of pathology.

A disadvantage of OCTA is that it cannot assess leakage since it is dyeless. Since dye leakage and staining do not occur in OCTA, the boundaries and areas of capillary drop out and neovascularization can be more precisely measured since there are no obscurations of the borders or edges of the neovascular network (figure 11.5).

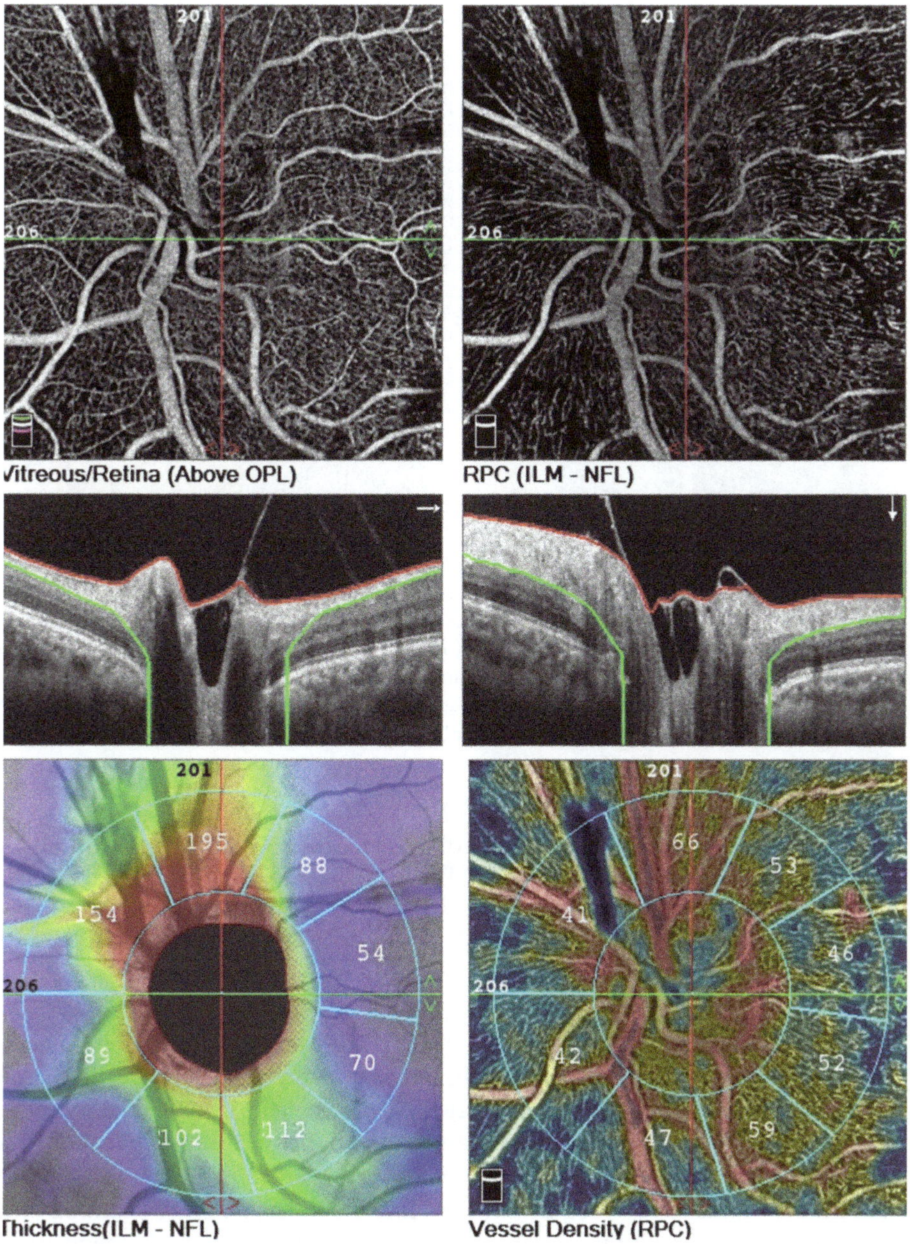

Figure 11.4. OCTA of the radial peripapillary capillaries (RPC) and includes the vessel density computation, a useful metric/parameter assessed in a glaucomatous eye.

Unlike conventional angiography, which is two-dimensional, making it challenging to distinguish vascular abnormalities within different layers, OCTA's three-dimensional nature allows for an entirely separate evaluation of retinal and choroidal circulation.

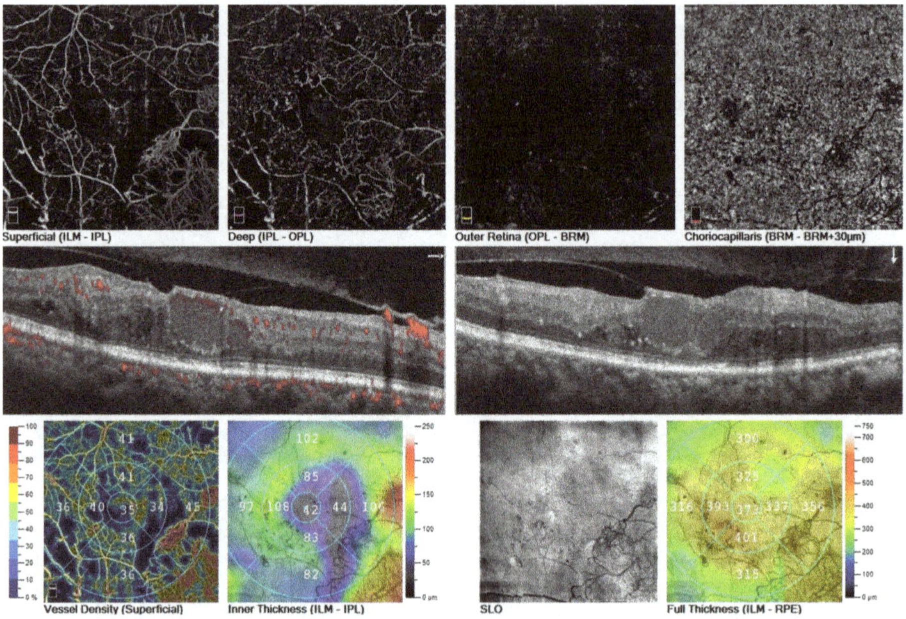

Figure 11.5. Significant foveal and macular ischemia and areas of capillary dropout are evident on both superficial and deep capillary plexus of this retina diagnosed with diabetic retinopathy. Also seen is a cluster of neovascularization elsewhere (NVE) in the lower right segment of the superficial and deep slabs. Since there is no dye leakage and therefore no hyperfluorescence to obscure, the neovascularization (NVE) outline is better visualized than in conventional FA. The vessel density computation shows a marked reduction.

FA cannot consistently image the radial peripapillary vascular plexus, which OCTA is very good at imaging, figure 11.4. FA underestimates the capillary density compared to histological measurements.

11.4.1 Limitations of OCTA

OCTA cannot image leakage as can be done in conventional angiography since it is dyeless imaging for only a short interval, in seconds. Therefore, OCTA cannot assess changes in vascular permeability or leakage. Also, widefield FA availability, which enables imaging of vasculature within the retina periphery, sets a high standard for OCTA, which images a limited retina area for now. Therefore, for these reasons, OCTA cannot replace FA. It requires faster scanners to enable the rescanning of the same retinal position multiple times, which is the fundamental requirement underlying the motion contrast principle of OCTA compared to FA. A much faster scan for the scanning of more expansive areas of retinal vasculature is required. Therefore, the speed of the scan needed to scan a 9 mm × 9 mm or 12 mm × 12 mm area of the retina is much faster than is necessary for a 3 mm × 3 mm retinal scan. The initial smaller size 3 mm × 3 mm and 6 mm × 6 mm scans were limitations, but 12 mm × 12 mm and expectations of widefield scans are more recent advances in OCTA.

In fast-flow situations, the saturation limit for the OCTA will be reached; therefore, fast flow can be detected by OCTA, but differences in flow will not be distinguishable because they are above the saturation limit.

11.4.2 A summary of OCTA artifacts

Shadow-graphic flow projection artifact makes interpretation of *en face* angiograms of deeper vascular beds more difficult. The projection artifacts occur due to the replication of superficial vessels on deeper layers. Also, motion artifacts can occur due to eye movements' effect on OCTA data acquisition [17]. Patients or persons with poorer fixation and uncontrolled eye movements are generally harder to scan. Real-time eye-tracking and post-processing software are used to remove these motion and projection artifacts [16]. Furthermore, to reduce noise, improve image quality, and vessel continuity, averaging of multiple *en face* OCTA images is done [18, 19].

Segmentation artifacts are encountered during imaging of diseases in which there is a significant alteration to the appearance and shape of the retinal layers. Examples of this clinical situation are macular edema, large pigment epithelial detachment, and bullous fluid collection in the subretinal or sub RPE space, as in figure 11.6.

Fading OCT and flow signal in large vessels due to the interferometric fringe washout effect is associated with rapid blood flow.

11.5 OCTA imaging of the normal retina and optic disc

The retina consists of neural tissue, vascular tissue within a connective tissue framework. OCT has revolutionized imaging, therefore understanding the retina in health and disease states. While OCT provided information on the retina structure, OCTA has given information on the retina's vascular network. Research using OCTA has provided much needed quantitative information or data on the retina's vasculature in a state of health and has also validated information on vasculature obtained from a histological specimen of the retina.

The retinal vascular network in the human eye axially is divided into four distinct capillary plexuses. While the vasculature within each plexus is densely linked, interconnecting vessels between these sub-networks are relatively few in comparison. With OCTA, the following layers have been confirmed and validated using confocal microscopy *ex vivo* [20]. From the anterior boundary of the retina to more posterior axial locations, the four distinct plexuses are nerve fiber layer vascular plexus (NFLVP), superficial vascular plexus (SVP), intermediate capillary plexus (ICP), and deep capillary plexus (DCP) [6]. While the three deeper layers can be observed and axially separated even in the retina periphery using OCTA at sufficient axial resolution, the NFLVP is most pronounced at locations where the nerve fiber layer holds a substantial width, this is in the peripapillary region as well as parafoveally. Therefore, the perifoveal vascular plexus is visualized using OCTA to assess glaucomatous optic neuropathy. To image the healthy and diseased eye properly, segmentation of the retinal layers is crucial in OCTA. It has been proposed that to allow for more accurate segmentation of retinal boundaries, the retinal layer interface between the inner plexiform layer (IPL) and inner nuclear layer (INL)

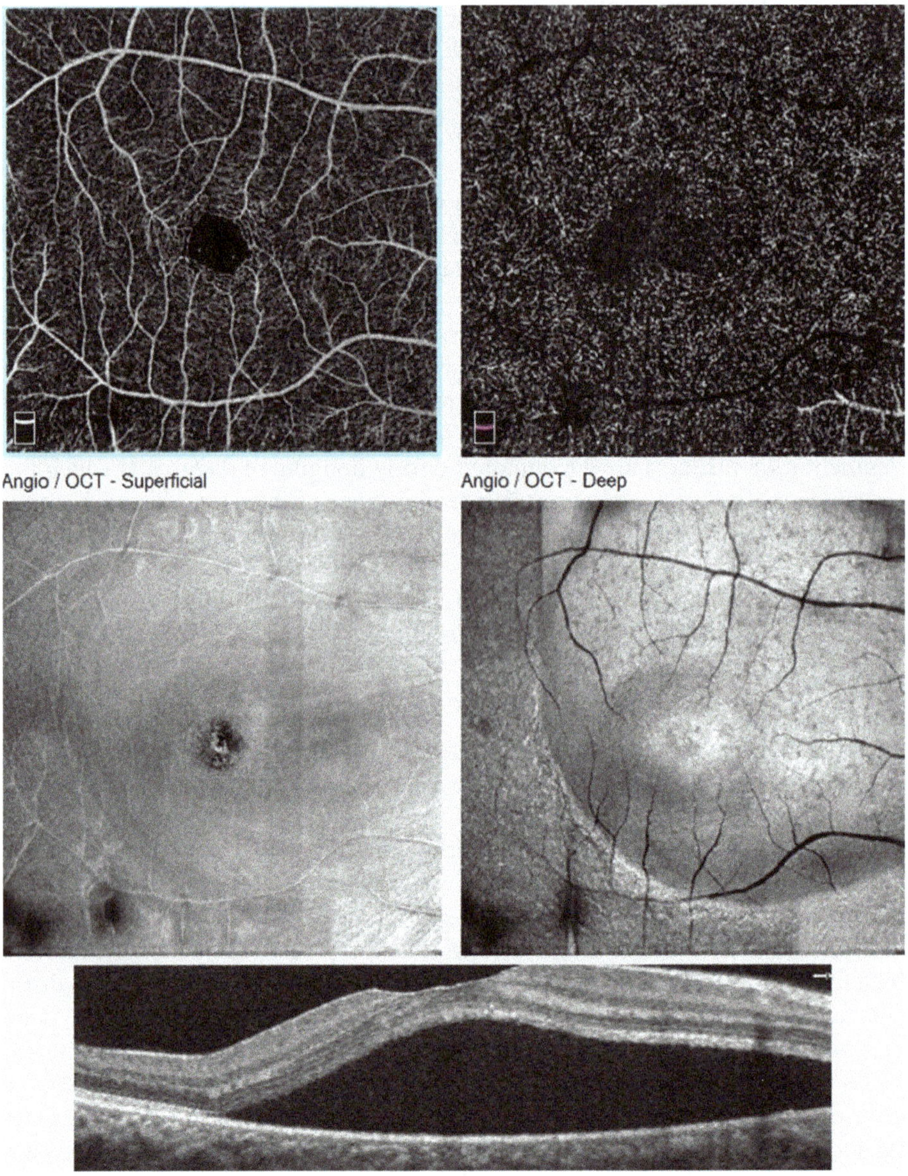

Figure 11.6. While the superficial vascular plexus is well imaged, the deep plexus cannot be easily imaged due to difficulty with automated segmentation by the OCTA machine. This difficulty in image segmentation is because of the large amount of subretinal fluid elevating the macula.

when shifted anteriorly and posteriorly by appropriate constant distances, represents a conveniently defined separating boundary for visualizing the three deeper plexuses independently within *en face* projections [6]. Currently, conflicting definitions of the axial location of borders between retinal vascular networks make the direct comparison of *en face* images from different OCTA devices difficult [21].

We will discuss the clinical application of OCTA next.

11.6 OCTA clinical application

11.6.1 Uveitis

OCTA finds usefulness in determining ischemia associated with posterior uveitis cases. There is an absence of perfusion seen on OCTA in such uveitic eyes having ischemia. Some uveitis cases, such as tuberculous choroiditis, are associated with choroidal neovascularization (CNV), which can be clearly defined using OCTA. Upon treatment with antiVEGF, sequential OCTA is used for monitoring response to treatment. Furthermore, there is associated macular edema in several uveitis cases evaluated using OCTA. The OCTA can find usefulness in assessing for non-perfusion related to the CME (figure 11.7), which is associated with some cases of posterior uveitis. Retinal angiomatosis proliferations can complicate TB granulomas, responsive to intravitreal antiVEGF [22]. Lastly, in uveitis involving the outer retina and choroid, patterns of flow voids in the choriocapillaris have been documented.

11.6.2 Diabetic retinopathy

OCTA has been studied extensively in evaluating DR alone and with DME. On OCTA, one can recognize the following microaneurysms, areas of impaired capillary perfusion (non-perfusion), defects in size and shape of the FAZ, neovascularization on the disc and retina, and intraretinal microvascular anomalies (IRMA). OCTA has been used in DR and DME clinical studies to quantify the FAZ area, vessel density, and perfusion density in the superficial capillary plexus

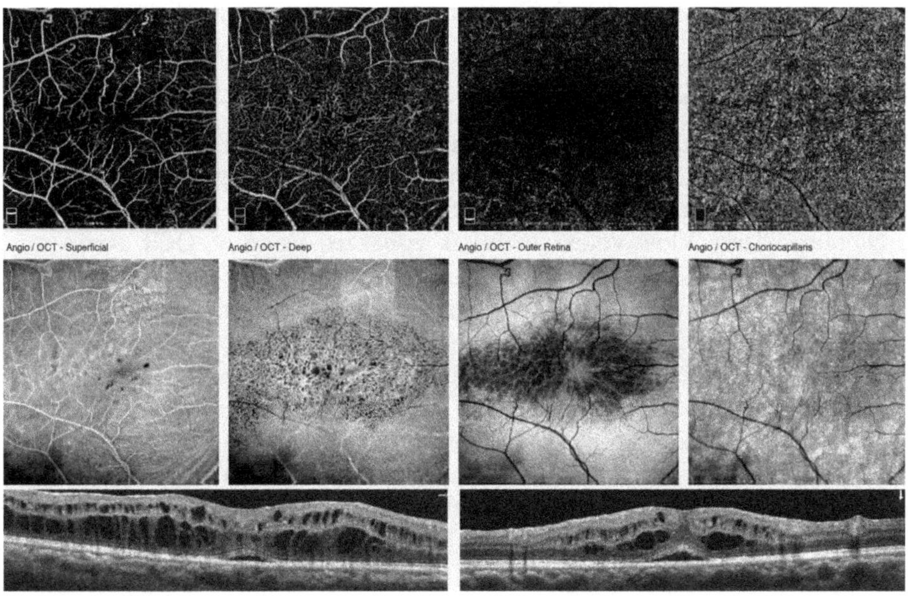

Figure 11.7. This represents the OCTA and OCT of a case with chronic, severe uveitis and cystoid macular edema. The superficial plexus does not show a distinct FAZ but suggests adequate perfusion. The deep plexus shows a much-reduced FAZ. The absence of foveal ischemia in the presence of macular edema offers a better visual prognosis.

[2, 23–25]. OCTA has demonstrated a clinically significant increase in the size of FAZ in eyes with progressively severe DR (figure 11.8(a)). OCTA is not associated with leakage and, therefore, can distinctively visualize and define neovascularization on the disc (NVD) and elsewhere (NVE). In this way, it is possible to differentiate NVD from collaterals. Also, OCTA is very useful in monitoring response to treatment [26]. It can be repeated several times in a day and at frequent clinic visits, unlike conventional FA, and therefore detects a longitudinal change in normal and abnormal retinal vasculature over time [26]. OCTA is also useful in identifying macular ischemia in the presence of macular edema, which helps offer a prognosis before commencing macular edema treatment. The finding of significant changes in FAZ and capillary vessel density is used to monitor eyes with DR, and these findings are more severe with the progression of DR severity. OCTA is useful in monitoring capillary non-perfusion areas in the superficial and deep plexus, as observed outside the fovea. Figure 11.8(b) shows the OCTA of an eye that has been treated with pan-retinal laser photocoagulation for active PDR and received several intravitreal antiVEGF injections for the treatment of DME. There is an enlarged FAZ seen in the superficial plexus and significant areas of reduced vessel density and non-perfusion in the superior perifoveal aspect of the deep plexus.

There are practical challenges experienced with OCTA imaging in the setting of DR. In DR, when OCTA imaging is performed in the presence of significant macular edema, image segmentation can be a challenge and result in errors using automated segmentation by the machines (figure 11.9). Image quality, which can be a challenge in such a clinical situation, may also affect the segmentation. Some OCTA machines offer manual segmentation, which is preferred in this clinical situation.

11.6.3 Macular degeneration

OCTA has found usefulness in age-related macular degeneration (AMD), wet and dry variants as well as in other forms or variants of macular degeneration. Advanced forms of dry AMD feature flow voids, which are areas of absent flow, are visualized in the choriocapillaris. Flow voids are better visualized using SS-OCT compared to spectral domain (SD) OCT [27–30]. Figure 11.10 shows some areas of flow void, which are seen as minute areas of void present within the choriocapillaris layer. The cross-sectional OCT of this same eye shows PED and reveals drusenoid deposits seen on *en face* OCT imaging. The superficial and deep plexus are within normal limits.

One of the several benefits of OCTA has been imaging neovascularization in AMD. OCTA provides useful, clear, depth-resolved images of CNV. It is used to study subclinical 'quiescent' CNV. Furthermore, OCTA has helped monitor and understand the response of CNV to treatment with intravitreal antiVEGF [31–33]. We have used OCTA to evaluate CNV in AMD, PCV (figures 11.11(a)–(d)), and in association with Angioid Streaks, (figure 11.12). Studies in myopic CNV have also been done using OCTA. OCTA gives valuable information on treatment outcome, as illustrated in figure 11.11(a–d), which shows CNV before and after treatment with intravitreal Aflibercept. A valuable use of OCTA is its ability to localize the depth of the CNV and therefore enable typing of the CNV. Figure 11.2 is a type 1 CNV.

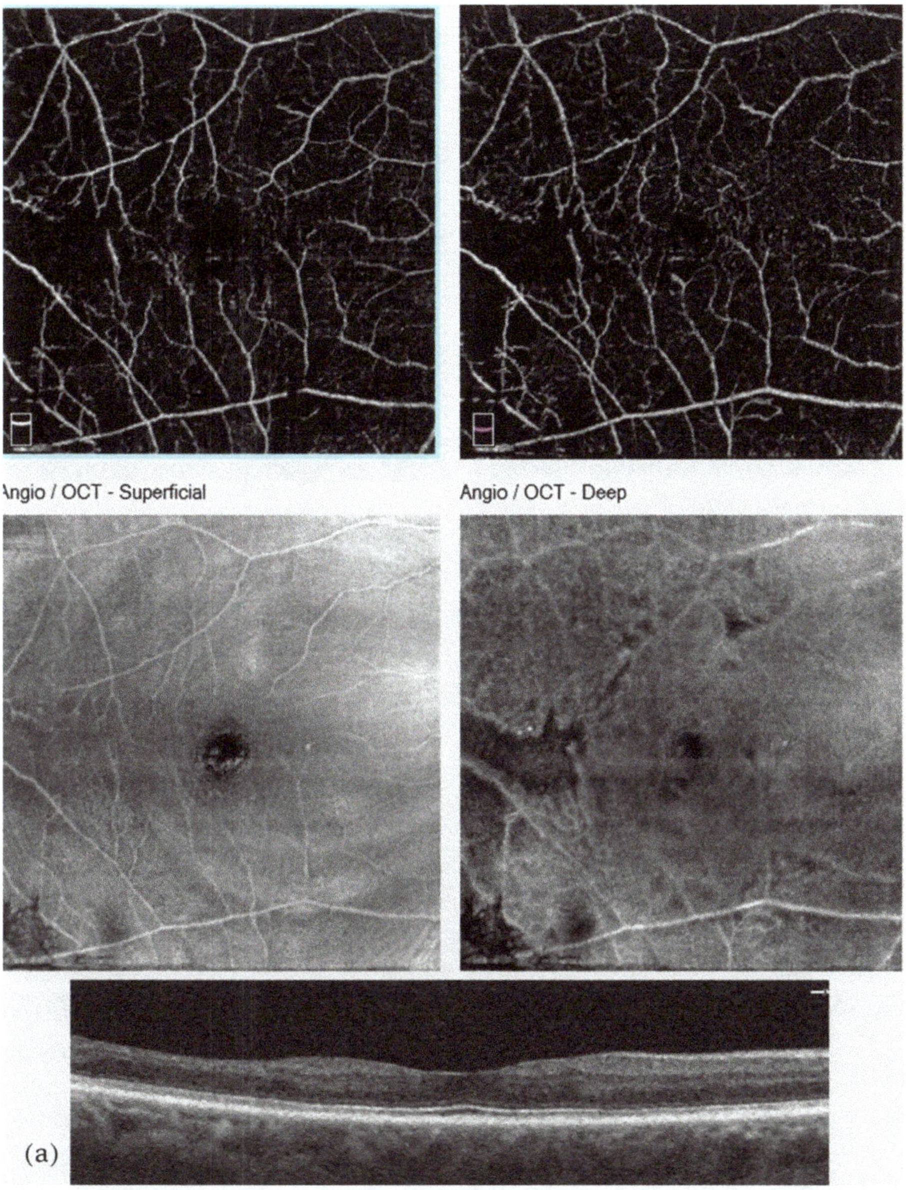

Figure 11.8. (a) Shows area of capillary drop out and non-perfusion, which extends into the FAZ in the superficial plexus; therefore, the FAZ is enlarged in this slab. A similar finding is noted in the deep capillary plexus, but the FAZ is smaller in size. The cross-sectional OCT shows a structural thinning of the inner retina in the area of non-perfusion. This is a case of diabetic retinopathy that presented with a featureless retina. (b) There is foveal perfusion, with normal vasculature in the inferior foveal area. The superior foveal vasculature and FAZ appear disorganized due to the cystic retina and hard exudates seen on *en face* and cross-sectional OCT.

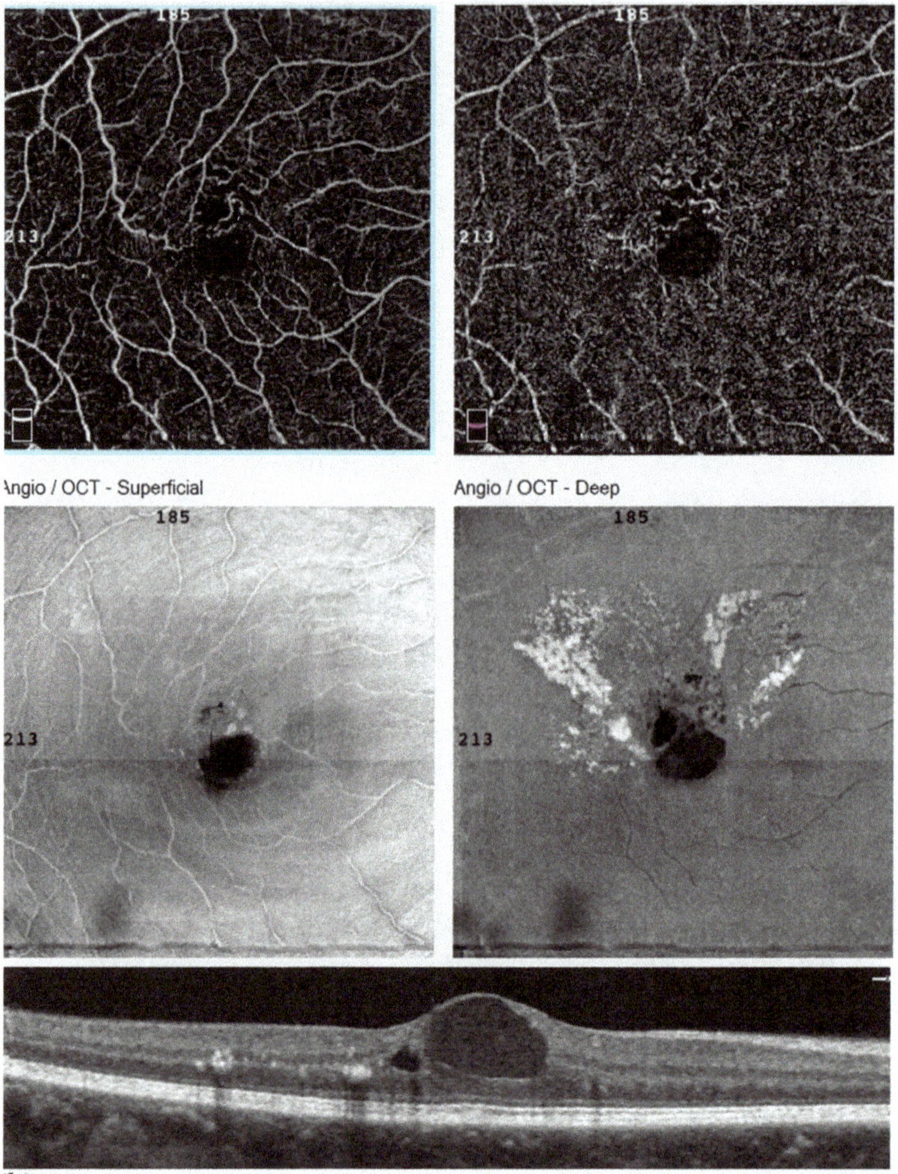

(b)

Figure 11.8. (Continued.)

Conversely, there is a presence of neovascularization in the outer retina and choriocapillaris scans in figure 11.11(a–d), indicative of a type 3 CNV.

11.6.4 Retinal vein occlusion

OCTA has found usefulness in RVO for assessing areas of retinal impaired perfusion, vascular changes such as dilatation and tortuosity, collateral vessels, cotton wool spots,

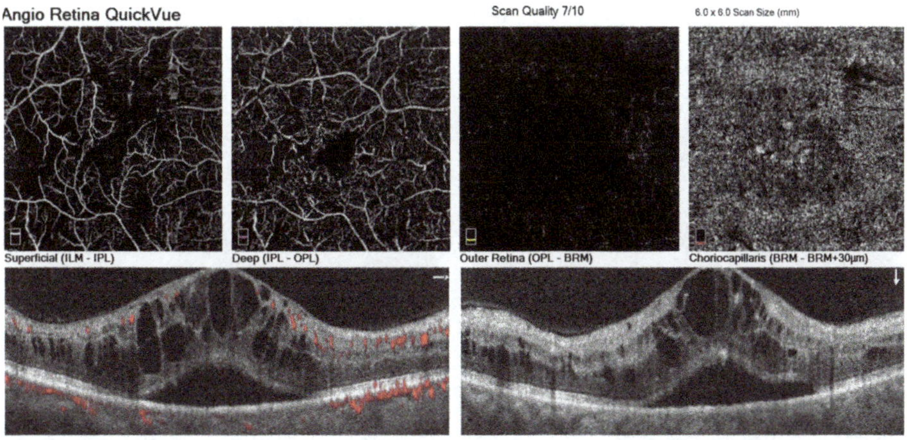

Figure 11.9. This is a clinical situation in which there is significant intraretinal and sub-retinal fluid, distortion of the macula seen on cross-sectional OCT. This presentation can result in challenges for automated machine segmentation and in segmentation artifacts. This segmentation difficulty can negatively impact the quality of the OCTA. The quality of this OCTA image is reduced to a score of 7. The superficial and deep plexus show macular non-perfusion and irregular and widened FAZ in diabetic retinopathy. Also, there are notable hyporeflective areas in the choriocapillaris layer; these are due to projection artifacts.

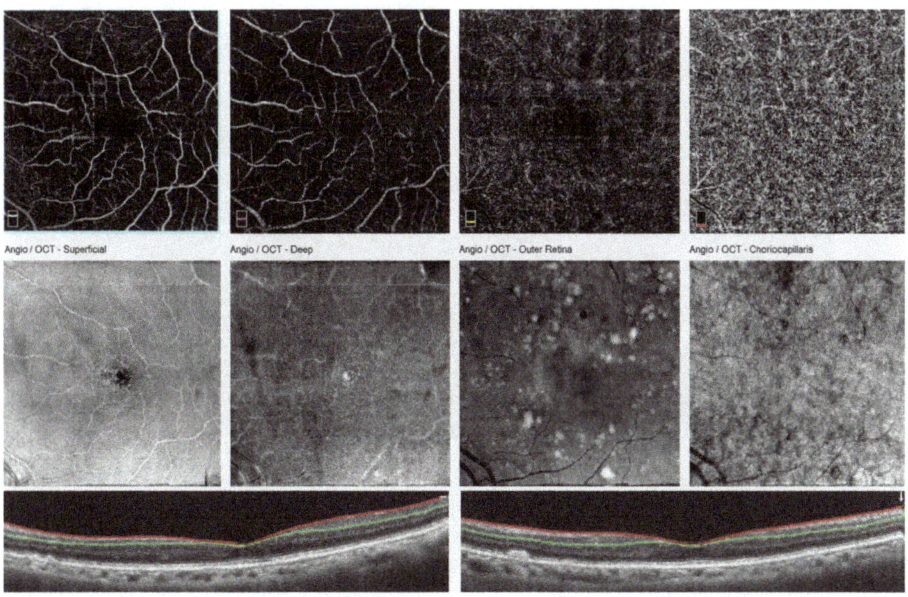

Figure 11.10. There are minute hyporeflective spots seen as areas of flow void within the choriocapillaris slab. This can be a feature of dry AMD. PED and drusen are noted on the cross-sectional and *en face* OCT images.

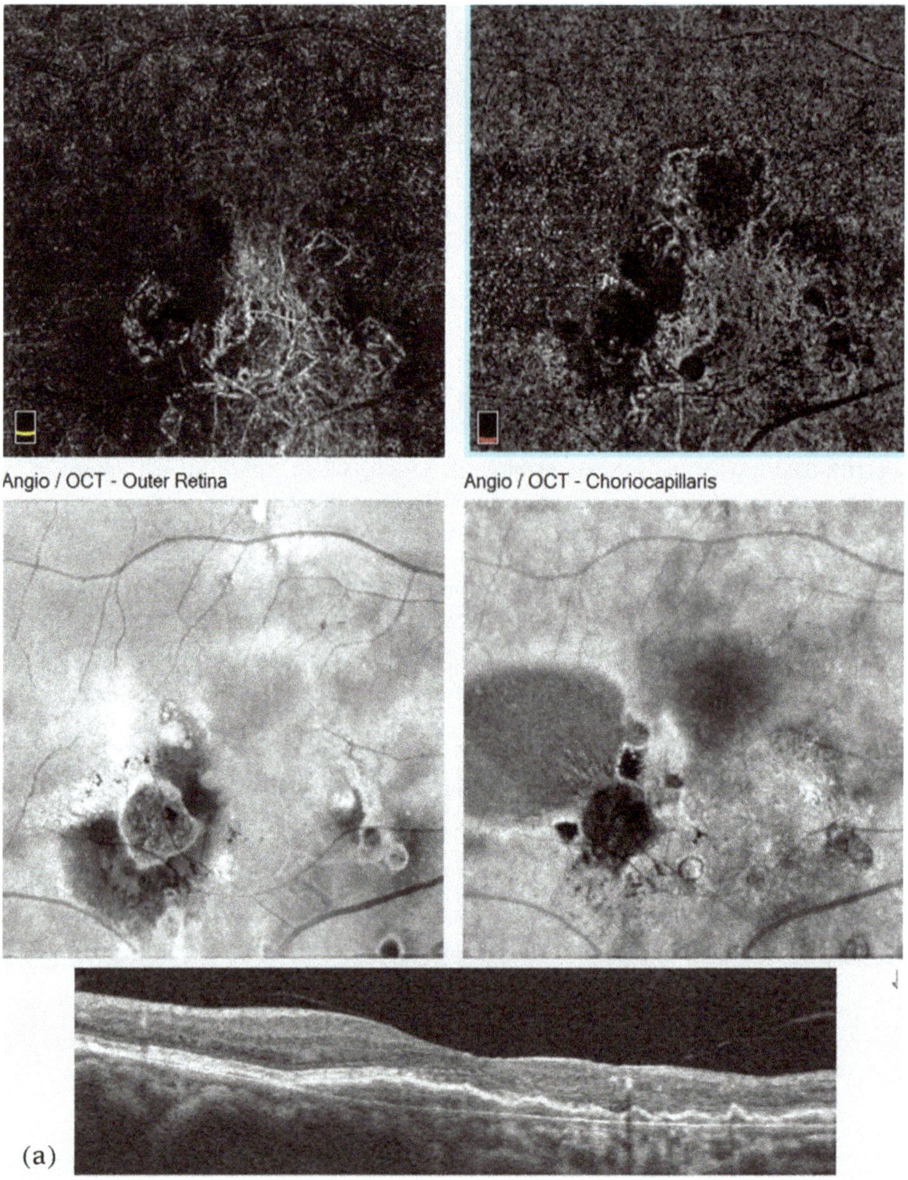

Figure 11.11. OCTA allows for monitoring the neovascular complex response to intravitreal injection of anti-vascular endothelial growth factors. (a) Neovascular complex at initial presentation; neovascular components are seen in the outer retina and choriocapillaris slabs. (b) Poor response to intravitreal injection of Bevacizumab; the neovascular complex was noted to increase in size compared to its size at initial presentation. (c) Improved response after switching treatment to monthly intravitreal Aflibercept; the components of the neovascular complex were noted to reduce in size significantly. (d) There is an increase in the neovascular complex's size following an increase of intravitreal Aflibercept treatment interval to every three months.

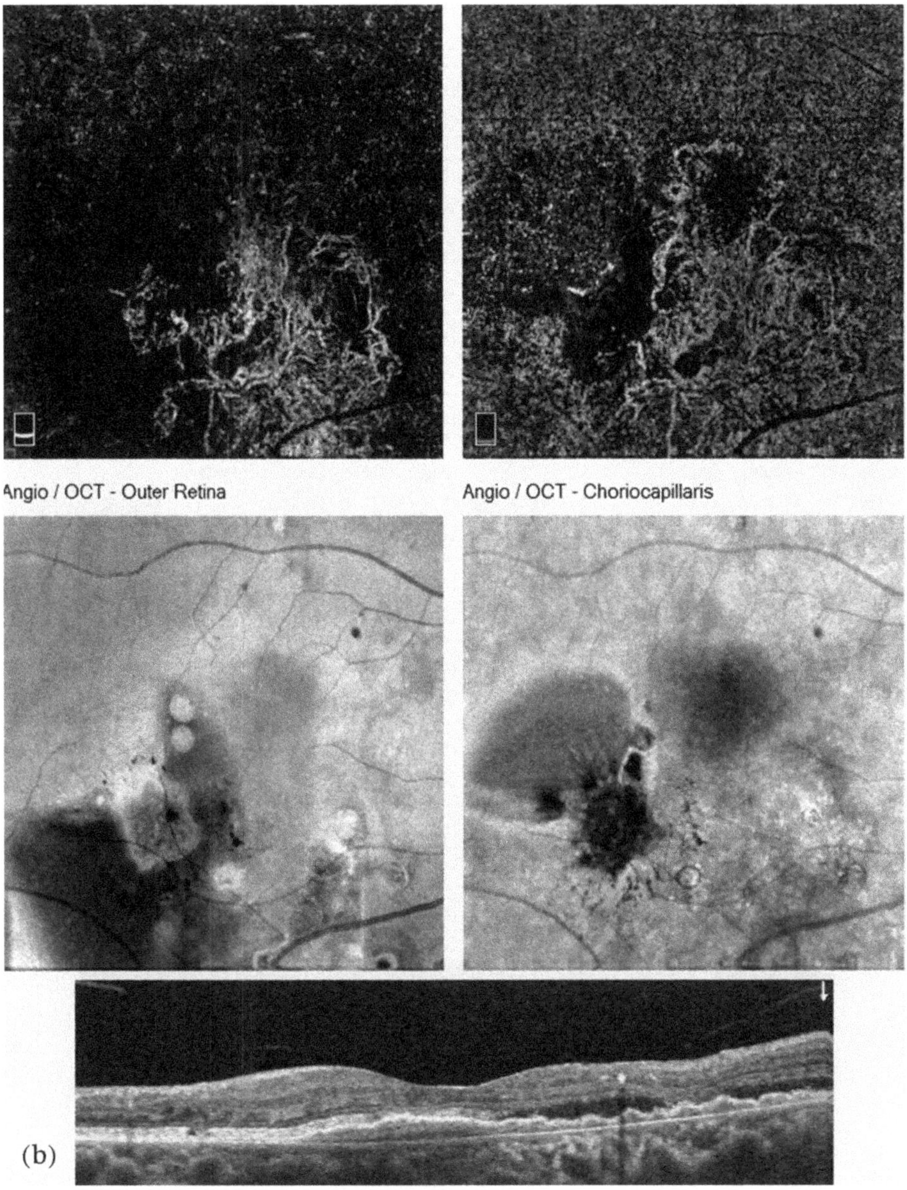

Figure 11.11. (Continued.)

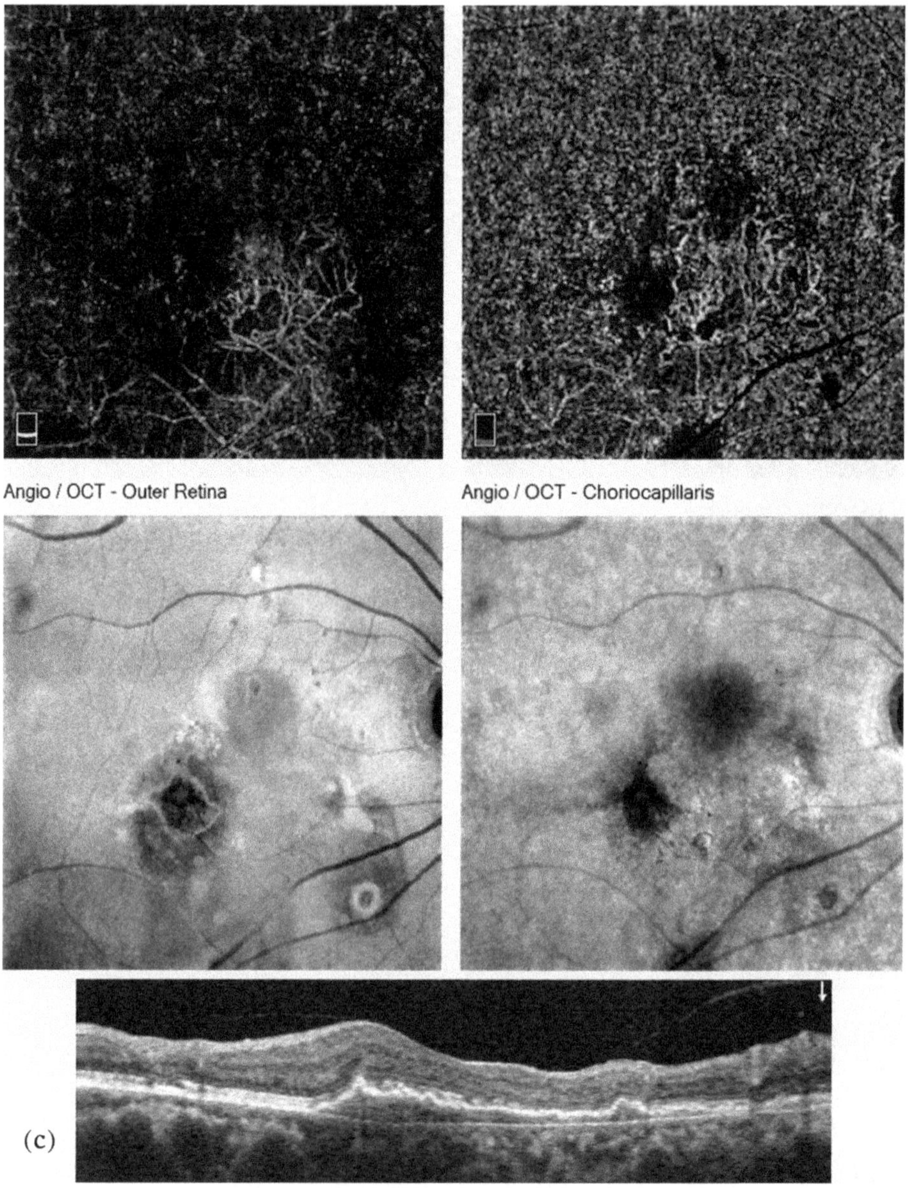

Figure 11.11. (Continued.)

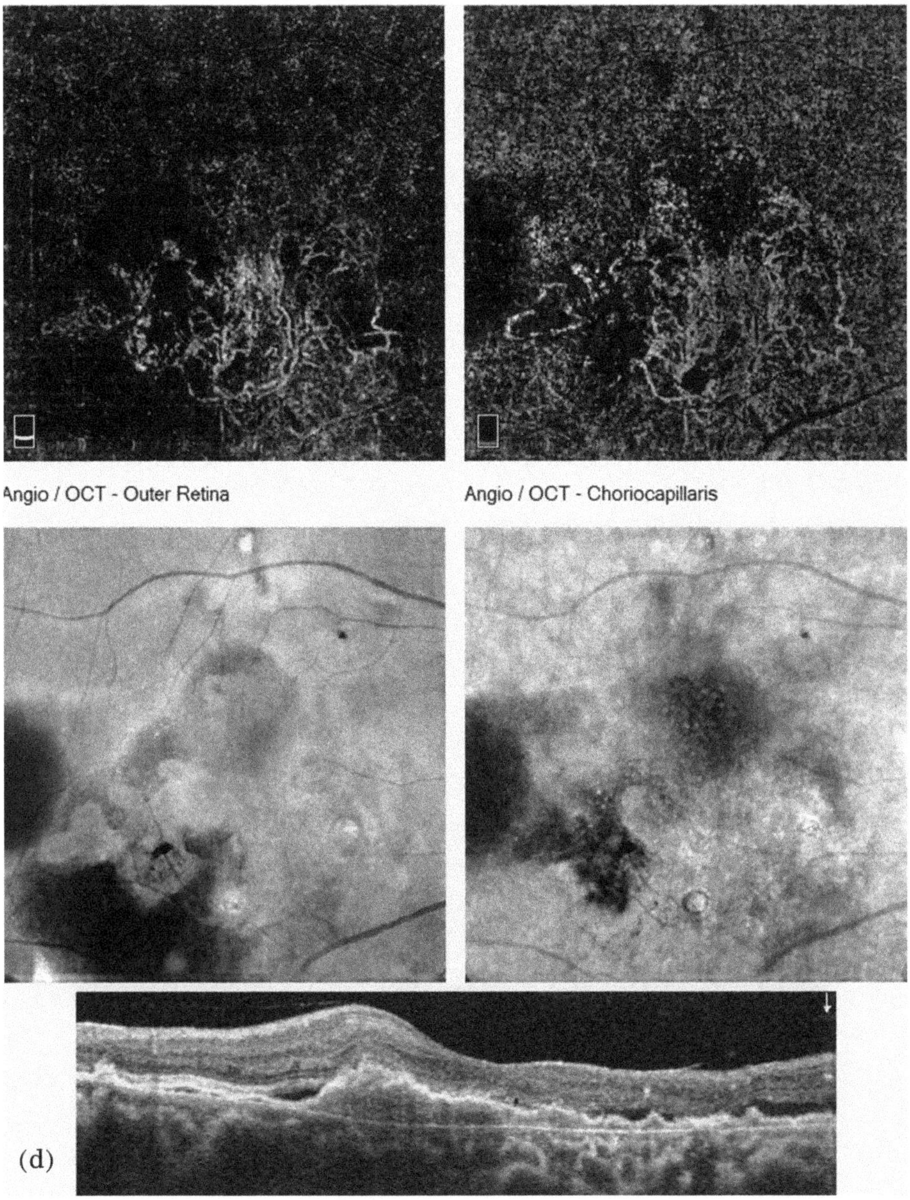

Figure 11.11. (Continued.)

retinal hemorrhages, increase in the areas of FAZ indicating foveal ischemia and neovascularization [34, 35]. The increasing severity of CRVO and BRVO is associated with decreased vessel density in both superficial and deep capillary plexus. Several RVO studies show a decrease in both superficial and deep capillary plexus density, FAZ enlargement, and microvascular abnormalities, as shown in (figures 11.13(a)–(c)). RVO is mostly associated with macular edema. As indicated earlier, macular edema, if

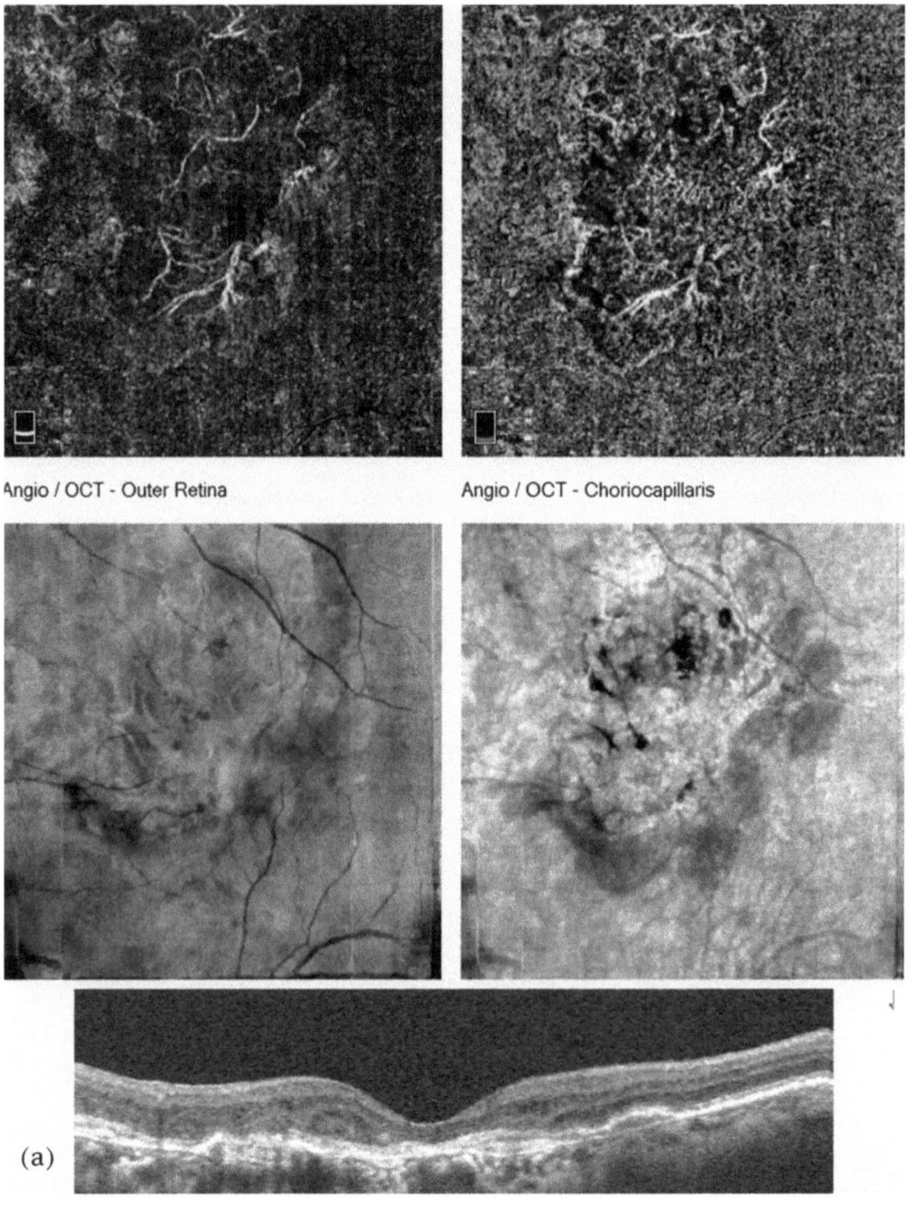

(a)

Figure 11.12. (a) Residual choroidal neovascular complex after multiple intravitreal Bevacizumab injections. The patient suffers from Angioid Streaks which can present with a subretinal neovascular membrane, macular hemorrhage, and scarring, as shown in the fundus photograph (b). Fundus photograph of the left eye showing macular scarring as a result of recurrent exudation and hemorrhage from CNVM, treated with multiple intravitreal antiVEGF. Notice the narrow irregular streaks radiating from the disc.

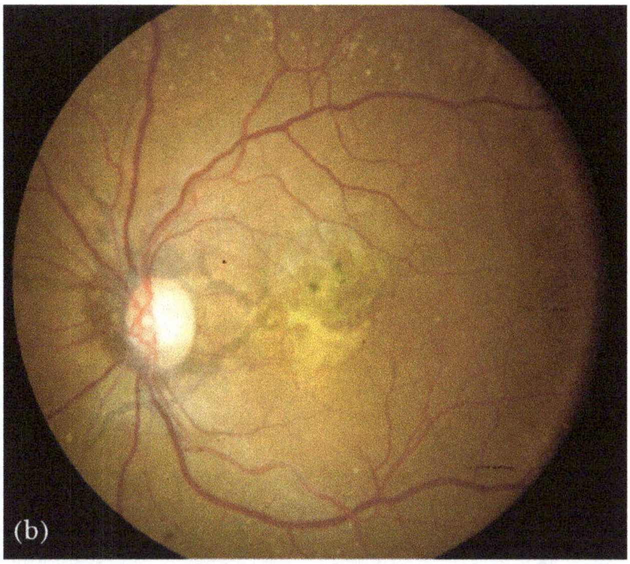

Figure 11.12. (Continued.)

present, affects OCTA image quality, and segmentation in the presence of macular edema becomes challenging. Figure 11.9 illustrates this difficulty in segmentation. Widefield FA provides valuable information concerning peripheral ischemia in RVO, which OCTA cannot do presently. It is hoped that SS-OCTA will offer peripheral imaging and equal widefield FA in the future. As in the case of DR, in the setting of enlarged FAZ and ME, it may indicate a negative prognostic outcome.

11.6.5 Retinal artery occlusion

RAO has been evaluated using OCTA. OCTA can demonstrate areas of vaso-occlusion and non-perfusion [36–38]. In one case report, non-perfusion on OCTA was similar to FA [39], figures 11.3(a) and (b). OCTA hasn't been evaluated in RAO as in RVO because of the lower incidence of RAO. Likely, the role of OCTA in evaluating RAO will evolve further. It presents a non-invasive method of acquiring frequent high-resolution images of the affected vasculature required in this disease, which is investigated for more effective treatment in those non-responsive to immediate anterior chamber paracentesis.

11.6.6 Vitreomacular interface disease

OCTA has been used for imaging pathology in the vitreoretinal interface. Epiretinal membrane (ERM) and macular holes (MHs) have been imaged using OCTA. One can see ERM with folds on the retina and tortuosity of the vessels shown in figure 11.14(a) due to the tractional forces on the retinal vasculature and inner retina. A change to less vessel tortuosity will be seen after surgery to remove the ERM. FAZ has been studied before and after macular hole repair and suggests a reduction in FAZ area after the successful macular hole closure [40, 41]. Figure 11.14(b) shows OCTA of MH.

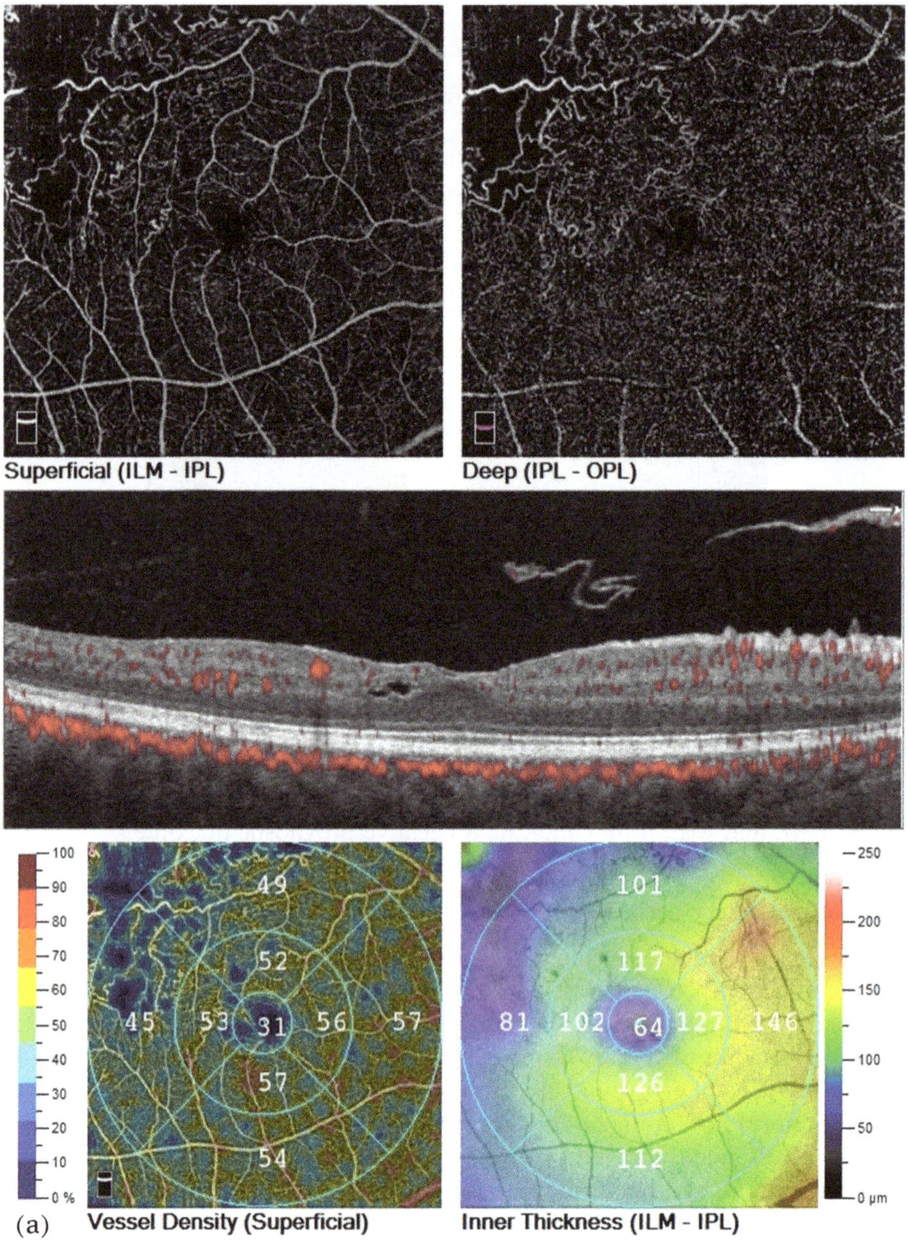

Figure 11.13. (a) Shows the OCTA of a right eye that has received intravitreal anti-vascular endothelial growth factor and scatter retinal laser (b) for treating superior-temporal branch vein occlusion. The superficial and deep segments of the OCTA show non-perfusion in the supero-temporal quadrant and tortuosity of the vessels in this area. Also, the vessel density in this region is reduced. Collateral vessels which can be appreciated in the fundus photograph (b), can also be seen in the superficial and deep slabs of the OCTA. Fundus photo of a right eye with superior branch vein occlusion. Notice the ghost vessels, collateral circulation and retinal laser scars. (c) Shows inferior branch retinal vein occlusion, OCTA demonstrating vascular non-perfusion area with inferior extension into the FAZ in the superficial slab.

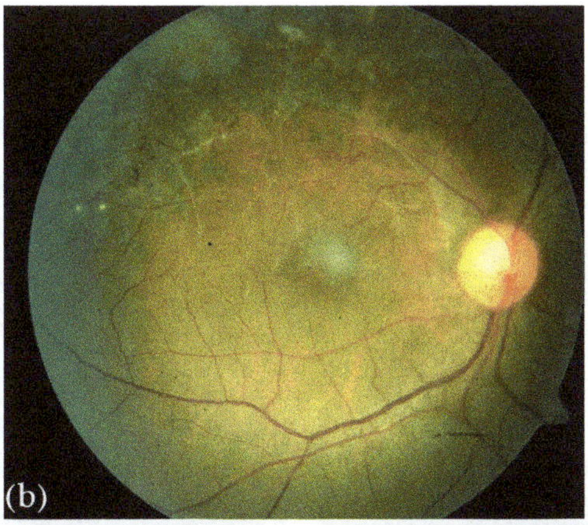

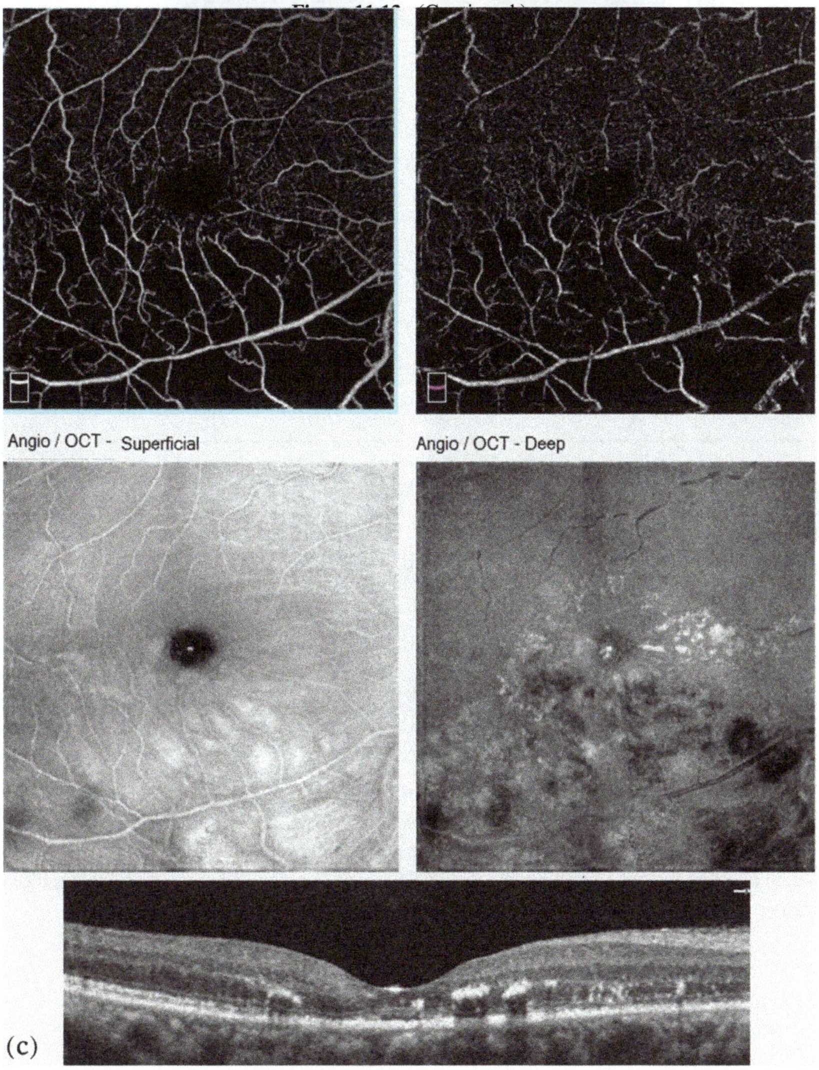

(c)

Figure 11.13. (Continued.)

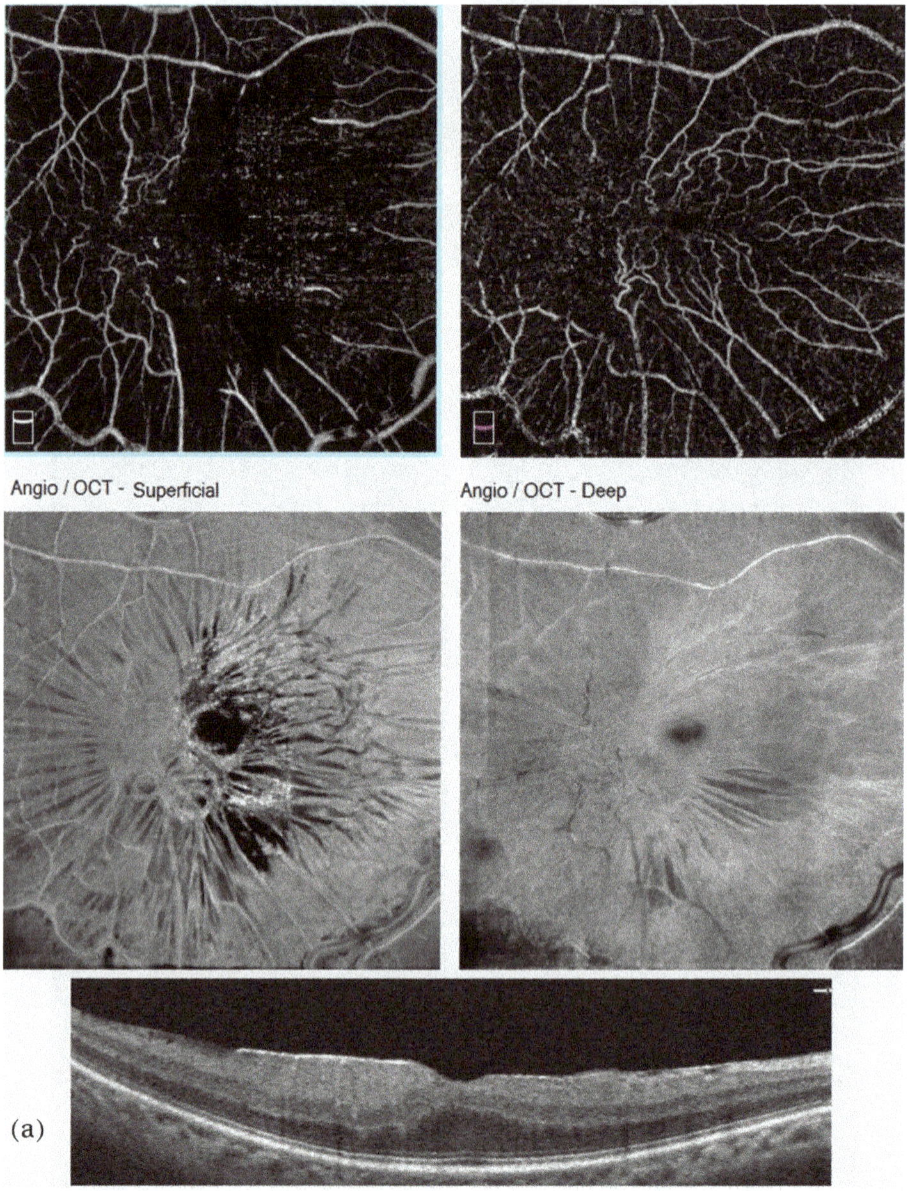

Figure 11.14. (a) Notice tortuosity and traction on the retinal vessels, which depicts the presence of ERM seen on the deep capillary plexus of the OCTA. The *en face* and cross-sectional OCT scans highlight the ERMs nicely. In this case, the FAZ in the deep plexus is irregular and appears smaller than usual because of the ERM-induced traction, further approximating the foveal vessels closer to one another. (b) Shows the OCTA of a full-thickness macular hole (FTMH) with normal-sized FAZ and macular perfusion.

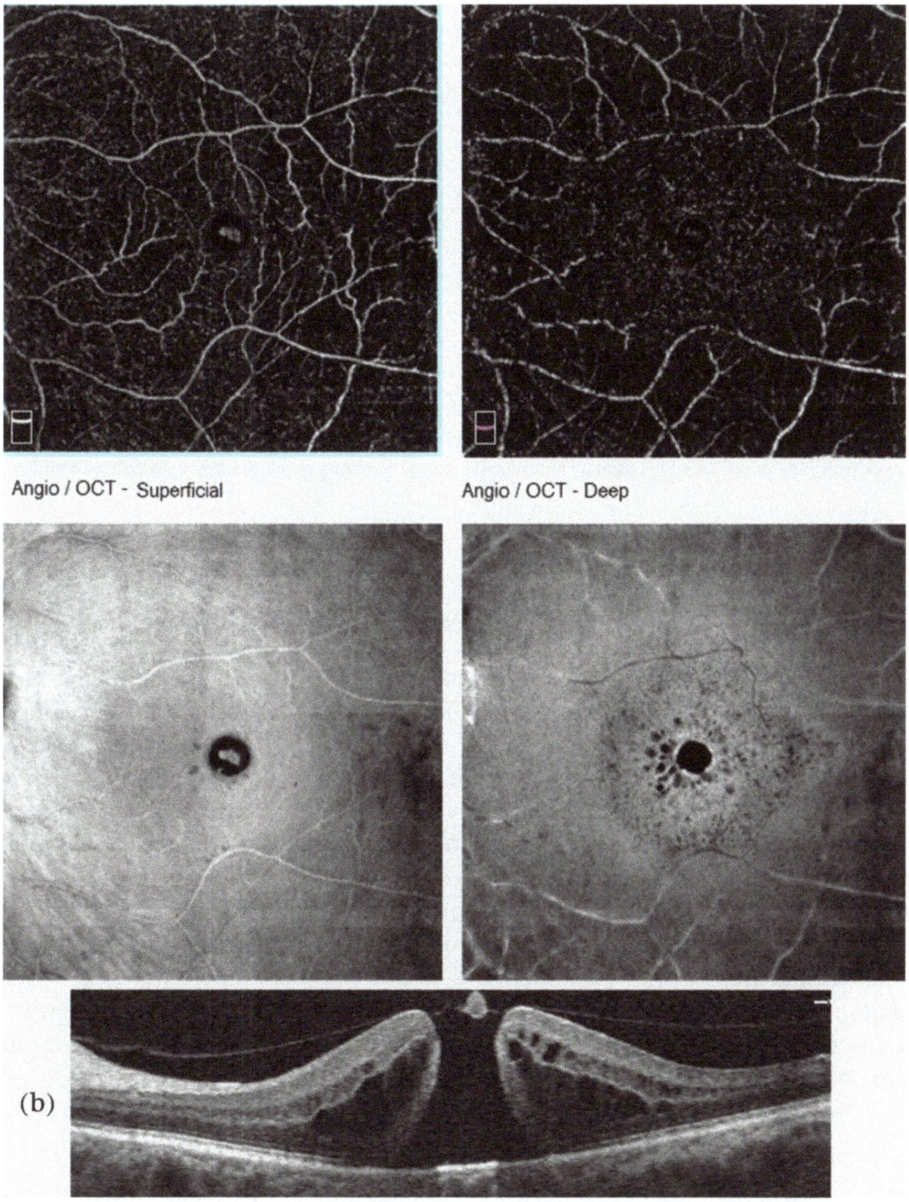

Figure 11.14. (Continued.)

11.6.7 Glaucomatous optic neuropathy (GON)

OCTA in established glaucoma showed a reduction in perfusion in the optic nerve head, peripapillary retina, which suggests that OCTA can be a useful addition in the investigations for the early detection and monitoring of glaucoma. Already imaging of the peripapillary vessel density using OCTA shown in figure 11.4, (which is not

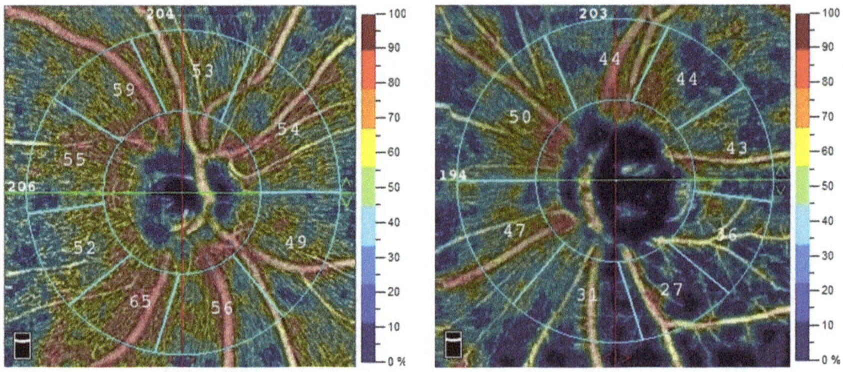

(a) (b)

Figure 11.15. (a) Shows the peripapillary vessel density of a normal-appearing right eye, and one should contrast this with the marked reduction in peripapillary vessel density of the left eye (b) of the same patient. This peripapillary vessel density is used as an objective metric for monitoring disease progression in glaucoma eyes.

imaged using FA) has been used to monitor glaucoma progression with high reproducibility and sensitivity for glaucoma progression [42–44].

The disc flow index and the optic disc's vessel density are used as two reproducible metrics obtainable using OCTA to evaluate GON [45]. A correlation has been observed between perfusion parameters and glaucoma stage severity, RNFL thickness, ganglion cell complex thickness, and visual field mean deviation in open-angle glaucoma patients. Other authors have confirmed this finding. Significantly reduced peripapillary flow index and peripapillary vessel density have been demonstrated in glaucomatous eyes compared to normal eyes, with vital diagnostic accuracy, repeatability, and reproducibility. Figures 11.15(a) and (b) depict an OCTA of a normal eye and a glaucomatous eye of the same patient. Notice a significant difference in the peripapillary vessel density between the expected normal and glaucomatous eye.

The peripapillary vessel density's diagnostic ability, particularly the inferotemporal sector, in both primary open-angle glaucoma and primary angle-closure glaucoma is of similar diagnostic accuracy to the RNFL.

11.6.8 Miscellaneous

OCTA has also been used to image other vascular and non-vascular retinal diseases, including sickle cell retinopathy [46, 47] (figure 11.16) and retinal dystrophies such as retinitis pigmentosa [48], and for monitoring the outcome of the surgery [49].

11.7 Discussion

Undoubtedly OCTA is innovation at its best and has provided a quantitative assessment of retinal and choroidal vasculature in a way that has not been done before. Since one can make image acquisition in seconds, it gives the possibility

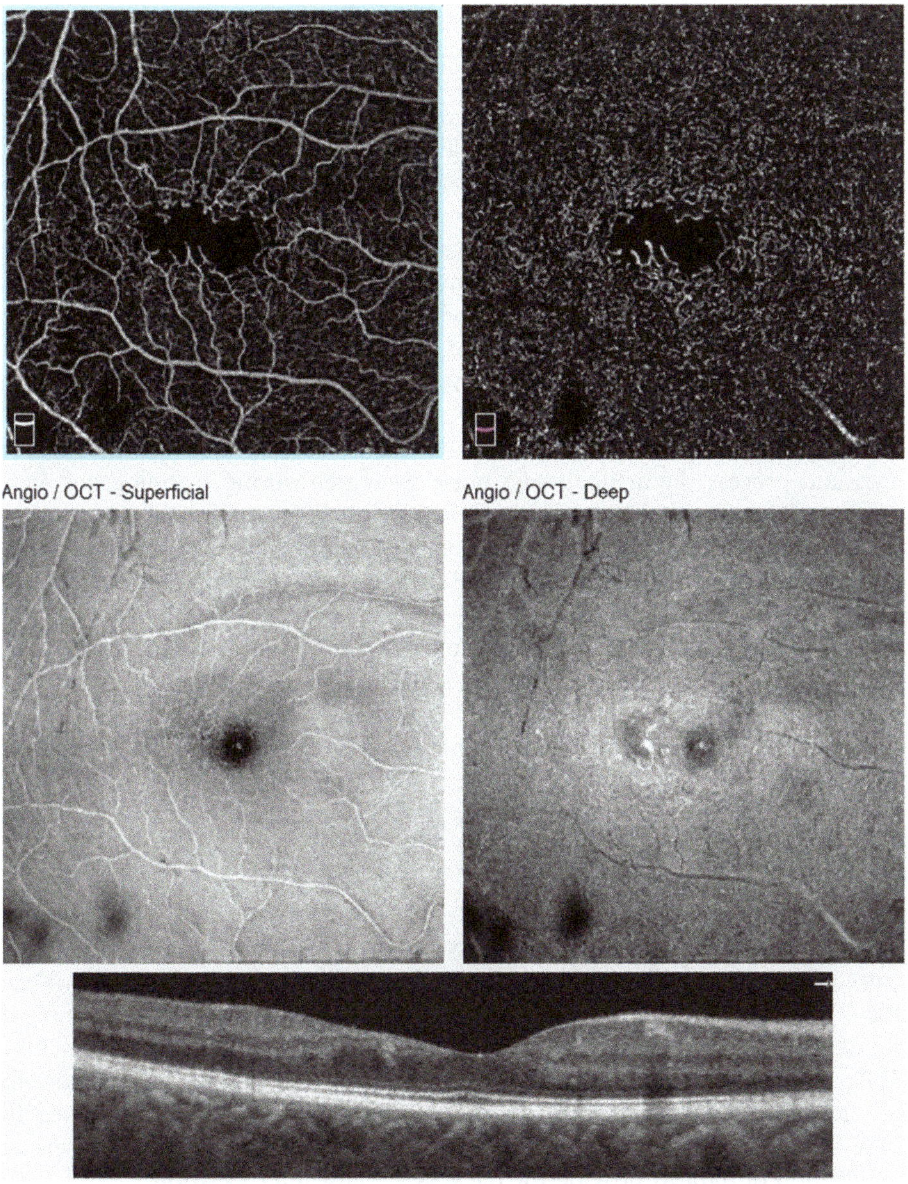

Figure 11.16. OCTA of an eye diagnosed to have proliferative sickle cell retinopathy (PSCR). This proliferative disease manifests mostly as a peripheral disease with the formation of peripheral sea fan neovascularization. However, an enlarged FAZ can be seen on OCTA, as demonstrated in this case. This could be indicative of perifoveal vaso-occlusion that may occur in some of these eyes.

for multiple imaging and, therefore a unique opportunity to understand vascular pathology and interventions in more detail. Yet, there are several drawbacks to this technology.

Currently, OCTA is not useful in the study of peripheral vascular pathology in the way that conventional FA has been. Peripheral vascular imaging using OCTA is a significant unmet need since several diseases, including retinal vein occlusions, sickle cell retinopathy, and other peripheral vascular pathology, are currently investigated using widefield angiography. Expected advances in OCT technology, with the availability of faster SS scans, may soon offer widefield OCTA. Physicians appear to prefer images acquired using SS-OCTA over SD-OCTA [22]. Preference for images using faster scanning OCTs means that the future of OCTA may tilt more towards the SS device.

OCTA cannot evaluate dynamic changes such as leakage, pooling, and staining, which are essential features seen with conventional angiography, FA, and ICG. It seems rational to expect that OCTA use will be limited in conditions in which these dynamic phenomena will be evaluated; preferably, FA will be used instead. Therefore, OCTA and FA can be complimentary in evaluating several retinal pathologies.

Commercially available OCTA machines have different software that resolves images and process the artifacts and noise generated during scanning and image acquisition. One can expect that there will be further improvements in current OCTA software to make the process even faster and image segmentation more accurate and precise. Efficient removal of artifacts will enable even better imaging and analysis of even deeper layers of the choroid and provide useful information concerning disease effect. Yet OCTA offers an enormous amount of data that needs resolving. One of the unmet needs in OCTA technology is the standardization of the data across different imaging platforms. As has been suggested, there is a need to standardize the flow within imaged vascular tissue, such as choroidal neovascular membranes (CNVs).

Furthermore, in the clinical application of OCTA technology, there is a need for OCTA metrics standardization. OCTA metrics or OCTA analytics refer to numerical parameters that describe the vascular network structure as obtained using OCTA. Some of the metrics that have been proposed for definition and standardization include perfusion density, vessel density, blood flow index, skeletonized density, vessel length density, and fractal dimensions, which are all being studied. However, several have not achieved standardized nomenclature yet. Available OCTA technology must necessarily achieve further improvements in OCTA image quality, accurate software segmentation, and standardization before usage in clinical trials for retinovascular diseases. Researchers have argued that OCTA should evolve into providing fast and reliable 3D representations and quantification of data to provide for the vessel image interpretation.

Despite these limitations of OCTA, it is expected that this technology's evolution will benefit the appreciation of retinovascular pathology and herald the use of ocular intravascular surgical techniques. The ability to provide three-dimensional image cubes will help the accuracy of localization of retinal blood vessels. As the evolution of OCT to intraoperative devices, i.e., intraoperative OCT (iOCT), it is likely OCTA will someday be used during retinal endovascular surgery. Combining OCTA and other imaging techniques such as adaptive optics (AO) will provide even greater image resolution and understanding of intravascular blood flow [50].

11.8 Conclusion

OCTA is a rapidly evolving technology that has gained acceptance within the ophthalmic community. It has benefited from advances in OCT, especially the availability of faster scans. Hardware and software refinements will likely see even more improvements in this technology. Standardization of software between the various available technologies will improve widespread utility and extrapolation of data from one source to another. Like most recent advances in ophthalmology, patient care will improve significantly over the coming years through OCTA's benefits.

References

[1] de Carlo T E, Romano A and Waheed N K *et al* 2015 A review of optical coherence tomography angiography (OCTA) *Int. J. Retin. Vitr. 1* **5**

[2] Spaide R F, Fujimoto J G, Waheed N K, Sadda S R and Staurenghi G 2018 Optical coherence tomography angiography *Prog. Retin. Eye Res.* **64** 1–55

[3] Fang P P, Harmening W M, Müller P L, Lindner M, Krohne T U and Holz F G 2016 Technische Grundlagen der OCT-Angiographie [Technical principles of OCT angiography] *Ophthalmologe.* **113** 6–13 (in German)

[4] Kashani A H, Chen C L, Gahm J K, Zheng F, Richter G M, Rosenfeld P J, Shi Y and Wang R K 2017 Optical coherence tomography angiography: A comprehensive review of current methods and clinical applications *Prog. Retin. Eye Res.* **60** 66–100

[5] Maram J, Srinivas S and Sadda S 2017 Evaluating ocular blood flow *Indian J. Ophthalmol* **65** 337–46

[6] Rocholz R, Corvi F, Weichsel J, Schmidt S and Staurenghi G 2019 OCT Angiography (OCTA) in Retinal Diagnostics ed J Bille *High Resolution Imaging in Microscopy and Ophthalmology* (Cham: Springer)

[7] Borrelli E, Parravano M, Sacconi R, Costanzo E, Querques L, Vella G, Bandello F and Querques G 2020 Guidelines on optical coherence tomography angiography imaging: 2020 focused update *Ophthalmol. Ther.* **9** 697–707

[8] Coscas G, Lupidi M and Coscas F 2016 Image analysis of optical coherence tomography Angiography *Dev. Ophthalmol* **56** 30–6

[9] Agemy S A *et al* 2015 Retinal vascular perfusion density mapping using optical coherence tomography angiography in normals and diabetic retinopathy patients *Retina* **35** 2353–63

[10] Spaide R F, Klancnik J M Jr and Cooney M J 2015 Retinal vascular layers imaged by fluorescein angiography and optical coherence tomography angiography *JAMA Ophthalmol* **133** 45–50

[11] Jia Y *et al* 2012 Split-spectrum amplitude-decorrelation angiography with optical coherence tomography *Opt. Express* **20** 4710–25

[12] Chalam K V and Sambhav K 2016 Optical coherence tomography angiography in retinal diseases *J. Ophthalmic Vis. Res* **11** 84–92

[13] Zhi Z, Qin W, Wang J, Wei W and Wang R K 2015 4D optical coherence tomography-based micro-angiography achieved by 1.6-MHz FDML swept source *Opt. Lett.* **40** 1779–82 001779

[14] Rosenfeld P J, Durbin M K, Roisman L, Zheng F, Miller A, Robbins G, Schaal K B and Gregori G 2016 ZEISS Angioplex™ spectral domain optical coherence tomography angiography: technical aspects *Dev. Ophthalmol* **56** 18–29

[15] Stanga P E, Tsamis E, Papayannis A, Stringa F, Cole T and Jalil A 2016 Swept-source optical coherence tomography Angio™ (Topcon Corp, Japan): technology review *Dev. Ophthalmol* **56** 13–7
[16] Liu L, Gao S S, Bailey S T, Huang D, Li D and Jia Y 2015 Automated choroidal neovascularization detection algorithm for optical coherence tomography angiography *Biomed. Opt. Express* **6** 3564–76
[17] Spaide R F, Fujimoto J G and Waheed N K 2015 Image artifacts in optical coherence tomography angiography *Retina* **35** 2163–80
[18] Uji A, Balasubramanian S and Lei J *et al* 2018 Multiple enface image averaging for enhanced optical coherence tomography angiography imaging *Acta Ophthalmol* **96** 820–7
[19] Uji A, Balasubramanian S, Lei J, Baghdasaryan E, Al-Sheikh M and Sadda S 2017 Impact of multiple en face image averaging on quantitative assessment from optical coherence tomography angiography images *Ophthalmology* **124** P944–52
[20] Chan G, Balaratnasingam C, Yu P K, Morgan W H, McAllister I L, Cringle S J and Yu D Y 2012 Quantitative morphometry of perifoveal capillary networks in the human retina *Invest. Ophthalmol. Vis. Sci.* **53** 5502–14
[21] Rocholz R, Teussink M M, Dolz-Marco R, Holzhey C, Dechent J, Tafreshi A and Schulz S 2018 SPECTRALIS optical coherence tomography angiography (OCTA): principles and clinical applications *Heidelberg Engineering Academy* https://academy.heidelbergengineering.com/course/view.php?id=505
[22] Agarwal A, Grewal D S, Jaffe G J, Stewart M W, Srivastava S and Gupta V 2018 Current role of optical coherence tomography angiography: expert panel discussion *Indian J. Ophthalmol* **66** 1696–9
[23] Tey K Y, Teo K, Tan A C S, Devarajan K, Tan B, Tan J, Schmetterer L and Ang M 2019 Optical coherence tomography angiography in diabetic retinopathy: a review of current applications *Eye Vis. (Lond.)* **6** 37
[24] Khadamy J, Abri Aghdam K and Falavarjani K G 2018 An update on optical coherence tomography angiography in diabetic retinopathy *J. Ophthalmic Vis. Res* **13** 487–97
[25] Gildea D 2019 The diagnostic value of optical coherence tomography angiography in diabetic retinopathy: a systematic review *Int. Ophthalmol* **39** 2413–33
[26] Falavarjani K G, Iafe N A, Hubschman J-P, Tsui I, Sadda S R and Sarraf D 2017 Optical coherence tomography angiography analysis of the foveal avascular zone and macular vessel density after anti-VEGF therapy in eyes with diabetic macular edema and retinal vein OcclusionOCTA After Intravitreal Injection *Invest. Ophthalmol. Vis. Sci.* **58** 30–4
[27] Arya M, Sabrosa A S, Duker J S and Waheed N K 2018 Choriocapillaris changes in dry age-related macular degeneration and geographic atrophy: a review *Eye Vis. (Lond.)* **5** 22
[28] Moreira-Neto C A, Moult E M, Fujimoto J G, Waheed N K and Ferrara D 2018 Choriocapillaris loss in advanced age-related macular degeneration *J. Ophthalmol* **2018** 8125267
[29] Waheed N K, Moult E M, Fujimoto J G and Rosenfeld P J 2016 Optical coherence tomography angiography of dry age-related macular degeneration *Dev. Ophthalmol* **56** 91–100
[30] Müller P L, Pfau M, Schmitz-Valckenberg S, Fleckenstein M and Holz F G 2020 OCT-angiography in geographic atrophy *Ophthalmologica* **244** 42–50

[31] Perrott-Reynolds R, Cann R, Cronbach N, Neo Y N, Ho V, McNally O, Madi H A, Cochran C and Chakravarthy U 2019 The diagnostic accuracy of OCT angiography in naive and treated neovascular age-related macular degeneration: a review *Eye (Lond)* **33** 274–82
[32] Malamos P, Tsolkas G, Kanakis M, Mylonas G, Karatzenis D, Oikonomopoulos N, Lakoumentas J and Georgalas I 2017 OCT-angiography for monitoring and managing neovascular age-related macular degeneration *Curr. Eye Res* **42** 1689–97
[33] Coscas G, Lupidi M, Coscas F, Français C, Cagini C and Souied E H 2015 Optical coherence tomography angiography during follow-up: qualitative and quantitative analysis of mixed type I and II choroidal neovascularization after vascular endothelial growth factor trap therapy *Ophthalmic Res* **54** 57–63
[34] Tsai G, Banaee T, Conti F F and Singh R P 2018 Optical coherence tomography angiography in eyes with retinal vein occlusion *J. Ophthalmic Vis. Res* **13** 315–32
[35] Moussa M, Leila M, Bessa A S, Lolah M, Abou Shousha M, El Hennawi H M and Hafez T A 2019 Grading of macular perfusion in retinal vein occlusion using en-face swept-source optical coherence tomography angiography: a retrospective observational case series *BMC Ophthalmol* **19** 127
[36] Baumal C R 2016 Optical coherence tomography angiography of retinal artery occlusion *Dev. Ophthalmol* **56** 122–31
[37] Bonini Filho M A *et al* 2015 Optical coherence tomography angiography in retinal artery occlusion *Retina* **35** 2339–46
[38] de Castro-Abeger A H, de Carlo T E, Duker J S and Baumal C R 2015 Optical coherence tomography angiography compared to fluorescein angiography in branch retinal artery occlusion *Ophthalmic Surg. Lasers Imaging Retin.* **46** 1052–4
[39] Okonkwo O N, Hassan A O, Akanbi T, Umeh V C and Ogunbekun O O 2021 Vitrectomy and manipulation of intraocular and arterial pressures for the treatment of nonarteritic central retinal artery occlusion *Taiwan J. Ophthalmol.* [Epub ahead of print]
[40] Wilczyński T, Heinke A, Niedzielska-Krycia A, Jorg D and Michalska-Małecka K 2019 Optical coherence tomography angiography features in patients with idiopathic full-thickness macular hole, before and after surgical treatment *Clin. Interv. Aging* **14** 505–14
[41] Kita Y, Inoue M, Kita R, Sano M, Orihara T, Itoh Y, Hirota K, Koto T and Hirakata A 2017 Changes in the size of the foveal avascular zone after vitrectomy with internal limiting membrane peeling for a macular hole *Jpn. J. Ophthalmol* **61** 465–71
[42] Chansangpetch S and Lin S C 2018 Optical coherence tomography angiography in glaucoma care *Curr. Eye Res* **43** 1067–82
[43] Kwon H J, Kwon J and Sung K R 2019 Additive role of optical coherence tomography angiography vessel density measurements in glaucoma diagnoses *Korean J. Ophthalmol* **33** 315–25
[44] Chung J K, Hwang Y H, Wi J M, Kim M and Jung J J 2017 Glaucoma diagnostic ability of the optical coherence tomography angiography vessel density parameters *Curr. Eye Res* **42** 1458–67
[45] Yarmohammadi A *et al* 2016 Optical coherence tomography angiography vessel density in healthy, glaucoma suspect, and glaucoma eyes *Invest. Ophthalmol. Vis. Sci.* **57** OCT451–9
[46] Alam M, Thapa D, Lim J I, Cao D and Yao X 2017 Quantitative characteristics of sickle cell retinopathy in optical coherence tomography angiography *Biomed. Opt. Express* **8** 1741–53

[47] Han I C, Linz M O, Liu T Y A, Zhang A Y, Tian J and Scott A W 2018 Correlation of ultra-widefield fluorescein angiography and OCT angiography in sickle cell retinopathy *Ophthalmol. Retin.* **2** 599–605
[48] Mastropasqua R *et al* 2020 Widefield swept source OCTA in retinitis pigmentosa *Diagnostics (Basel)* 10 50
[49] Okonkwo O N, Sibanda D, Akanbi T and Hassan A O 2020 Natural history of a large bubble of migratory submacular perfluorocarbon *Int. Med. Case Rep. J.* **13** 477–86
[50] Camino A, Zang P, Athwal A, Ni S, Jia Y, Huang D and Jian Y 2020 Sensorless adaptive-optics optical coherence tomographic angiography *Biomed. Opt. Express* **11** 3952–67

IOP Publishing

Photo Acoustic and Optical Coherence Tomography Imaging, Volume 3
Angiography: an application in vessel imaging
Ayman El-Baz and Jasjit S Suri

Chapter 12

A comprehensive survey on computer-aided diagnostic systems in diabetic retinopathy screening

Meysam Tavakoli and Patrick Kelley

Diabetes mellitus (DM) can lead to significant microvasculature disruptions that eventually cause diabetic retinopathy (DR), or complications in the eye due to diabetes. If left unchecked, this disease can increase over time and eventually cause complete vision loss. The general method to detect such optical developments is through examining the vessels, optic nerve head, microaneurysms, haemorrhage, exudates, etc from retinal images. Ultimately this is limited by the number of experienced ophthalmologists and the vastly growing number of DM cases. To enable earlier and efficient DR diagnosis, the field of ophthalmology requires robust computer aided diagnosis (CAD) systems. Our review is intended for anyone, from student to established researcher, who wants to understand what can be accomplished with CAD systems and their algorithms to modeling and where the field of retinal image processing in computer vision and pattern recognition is headed. For someone just getting started, we place a special emphasis on the logic, strengths and shortcomings of different databases and algorithms frameworks with a focus on very recent approaches.

Important acronyms

- **Diabetes mellitus**: DM
- **Diabetic retinopathy**: DR
- **Non-proliferation diabetic retinopathy**: NPDR
- **Proliferation diabetic retinopathy**: PDR
- **Microaneurysm**: MA
- **Haemorrhage**: HE

doi:10.1088/978-0-7503-2060-3ch12

- **Exudate**: EXs
- **Cotton wool spots**: CWS
- **Neovascularization**: NV
- **Macular edema**: ME
- **Field-of-view**: FOV
- **Computer-aided diagnosis**: CAD
- **Optic nerve head**: ONH
- **Receiver operating characteristic**: ROC
- **Area under the curve**: AUC
- **k-nearest neighbor**: kNN
- **Support vector machine**: SVM
- **Neural network**: NN
- **Artificial intelligence**: AI
- **Machine learning**: ML
- **Deep learning**: DL
- **Convolution neural network**: CNN
- **Recurrent neural network**: RNN
- **Autoencoder**: AE

12.1 Introduction

The rapidly rising diabetes mellitus (DM) is one of the major issues of our recent health care [1]. The number of people affected by DM continues to increase at an alarming rate [2, 3]. The World Health Organization estimates for the next 25 years the number of diabetic people to grow from 130 to 350 million [4]. In the same direction, in the time between 2000 and 2030 another source anticipates the spread of the DM population worldwide to increase from 2.8% to 4.4% [5]. The real catch is that just half of diabetic people are diagnosed with the disease [6]. From a medical aspect viewpoint, DM leads to serious late complications such as macro and micro vascular changes which cause diabetic retinopathy (DR), heart disease, and renal problems [7, 8]. In DM, because of insulin insufficiency there is incapacitated metabolism of glucose; this in turn leads to damage and complications in blood vessels [9]. According to a study in the United States, DM is among the five lethal diseases, and yet there is no treatment for it. In the same study, the total yearly expense to diagnose and treat DM in 2002 was estimated to be $132 billion [10]. In fact, unmanaged DM and its complication leads to different malfunctions and death.

DR is a serious issue stemming from DM that affects the eye and is a main cause of vision loss and blindness [11, 12]. DR is the resulting condition specifically for these complications and damaged vessels lying in the tissue behind the retina [13]. Unfortunately, so many cases of diabetes go undiagnosed and the connection between diabetes and eye vision isn't common knowledge; patients with DM are 25 times more likely to be at risk of vision loss when compared with non-diabetic ones [14]. DR is often overlooked and detected too late, since it is a silent disease, and it may be diagnosed by the patient when the vascular abnormalities in the retina have progressed to a stage when its cure is complicated and sometimes unfeasible [15, 16].

Therefore, the most successful therapy for DR can be managed just in its early phases. Consequently, early detection of DR through systematic screening is of critical importance.

It is believed that the screening of diabetic people for the evaluation of DR potentially reduces the risk of blindness in these people by 50% [17]. The problem is, however, that this would produce immense costs, due to the large number of examinations. Moreover, there are not enough specialists, especially in rural areas, to do these examinations [18]. Physical screening of DR to address the large and mounting number of DR cases is therefore not possible. Instead, automated systems based on image processing methods may help to overcome this issue [19]. It would be more beneficial and helpful if the initial steps of analyzing the retinal images and detecting all retinal lesions in the images can be computerized through a robust computer automated system. In this way, the automated system would allow for speedier identification of the telltale signs of DR from the early onset of retinal lesions and notification of the ophthalmologist for further assessment. This would allow more patients to be screened per year and for ophthalmologists to better manage and direct their time towards those priority patients who would have otherwise been overlooked or delayed in treating [20].

12.1.1 Clinical signs of diabetic retinopathy

Now, before we start DR diagnosis the big question is *what are the revealing indications of DR?* There are several abnormalities in the retina that are manifestations of DR which are concisely described below.

1. Microaneurysms (MAs): They are the first visible sign of DR. They appear as small and round shape dots near tiny retinal blood vessels in retinal images [21]. The size of MAs usually ranges from 10 μm to 125 μm in diameter [19, 22].
2. Haemorrhages (HEs): A more severe level of DR is caused by retinal haemorrhages or leakage of weak capillaries and ruptured MAs into the surface of the retina [23]. They are defined as a red spot with different shapes, such as 'dot', 'blot' and 'flame' [21, 23] and irregular margin and uneven density [24].
3. Hard exudates (EXs): When the lipoproteins and other proteins are leaking because of the degradation or breakdown of the blood–retina barrier through abnormal retinal vessels, hard EXs appear [25]. They become visible as small yellowish-white patches with sharp borders [24]. They are often collocated in bulks or semicircular hoops and positioned in the exterior layer of the retina [24].
4. Cotton wool spots (CWS): Also called soft exudates, they occur due to occlusion, or blockage of blood supply, of arterioles and cause disruption and damage to areas of tissue [26]. The ischemia of the retinal layer of the nerve fiber caused by reduced blood flow to the retina changes the axoplasmic flow and leads to accumulation of axoplasmic residue in the retinal ganglion cell axons. The accumulation of this axoplasmic residue arises as fleecy yellowish-white lesions in the retinal nerve fiber layer known as soft exudates or CWS [26, 27].

5. Neovascularizations (NVs): The new blood vessels start to grow on the inner surface of the retina primarily because of the lack of oxygen. These new blood vessels push into the surrounding retinal space and are feeble and very often bleed into the retinal cavity, effecting the eyesight [10, 28].
6. Macular edema (ME): It is recognized as the swelling beneath the macula, an area at the center of the retina. It results in penetration of abnormal capillaries in the retinal because of the fluid leakage or solutes around the macula [29, 30]. Eventually, it influences the central the eyesight [31].

Figures 12.1 and 12.2 show typical normal retinal images with their basic landmarks, and abnormal images with different signs and stages of DR.

12.1.2 Stages of diabetic retinopathy

According to the existence of the aforementioned clinical signs, and the severity/density of these signs [34], DR is categorized into four kinds including: mild, moderate, severe non-proliferative DR (NPDR), and PDR [24, 35, 36]. With respect to number of MAs and HEs, the NPDR is introduced as normal, mild, moderate, and severe DR [37].

- **Normal**: If no DR sign is observed.
- **Mild**: Here, only MAs are presented.
- **Moderate**: In this case, both MAs, and HEs exist and their numbers are less than 20 in each quadrant, and we see soft EXs as well.
- **Severe**: Here, in each quadrant there are more than 20 MAs, and HEs. Also, in more than two quadrants there are some venous beadings and different shapes of hard EXs.

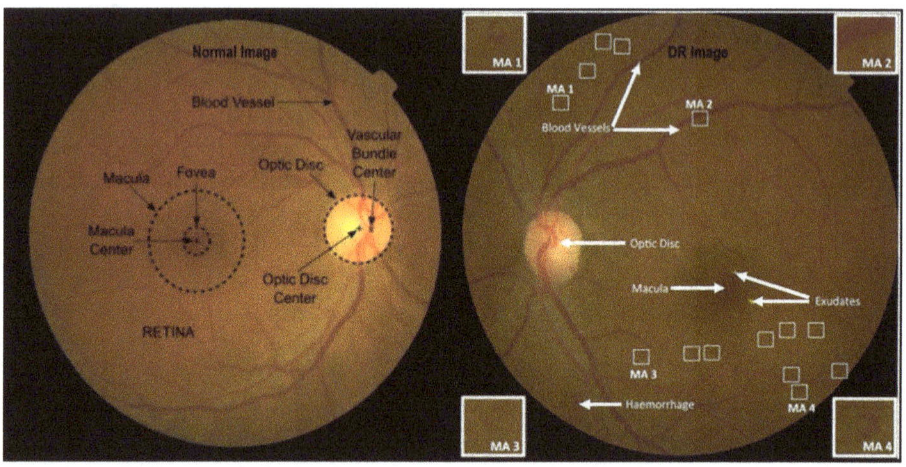

Figure 12.1. A normal retinal image versus a DR one. Sample retinal images for both normal and abnormal ones. On the left image, all available anatomical landmarks have been shown including fovea, macula, blood vessels, optic nerve head [32]. On the right one: all pathologic signs of DR including MAs, HEs, and EXs are visible [33].

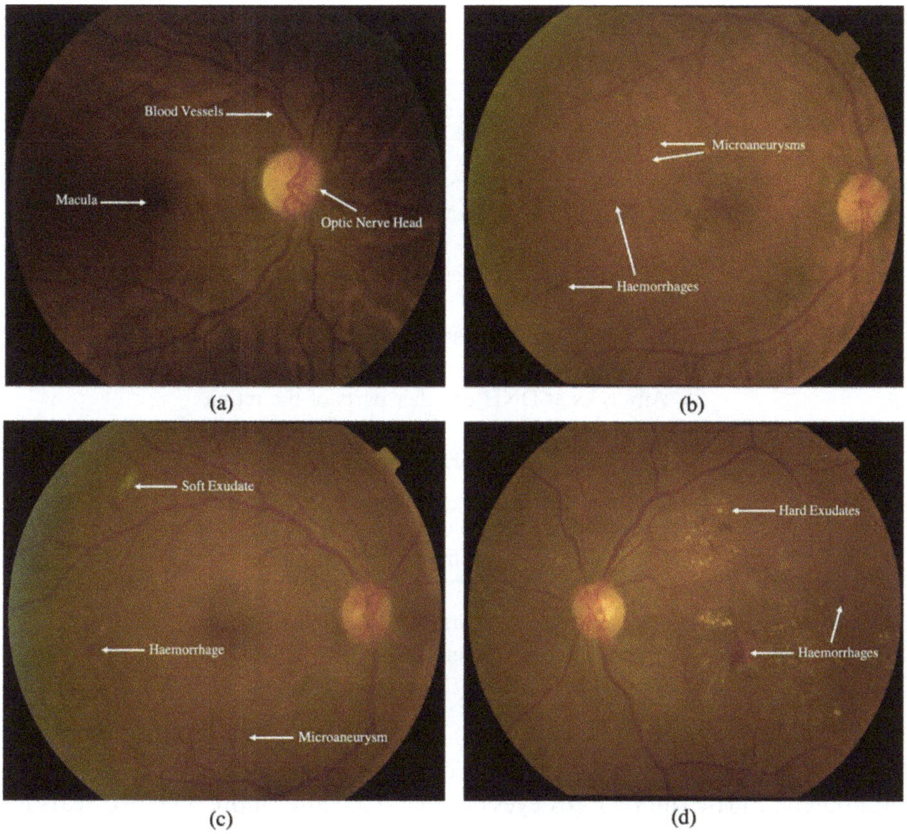

Figure 12.2. Different types of DR. (a) Normal; (b) mild NPDR; (c) moderate NPDR; and (d) severe NPDR.

Figure 12.2 shows the retinal images of these categories and related signs. In the PDR, we see the initiation of new vessels, which are abnormal vessels (NVs). It is the severe phase of DR, in which this new abnormal vessel growth, caused by poor circulation especially from formations of MAs, protrudes into the retina and even into the vitreous, the gel-like medium that gives the eye its spherical shape. Thus, both MAs and NVs are two important clinical lesions [38] and fluid in DR is classified as EXs and non-EXs [39].

There are several organizations for DR grading such as American Academy of Ophthalmology, whose classification of DR was established by curing of DR at its early stage [40], and a protocol introduced by the Scottish DR grading system [41]. Table 12.1 presents the Scottish protocol.

MAs, in general, are a very good indication of possibly worsening conditions of DR. From a statistics point of view, the NPDR type with MAs has a 6.2% chance to expand into PDR during a year [1]. An increased number of MAs is a primary characteristic of DR progression. Moreover, with development of ischemia, there is an increased chance of PDR progress during the first year. The first year risk growth increases from 11.3% (lower) to 54.8% (advanced) phase [1, 42]. In the NVs stage,

Table 12.1. DR Scottish grading protocol [41].

Grade	Characteristics
No DR	Without abnormalities
Mild NPDR	Just MAs appear
Moderate NPDR	There are more than 5 MAs but less than severe NPDR
Severe NPDR	• In each quadrant, there are more than 20 HEs • In two quadrants, there is venous beading • Intra-retinal microvessel abnormalities
PDR	• Any NVs at ONH or other parts of the retina • Vitreous/ pre-retinal HE
No ME	There are no EXs or thickening of retina in posterior pole
Mild ME	There are EXs or thickening of retina at posterior pole, >1 disc diameters from fovea
Moderate ME	Same as mild ME's signs plus 1 disc diameters, without affecting fovea
Severe ME	There are EXs or thickening of retina with affecting fovea

diabetics have a 25.6% to 36.9% chance of vision loss and blindness, if not cured appropriately. Furthermore, PDR eyes not therapied after more than two years have a chance of vision loss and blindness at 7.0% and if for more than four years it is not cured, the chance of vision loss goes up to 20.9%. On the other hand, this vision loss and blindness decreases to 3.2% during two years and 7.4% during four years of therapy [42]. Diabetic people with mild DR do not need specific treatment. In fact, they need to manage their DM and the related risk factors such as hypertension, anaemia, and renal malfunction. They need to be closely monitored, to decrease the possibility of progressing to higher stages of DR [18]. In the severe and advanced phase of DR, like NV, therapy is limited [43].

12.1.3 Retinal imaging and databases

Now we are aware of the signs and stages of DR, and the way diagnosis and determination of the stage of DR involves processing retinal images. Retinal photography is a procedure where three-dimensional retinal landmarks are projected on the two-dimensional image. The created image shows the amount of reflected light [44]. In the camera that creates the fundus image, the optical layout is equivalent to an ophthalmoscope, which is used to observe the inner part of the eye. The camera sees the retinal area with 2.5 times magnification at the angle between 30° and 50°. More information can be found in [1]. In retinal photography for DR screening, the below modalities/techniques are utilized [44].

(a) Red-free retinal imaging: The image is taken by applying the amount of reflected light at a particular waveband.
(b) Color retinal image: This image is based on the red–green–blue (RGB) spectrum and the camera detector sensitivity to the light.
(c) Fluorescein angiography retinal image: The created image is based on the amount of photons that are emitted by fluorescein dye. The dye is injected through the blood stream [44].

Automated algorithms, in the process of testing and improving upon themselves, require a set of exemplary data with the proper classifications. In our case, these would be retinal images from reputable databases, many of which have been classified by professional ophthalmologists. Most of the provided retinal images in these databases are used in application of screening and testing the CAD methods related to detection of MAs, HEs, exudates and other DR features. Outlined below, many of these online research databases are included. Table 12.2 summarizes all databases for detection of DR.

- **DIARTDB**: The dataset was created by Kauppi *et al* [45] and is freely available for research from http://www.it.lut.fi/project/imageret/. Images were taken with a 50° field-of-view (FOV). It has two classes, DIARTDB0, including 130 color retinal images (20 images are normal without sign of DR and 110 contain symptoms of the DR). These signs include hard and soft EXs, MAs, HEs and NV. The second class is called DIARTDB1, including 89 images, of which five are normal and 84 contain at least mild NPDR signs (MAs).
- **HRF**: It is presented as the **H**igh-**R**esolution **F**undus retinal image dataset. The images were captured with different image acquisition settings and 45° FOV. There are images in this database for each category: healthy, glaucomatous, and DR. Information in the database was created by a group of experts from ophthalmology clinics [46].
- **STARE**: It consists of 400 images initially applied by Hoover *et al* [47]. These images were captured at 35° FOV.
- **KAGGLE**: The images were prepared by Kaggle [37] as a series of high resolution pictures. The images were captured under different spatial conditions and were indexed by a well-experienced ophthalmologist with respect to without DR to mild, moderate, severe and proliferative DR.
- **ROC**: **R**etinopathy **O**nline **C**hallenge is a dataset established for enabling research groups to establish a CAD system for diagnosis of DR. It aimed to create a platform for algorithms evaluation [48].
- **DRIVE**: It has been developed from a DR screening system. The database included 40 retinal images in which 33 images are without symptoms of DR and seven showing signs of DR. The database was taken using a non-mydriatic 3CCD fundus camera with a 45° FOV [49].
- **MESSIDOR**: It consists of 1200 eye retinal color images. The images were captured by a color non-mydriatic retinography with a 45° FOV with different resolutions [50].

Table 12.2. More details for the databases used for DR detection system.

Database	Numbers	Resolution	Format	Task
MESSIDOR [50]	1200	1440 × 960, 2240 × 1488, 2304 × 1536	TIFF	DR grading Risk of DME
Kaggle [54]	80 000	—	JPEG	No sign of DR-Mild Moderate-Severe-PDR
E-ophtha [55]	148 MAs, 233 normal, 47 EXs, 35 no signs of DR	2544 × 1696 1440 × 960	JPEG	MA small HE detection EX detection
DRIVE [49]	33 normal 7 mild DR stage	584 × 565	TIFF	Vessels extraction
STARE [47]	402	605 × 700	PPM	13 retinal diseases Vessels segmentation-ONH
DIARETDB1 [45]	5 no sign 84 NPDR symptom	1500 × 1152	PNG	MAs-HEs
CHASE [37, 56]	28	1280 × 960	JPEG	Vessels extraction
MUMS-DB [19, 53]	120		PNG	MAs
ARIA [57]	16 normal 92 AMD-59 DR	768 × 576	TIFF	ONH-Fovea-Vessel
SEED-DB [58]	192 normal 43 glaucomatous	3504 × 2336	—	ONH
EyePACS [59]	9963	—	—	Referable DR MA
ORIGA [60]	482 normal 168 glaucomatous	720 × 576	—	ONH

- **E-OPTHIA**: It was created because of a DR screening project which developed a color dataset for DR, called E-ophtha. This dataset was made of fundus images with a variety of signs such as EXs and MAs manually labelled by well-experienced ophthalmologists [51]. It contained 47 images with EXs, 35 with no sign, and 281 images with MAs or small HEs.

Table 12.3. Some CAD systems and databases (Blue = B, Green = G, Red = R).

Author	Image plan	Method	Database
Pourreza *et al* [53]	B&G	Radon transformation	DRIVE, STARE, MUMS-DB
Esmaeili [61]	R&G	Morphology	STARE, DRIVE, DIARETDB1
Qureshi *et al* [62]	G	Entropy and Hough	DIARETDB0, DIARETDB1, DRIVE
Welfer *et al* [63]	LUV	Mathematical Morphology	DRIVE, DIARETDB1
Harangi and Hajdu [64]	Grey	Probabilistic model and augmented Naive based	DIARETDB0, DIARETDB1, DRIVE, MESSIDOR
Akyol *et al* [65]	G	BLP	DIARETDB1, DRIVE,ROC
Dai *et al* [66]	Grey	Hough, PCA	MESSIDOR, DRIONS
Alshayeji *et al* [68]	Grey	Adaptive edge detection,	DERIVE, DMED, STARE, DIARETDB1
Fraz *et al* [68]	G	Watershed transformation	DRIVE, Shifa, CHASEDB1, DIARETDB1

- **DRIONS**: An evaluating dataset for ONH segmentation that included 110 retinal images. The fundus images were captured with a color analogical camera [37].
- **CHASEDB1**: It is a dataset available at blogs.kingston.ac.uk/retinal/chasedb1 for free. It includes 28 fundus images. An ophthalmologist and human experts were asked to mark the vessels in all these images and these hand-labelled vessels were used as the ground truth (or gold standard) for this dataset [52].
- **HEI-MED**: The database is a test database for evaluating algorithms in the EXs and macular edema detection. It contained 169 images which were segmented manually by a highly skilled ophthalmologist [30].
- Others: These are other datasets established by individual researcher groups such as MUMS-DB [19, 53] DMED, ARIA, Eyepack [1]

Table 12.3 summarizes some CAD systems using different image database analysis for DR detection. These approaches are distinguished based on features, processing, and classification of DR.

12.2 Computer-aided diagnosis and diabetic retinopathy

As previously mentioned, manual detection and diagnosis is an exhaustive task in both cost and lack of experts [16] when screening DR. It would be more cost effective and helpful if the initial task of analyzing the retinal images could be automated. Automated approaches address these issues by decreasing the time, expense, and effort considerably [69]. Moreover, since image processing techniques are growing in all areas of medical science, by assisting them, especially in advanced ophthalmology, we can do automated screening [39].

For this reason, automated DR detection and classification utilizes computer aided diagnostic (CAD) systems. These computer-based systems have the ability to detect any change in normal and abnormal images and systematize these variations to form a feature space. At the end, the mixture of these features introduces type and stage of DR. Different CAD systems have been presented as the state-of-the-art for early DR detection and related lesions [12, 18, 19, 70–77].

Normally, each CAD system is the sequence of two different approaches, i.e., feature detection and classification [37]. Since there are many different approaches in CAD systems, it is also necessary to evaluate their robustness and accuracy. There are many different standard methods used in assessing the effectiveness of CAD systems. A common one is using receiver operating characteristic (ROC) analysis and concept of area under the curve (AUC) analysis. The AUC of ROC curve is a measurement of performance for an automated method, for example, extraction problem, at different thresholds values. ROC is a probability curve and AUC shows degree or rate of distinguishability. In fact, AUC shows us how much the method is able to distinguish between classes. The higher the AUC, the better is the method in its prediction. By analogy, the higher the AUC, the better the method is at distinguishing between images with DR and no DR. ROC curves illustrate the tradeoff between sensitivity and specificity for a range of thresholds and enable the identification of an optimal value [78].

Here, the analytical definition when using ROC is assessed according to the true positive fraction (TPF), given by sensitivity, and the false positive fraction (FPF) (1-specificity) [16]. Moreover, the accuracy is specified as a quantification including the ratio of well-classified pixels. The final outputs for the automated approach as compared to the ground truth or gold standard are calculated for each image. These metrics are defined as:

$$\begin{aligned} Sensitivity(Se) &= \frac{TP}{TP + FN} \\ Specificity(Sp) &= \frac{TN}{TN + FP} \\ Accuracy(Acc) &= \frac{TP + TN}{TP + FN + TN + FP} \end{aligned} \tag{12.1}$$

where TP is true positive, TN is true negative, FP is false positive and FN is false negative same as in [79, 80].

A look into the results for both ROC and AUC (equation (12.1)) obtained by other CAD systems in DR screening. Table 12.4 summarizes some of the algorithms at different image analysis stages for detection of DR.

Briefly, both Tavakoli *et al* [19] and Philip *et al* [36] introduced a DR screening system based on the comparison of automated system with respect to the gold standard which was manual detection by an ophthalmologist. Their approach

Table 12.4. Some of the CAD systems with different image analysis stages.

CAD system	Methodology	Authors
ONH detection and segmentation		Qureshi *et al* [62], Welfer *et al* [63], Harangi and Hajdu [64], Akyol *et al* [65], Dai *et al* [66], Alshayeji [67], Basit and Fraz [68], Bharkad [81], Lesay *et al* [82], Zou *et al* [83], Xiong and Li [84], Sarathi *et al* [85], Lu [86], Lowel *et al* [87], Hoover and Goldbaum [88]
Vessel segmentation	Thresholding approach Tracking approach Machine classifiers	Hoover *et al* [47], Dash and Bhoi [89], Zhu *et al* [90], Neto *et a l*[91], Xu and Luo [92], Nguyen *et al* [93], Zhang *et al* [94], Fraz *et al* [95], Marin *et al* [96], Akram *et al* [97], Vermeer *et al* [98], Chakraborti *et al* [99], Wu *et al* [100], Tavakoli *et al* [101], Roychowdhury *et al* [102], Tolias and Panas [103], Azzopardi *et al* [104], Zana *et al* [4, 105], Soares *et al* [106]
Lesion detection	Red lesions (MAs, HEs)	Quellec *et al* [107], Walter *et al* [22], Tavakoli *et al* [19], Wang *et al* [108], Srivastava *et al* [109], Seoud *et al* [23], Pereira *et al* [112], Dai *et al* [110], van Grinsven *et al* [111]
	Hard and soft exudates	Liu *et al* [113], Zhang *et al* [115], Orlando *et al* [114], Youssef and Solouma [116], Fraz *et al* [117], Amin *et al* [118], Sopharak *et al* [119], Sanchez *et al* [120]
DR detection and classification	DR, Non DR NPDR, PDR	Tavakoli *et al* [19], Koh *et al* [121], Gardner *et al*[122], Rubini and Kunthavai [123], Gulshan *et al* [124], Kose *et al* [125], Yang *et al* [126], Sisodia *et al* [127], Kumar *et al* [128], Gupta *et al*[129], Leontidis [82], Sinthanayothin *et al* [130], Usher *et al* [131], Fleming *et al* [17, 132], Perumalsamy *et al* [133], Winder *et al* [134], Mookiah *et al* [135]

detected DR images with an accuracy of more than 90%. Sinthanayothin *et al* [130] established a trustable automatic DR online screening system purposed to simplify ophthalmologists' work in their country. They got sensitivity and specificity for their screening program to classify the DR as 80.2% and 70.7%, respectively. Usher *et al* [131] employed a DR screening tool which detected DR with both sensitivity and specificity of 95.1% and 46.3%, respectively. Niemeijer *et al* [70] introduced a DR detection algorithm based on information fusion approach with AUC of 0.88. Quellec *et al* [71] established an automated DR screening system for both MAs and age-related macular degeneration detection, and their method reached an AUC of 0.93. Fleming *et al* [132] introduced a CAD system to detect blot HEs that provided sensitivity and specificity of 98.60% of 95.50%, respectively. Fleming *et al* [17] also established another CAD system to detect MAs which got a sensitivity and specificity of 85.4% and 83.1%. Perumalsamy *et al* [133] employed a CAD system and obtained an accuracy of 81.3% by evaluating the performance with ophthalmologists' idea. Abramoff *et al* [136] established a web-based DR detection system with a specific protocol and short questionnaire, measurement of visual eyesight, and four retinal images. The accuracy of their protocol reached 93%. Abramoff *et al* [137] also presented an automated DR detection system which acquired an AUC of 0.84 of DR detection.

There are also some authors who reviewed most of the CAD methods, and their applications [1, 39, 44, 134, 138] for automated screening of DR. Winder *et al* [134] discussed different segmentation approaches of optic nerve head (ONH), retinal vessels, macula, and DR detection. Teng *et al* [138] for retinal landmarks, and lesions segmentation, and image registration reviewed different methods. Abramoff *et al* [44] surveyed various retinal imaging modalities, vessel segmentation, lesion, and ONH detection. Patton *et al* [39] discussed image preprocessing, and registration methods, besides landmarks and lesions segmentation. Mookiah *et al* [1] comprehensively reviewed the approaches to detect and extract the retinal lesions and landmarks such as ONH, vessels, EXs, MAs, and HEs.

From a classification viewpoint, color characteristics were utilized on a Bayesian classifier to group each pixel into either lesion or non-lesion categories [139]. In this study, authors obtained 100% accuracy in detection of all EXs in the retinal images, and accuracy of 70% in classification of normal retinal images. By using image processing techniques and multilayer neural network (NN), DR and normal retina were categorized with sensitivity and a specificity of 80.2% and 70.7%, respectively [21]. Automated detection of NPDR, based on MAs, HEs, and EXs was studied in [135, 140]. The approach was able to correctly detect the NPDR stage with an accuracy of 81.7%. For DR screening again above, lesions were used [141] and they obtained the sensitivity and specificity of 74.8% and 82.7%, respectively. An early automated DR detection system was developed by Kahai *et al* based on a decision support system [142]. In their method, Bayes optimality criteria were used to identify DR with a sensitivity and specificity of 100% and 67%. Different stages of DR were grouped using both area and geometry of the combination of the vessels with a feedforward NN [143]. The average classification efficiency for this method was 84% with both sensitivity and specificity of 90% and 100%. Nayak *et al* employed EXs and vessel area along with parameters related to its texture which was coupled with NN to classify NPDR, PDR, and normal [5]. They got accuracy of 93%, sensitivity 99%, and specificity of 100%. The feature of support vector machine (SVM) classifier was used by Acharya *et al* [144] to classify the retinal image in to normal, mild, moderate, severe and PDR categories. They showed a classification accuracy of 82% with sensitivity of 82%, and specificity of 88%. Nicolai *et al* established an automated lesion detection algorithm for DR screening purpose, which detected 90.1% of diabetics and 81.3% of those without DR [145]. Their system showed both sensitivity and a specificity of 93.1% and 71.6%.

12.3 Retinal landmarks detection

A standard CAD screening system needs to extract the retinal landmarks such as fovea, ONH, and vessels, and additionally lesions such as MAs, HEs, EXs, CWS, and NVs. Several studies have worked on various detection and segmentation approaches to identify these anatomical regions. Here, we briefly discuss them.

12.3.1 Localization and segmentation of optic nerve head

The ONH can reveal a critical eye disease like glaucoma [146]. The shape of the ONH appears as semicircle or ellipse in the retinal image. The ONH is also applied as a reference landmark to identify other retinal regions and structures such as the macula and fovea [87, 147–149]. Moreover, the ONH localization assists in determining central retinal arteries and veins [87, 150]. Many vessel tracking algorithms use the ONH center as the starting point, since the vessels diverge from the ONH center. The ONH localization is carried out by detecting its center or by drawing a circle around the ONH region [151–155]. ONH segmentation is introduced by specifying the boundary of ONH. Thus, identification and extraction of ONH is highly important [134, 146, 156–160].

The localization of the ONH is serious to differentiate it from the other lesions. The ONH identification approaches, as explained in literature mainly estimate its center or surround it by a circular mask. In both cases, the localization is quite an elaborate task because of the presence of thick vessels, inaccurate boundaries, and lesions such as EXs [134, 146]. The methods according to intensity changes are easy to use, fast and robust for images with no symptoms. These methods may lose out when the ONH is concealed by blood vessels and lesions such as EXs, CWS, or other similar intensity artifacts [161].

12.3.1.1 Detection of optic nerve head center

The ONH color and shape can be utilized to detect it. Normally, in some studies the ONH is localized by its relations to bright pixel clusters. Sinthanayothin *et al* [162] used a window size 80 × 80 pixel as the size of a typical ONH and localized the ONH by observing the maximum change in its intensity to its neighboring pixels. They chose the point with highest pixel variance to detect the center of the ONH and reported a sensitivity and specificity of 99.1%. Lowel *et al* [87] developed a correlation filter to identify the ONH in the presence of EXs and other artifacts and correctly detected it in 83% of cases. Hoover and Goldbaum [88] presented an approach using the concept of a fuzzy approach to locate the ONH. Their approach obtained 89% correct ONH detection. Walter *et al* [163] introduced a method to detect ONH based on its illumination, distance, and watershed transformation in different color space, HSL (hue, saturation, luminance), with accuracy of more than 85%. Tavakoli *et al* applied three different vessel segmentation algorithms, Laplacian-of-Gaussian, Canny, and Matched filter edge detectors to detect the ONH in both the images without DR signs and in the presence of DR [164].

The Hough and similarly the radon transform were applied by many authors to identify the ONH [53, 87, 165–167]. The method is based on integral transformation, or mapping the values of the line integral of an images' pixels into another domain. Ultimately, the ONH was identified by effective projection profiles in radon space and the boundary parts with maximum vote [53].

Principal component analysis (PCA) was another method for identifying the ONH [162, 168, 169]. In fact, this method turns the correlated variable into uncorrelated ones, which are known as principle parts [1]. The first principal component demonstrates the maximum information existing in the image. Li and Chutatape [169] detected the ONH using the concept of PCA with accuracy of 99%. Another method that applied the convergence of vessels to localize the ONH was proposed by Foracchia *et al* [170]. Youssif *et al* [171] applied matching the expected directional structure of the vessels for ONH detection. Their method started by normalizing luminosity and contrast throughout the image using illumination equalization and adaptive histogram equalization methods, respectively. Lu and Lim [172] utilized a different approach to place the ONH based on its bright appearance in a color retinal image. They used a series of concentric lines with various directions and assessed the image variation along the different directions. Figure 12.3 shows the retinal images of these categories and related signs.

12.3.1.2 Segmentation of optic nerve head

In the case of segmentation methods category, the brightness was applied by Li and Chutatape [173, 174] to detect the ONH candidates for their model-based technique.

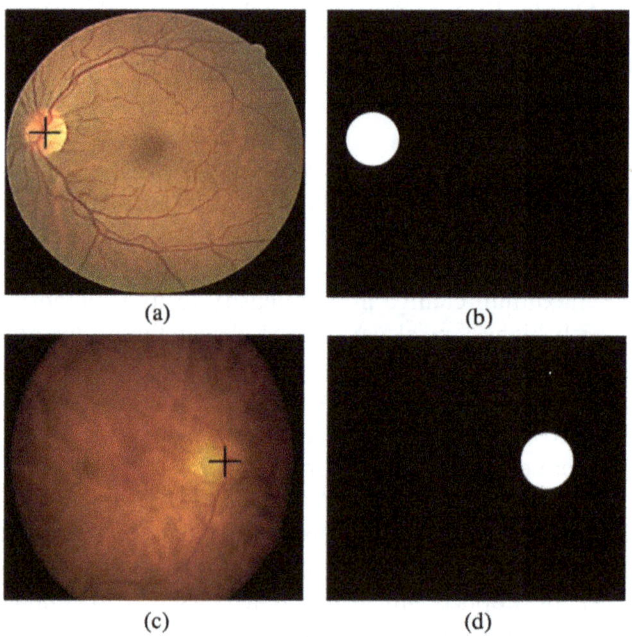

Figure 12.3. ONH detection. (a) and (c) retinal images from MUMS-DB and detected center of the ONH; (b) and (d) ONH mask segmentation.

They established an active shape model, which consisted of building a model with training cases and iteratively matching the landmark points on the disc edges and main vessels inside the disc [169]. A model-based approach was introduced effectively based on active contour model by Osareh *et al* [175, 176] to approximately detect the ONH. Lowell *et al* carefully created an ONH template, which was then correlated to the intensity component of the retinal image using the full Pearson-R correlation [87]. Another model-based algorithm was introduced by Xu *et al* [177]. They applied the deformable contour model technique that included ONH margin. Their method used a clustering-based classification of contour points and customized contour evolution step integrated in original snake formulation. The approach introduced by Wong *et al* [178] was based on the level-set method followed by ellipse matching to smooth the boundary of the ONH. They proposed: (1) the ONH location according to the previously obtained location by means of histogram analysis and initial contour definition, and (2) subsequently, a modified version of the conventional levelset method used for ONH boundary extraction from the red channel. This contour was finally matched by an ellipse.

On the other hand, Lu [86] introduced a circular transformation to take ONH and its boundary simultaneously. Lalonde *et al* presented Hausdorff-based template matching together with pyramidal decomposition and confidence assignment [179]. Pyramidal decomposition acts mainly for detection of large areas with bright pixels, with probable existence of the ONH, but it can be disturbed by large areas of bright pixels that may be near the image borders due to non-uniform illumination. In order to solve this problem, Frank ter Haar [180] used illumination equalization in the green channel of the image, and then a resolution pyramid using a simple Haar-based discrete wavelet transform was formed. Further, he applied the Hough transform and proposed two methods using a vascular branch constructed from a binary vessel image. The last technique of Frank ter Haar was based on fitting the vessel orientations on a directional model [180]. The ONH identification methods not only apply the characteristics of the ONH, but also extract the location and orientation of vessels [181–184]. Niemeijer *et al* [184] presented the use of local vessel geometry and image intensity features to find the correct positions in the image. A k-nearest neighbour (kNN) regressor was used to accomplish the integration. Tobin *et al* [181] used an algorithm that highly depended on vessel-related ONH characteristics. They applied a Bayesian classifier to categorize each pixel in images as with/without ONH. Abramoff and Niemeijer [185] applied same properties for the ONH detection, and then a kNN regression was used to estimate the ONH location. The approach developed by Abramoff *et al* [186] according to a pixel classification using the feature analysis and nearest neighbor technique. The final results of their algorithm came with the categorization of pixels to a group that belonged to the ONH and non-ONH. In a recent study, Hsiao *et al* placed ONH by illumination correction operation, and contour segmentation that was completed by a supervised gradient vector flow snake [187]. To identify ONH location candidates, first template matching was used by Yu *et al.* Next, they applied vessel features on the ONH to place the location of it. They combined region and local gradient information to the segmentation of the ONH boundary and used morphological filtering to remove retinal vessels [188].

Moreover, some approaches made use of the fact that all major vessels diverge from the ONH [170, 171, 180, 189]. Most of the studies identified the ONH with respect to its shape and color. Other algorithms recognized the ONH according to tracking vessels toward their origin. In table 12.5, we have provided the details of more methods in ONH detection.

12.3.2 Vessel segmentation

There is a substantial effort reported in the state of the art for the segmentation of retinal vessels in fundus images [56, 91, 94, 95, 97, 190–195]. Overall, the vessel segmentation approaches are classified into five categories: *tracking*, *mathematical morphology*, *matching filter*, *model-based thresholding*, and *supervised classification algorithms* [1]. Table 12.6 summarizes all these methods and we briefly discuss them below.

Table 12.5. Summary of some methods and their results of ONH detection.

Authors	ONH detection method	DRIVE	STARE
Sinthabayothin *et al* [162]	Highest average variation	99.1%	42%
Walter and Klein [163]	Largest brightest connected object	100%	58%
Li and Chutatape [173, 174]	Brightness guided, model base	99%	—
Osareh *et al* [175, 176]	Averaged ONH model based	100%	58%
Lowell *et al* [87]	ONH Laplacian of Gaussian template	99%	—
Lalonde *et al* [156]	Hausdorff based with pyramidal decomposition and confidence assignment	100%	71.6%
Frank ter Haar [180]	Resolution pyramid using Haar based wavelet transform	89%	70.4%
Frank ter Haar [180]	Hough used only to pixels on or close to the retinal vasculature	96.3%	71.6%
Frank ter Haar [180]	Pyramidal decomposition of both the vascular and green plan	93.2%	54.3%
Frank ter Haar [180]	Hough and fuzzy convergence	97.4%	65.4%
Tobin *et al* [181]	Vasculature related ONH features and a Bayesian classifier	81%	—
Abramoff and Niemeijer [184, 185]	Vasculature related ONH properties and a kNN regression	99.9%	—
Youssif *et al* [171]	Matching the expected directional pattern of the retinal blood vessels	100%	98.8%
Hsiao *et al* [187]	Illumination correction, and contour segmentation is completed by a supervised gradient vector flow snake	100%	90%
Tavakoli *et al* [164, 167]	Different vessel segmentation methods	90%	85%
Pourreza *et al* [53]	Radon Transform	100%	94%

Table 12.6. Some vessel segmentation methods.

Authors	Method	Classifier and databases	Evaluation
Xu and Luo [92]	Sobel filter for ONH, adaptive threshold	Thick and thin vessels, SVM, DRIVE	Acc. = 93.36% Sen. = 85.57%
Hassan *et al* [196]	Hidden Markov model	Vessels, DRIVE	Acc. = 95.7% Sen. = 81.0% Sp. = 97.0%
Dash and Bhoi [89]	Binary difference and enhanced image	DERIVE, CHASEDB1	Acc. = 95.5%,
Nguyen *et al* [93]	Linear combination of line responses	SVM, STARE, DRIVE, REVIEW	Acc. = 93.24%, 94.07%,
Fraz *et al* [56, 95]	Mathematical morphology, aggregation	DRIVE, STARE MESSIDOR	Acc. = 95.52% Sp. = 97.23%
Marin *et al* [96]	Grey level features	ANN, DRIVE, STARE	Acc. = 95.26%, Sen. = 69. 44%sp. = 98.19%
Aslani and Sarnel [197]	Hybrid features	DRIVE, STARE	Acc. = 95.13%, Sen. = 77. 96% sp. = 97.17%
Lupascu and Trucco [198]	AdaBoost	AdaBoost classifier, DRIVE	Acc. = 95.69% Sen. = 55.74% sp. = 99.36%
Rodrigues *et al* [159]	Histogram analysis, wavelet Hessian based and Gaussian filter	DRIVE, HRF	Acc. = 94.65%,
Tavakoli *et al* [101]	Radon transform	MUMS-DB	Acc. = 90%, Sen. = 85%
Zhang *et al* [94]	kNN	DRIVE	Acc. = 95.05% Sen. = 78.12% Sp. = 96.62%
Zhu *et al* [90]	Vessel profile, Gaussian, morphology	ELM, DRIVE, RIS	Acc. = 96.07, Sen. = 60.18, Sp. = 98.68%
Neto *et al* [91]	Binary morphological reconstruction	DRIVE, STARE	Acc. = 86.16%, 87.87% Sen. = 79.42, 76.96% Sp. = 95.37%, 96.31%
Vega *et al* [199]	Lattice NN and dendritic processing	ANN	Acc. = 94.83%, Sen. = 96.71%, Sp. = 70.19%
Waheed *et al* [200]	Connected component, binary threshold,	Localized Fisher discriminant analysis, DRIVE, STARE	Acc. = 95.81%, 96.16%,
Zilly *et al* [201]	AdaBoost	MESSIDOR, DRISHTIGS	—

A comprehensive review on existing methods in retinal vessel segmentation and available public datasets is presented in [33, 52]. In general, vessel segmentation algorithms can be classified into two broad categories: unsupervised and supervised methods [202]. A comparative survey of these two classes has been presented in [102]. Unsupervised methods can be further classified into techniques based on matched filtering [47, 99], morphological processing [203], vessel tracking, multi-scale analysis [204, 205], line detectors [93] and model-based algorithms [98, 206]. Supervised segmentation methods are based on pixel classification such as kNN [49], Gaussian mixture models [106], SVM [207], NN [207], decision trees [194], and AdaBoost [96, 198]. They utilize ground truth data for the classification of vessels, based on given features.

The principle of matched filter detection in unsupervised methods was proposed in [47]. With this technique, the authors used a 2D Gaussian-shaped template in all directions to search for vessel components. The kernel was rotated through a range of angles (usually 8 or 12) to detect vessels with different orientations. The resulting image is thresholded to produce a binary representation of the retinal vasculature. Another matched filtering approach was used for retinal vessel segmentation [208]. The application of local thresholding strategies, edge detection, matched-filtering and region growth was developed in [208]. Chakraborti *et al* [99] applied an unsupervised segmentation approach that combined a 'vesselness' filter and matched filter using an orientation histogram. Vermeer *et al* [98] achieved vessel detection by thresholding, after convolving the image with a 2D Laplacian kernel. A general framework of adaptive local thresholding based on the combination of applying a multithreshold scheme and classification procedure to prove all binary result objects were applied by Jiang *et al* [209]. Zana *et al* [4] proposed a morphology-based approach using an algorithm that combined morphological filters and cross-curvature evaluation to segment vessel-like patterns in retinal images. Other mathematical morphology approaches are described in [105, 203, 210]. Heneghan *et al* [210] combined morphological transformations with curvature information and matched filtering for centerline detection. In [100] vessels are extracted using multiple structure elements followed by connected component analysis. Martinez-Perez *et al* [204] introduced an automated algorithm for retinal images based on a multi-scale feature detection. The local maxima of the gradient magnitude and the maximum principal curvature of the Hessian tensor were used in a multi-pass region growing process. This procedure accurately segmented the retinal vasculatures by using both feature and spatial information. Zhang *et al* [205] employed innovative rotating multi-scale second-order Gaussian derivative filters which are referred to as orientation scores for the enhancement and segmentation of blood vessels. Roychowdhury *et al* proposed an iterative unsupervised retinal vessel segmentation algorithm that used an adaptive threshold and a region growing method with a stopping criterion to terminate the iteration [211].

Nguyen *et al* proposed a vessel detection method based on line detection [93]. This approach was based on the fact that changing the length of a line detector makes line detectors with a variety of scales. The final vessel segmentation results were achieved by combining line responses at varying scales. The combination of shifted filter responses for detection of bar-shaped structures in retinal images was presented by Azzopardi *et al* [104]. Their approach was rotation invariant, where the selection of the orientations was chosen from given structures similar to vessels which suffered from difficult crossing cases. Lam and Yan [206] used the Laplacian operator to segment blood vessels and detect centerlines from the normalized gradient vector field, pruning noisy objects. Vessel continuity employs measures of width and orientation, iteratively calculated in a local region near the current point, in order to track along the length of a vessel [100, 103]. Tavakoli *et al* [101] presented the combination of radon transformation and multi-overlapping windows to segment the retinal vessels. Since their approach relied on the concept of radon, which is integral based, they suppressed the intrinsic noise of images by using this transformation. Figure 12.4 shows a sample example for the results of vessel segmentation.

Supervised approaches use a pixel classification method, referred to as a primitive-based approach by Staal *et al* [212]. This approach was based on the detection of ridges, used as primitives for introducing linear segments, called line elements. Sinthanayothin *et al* [162] identified retinal vessels using an NN whose inputs were deduced from PCA of the image and edge extraction of the first principal component. An NN and backpropagation multilayer method was proposed by Gardner *et al* [122] for vessel tree detection following histogram equalization. The kNN classifier was implemented by Niemeijer *et al* [49]. In their algorithm, from the green plane of the RGB image, a vector of 31-component pixel features was created with the Gaussian and its derivatives up to order 2 at five different scales. Soares *et al* [106] applied a Gaussian mixture model Bayesian classifier. By using the Gabor wavelet transform, multiscale analysis was performed on the image. Ricci and

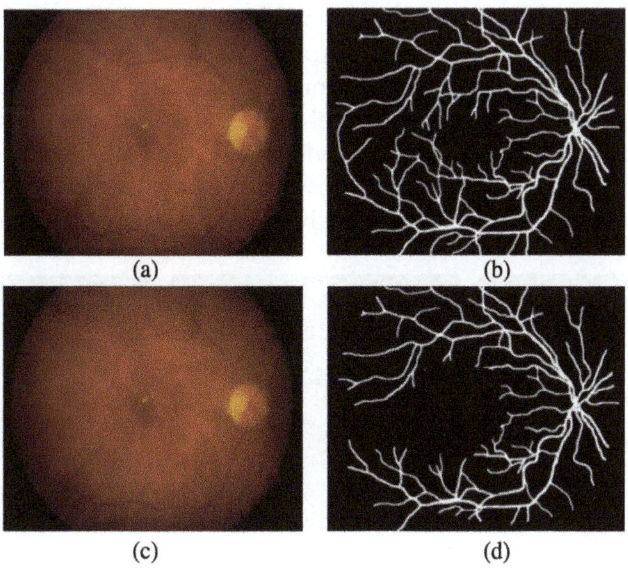

Figure 12.4. Vessel segmentation. (a) and (c) Retinal images from MUMS-DB; (b) and (d) vessel segmentation results. For more details see [101].

Perfetti [207] utilized an SVM for classifying pixels as vessel or non-vessel. In their algorithm, along with the gray-level of the target pixel they applied two orthogonal line detectors to create the feature vector. One of the interesting supervised methods was implemented by Fraz *et al* [56] which was an ensemble classification system of boosted and bagged decision trees. The drawback of supervised classifications is its necessity for a sufficient number of manually annotated training image sets. Moreover, it is not easy to generalize the trained models to meet the needs of different datasets. More recent studies have successfully applied the concept of deep learning to the segmentation of the retinal vasculature [90, 213–222].

12.3.3 Localization of the fovea

The fovea is a small region approximately at the center of the retina and responsible for central and high resolution vision. It is placed around 45 times of ONH radius to the temporal direction of ONH [151, 162, 169, 184, 223–228]. The visual cells placed in the fovea are closely packed and cause the highest sharpness of vision [229]. There are not any blood vessels in the fovea region to block the crossing of light to the fovea cells. The fovea detection approaches are divided into two classes: hybrid and constraint based approaches [1].

12.3.3.1 Hybrid-based method

Sinthanayothin *et al* [162] presented a method for fovea localization using intensity correlation. At the beginning, the fovea was identified with intensity correlation and from HSI transformation the peak was selected. The selected peak located a dark area that was considered as the fovea. The disadvantage of their method was that when the fovea was not at the center it failed. In another method, using location of the fovea, a polar retina coordinate system was established. Here, the fovea accurately was located using its appearance and geometry according to other landmarks [169]. Niemeijer *et al* [184] proposed a method to identify the fovea based on a single point distribution model. The method included all signs including global and regional, to identify this model. These signs were extracted from the related vessel position and its width. Tavakoli *et al* [147] proposed an approach based on vessel segmentation and parabola fitting to identify the fovea. Their

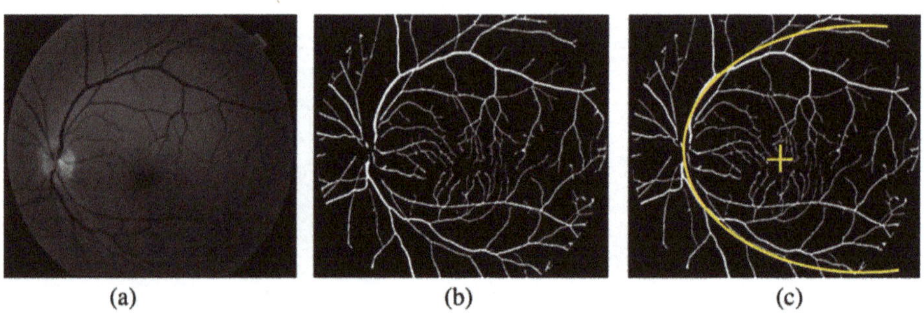

Figure 12.5. Fovea detection. (a) Retinal images from MUMS-DB; (b) vessel segmentation results; and (c) detected fovea. For more details see [147].

algorithm included four steps: multi-overlapping windows, local radon transform, vessel validation, and parabolic fitting. Figure 12.5 shows an example for the results of fovea detection.

12.3.3.2 Positional constraints-based

The positional restrictions [162, 169, 183, 184, 223–225] were applied to localize the fovea and determine its position. The fovea identification was obtained using the location of other retinal regions. The main part of this approach [181] specifies the horizontal retinal ridge that is passing across the ONH and fovea, which divides the upper and lower retinal regions. Further, the region of fovea was extracted using a fixed distance of 5 ONH radius from the ONH center. Table 12.7 summarizes some related methods with brief details.

12.4 Detection of retinal lesions

An acceptable CAD system requires identifying lesions such as MAs, HEs, EXs, CWS, and NVs as perfect as possible to finally result in distinguishing between DR and no DR images. Several studies have worked on various detection and segmentation approaches to detect these lesions. Here, we briefly discuss them.

12.4.1 Microaneurysm and haemorrhage detection

The most reliable method for MA detection is from fluorescein angiography retinal images [19]. However, the fluorescein angiography images are costly, invasive and not suitable for everyone [22]. Moreover, this approach is time-consuming. Introductory global image processing approaches were applied for automated detection. There are a number of methods for the detection of both MAs and HEs. These methods can be generally categorized into two general groups including unsupervised and supervised approaches [17, 22, 107, 108, 110, 111, 226, 233–247]. Many studies have searched both MAs and HEs detection from the retinal images by using proper thresholding after masking the landmarks such as ONH, fovea, and vessel. See figure 12.6, from [2], for more details. After the thresholding,

Table 12.7. Details of some fovea detection methods.

Author	Approach	No. of images	Accuracy
Sinthabayothin *et al* [162]	NN and template-based	112	80.4%
Li and Chutatape [174]	Modified active shape model	35	100%
Tavakoli *et al* [147]	Fitting the main vessels with parabola function	140	90%
Tobin *et al* [181]	ONH and fovea localization	345	92.5%
Seckar *et al* [231]	Hough transform and morphological function	44	100%
Fleming *et al* [183]	Vessel segmentation with semiellipse fitting	1056	96.6%
Niemeijer *et al* [184]	Point distribution model and kNN	500	96.8%
Welfer *et al* [232]	ROI selection and morphological operator	126	96.1%

mathematical morphology was used to distinguish MAs and other structures, such as bifurcations and small vessels. From an unsupervised viewpoint, the MAs detection approaches are classified in to different classes: morphology, region growing, wavelet, and hybrid approaches [1]. Among these groups, the HEs detection is mostly based on mathematical morphology combined with pixel classification. In this section, we briefly explain the related studies.

Walter *et al* [22] presented a four-part method for MAs detection. At the beginning, the preprocessing step used green channel images. Top-hat transformation and global thresholding were used to identify MAs. Next, fifteen characteristics were used to differentiate true MAs from false ones using kNN, Gaussian and kernel density classifiers. Hatanaka *et al* [248] proposed a brightness correction approach for automated detection of HEs. The correction was applied in hue saturation value (HSV) space.

In the morphological approach, different adjustments often increase the accuracy of MAs detection [17, 240]. Despite this type of hand-crafted method being typically fast, easy, and user- friendly, its ability is constrained by its builder. In better words, some main hidden structures and uncovered patterns might be ignored by the builder and cause false detection [110]. The morphological operations such as closing [22], or using top-hat transformation [19] have been applied for MAs detection due to their relatively uniform circular shape and limited size range. However, besides the above issue, most of the mathematical morphology methods mainly depend on the choosing of optimal structuring elements; changing their size and shape decreases the efficiency of these algorithms.

Fleming *et al* [17] developed contrast normalization, watershed and region growing approaches to detect MAs in retinal images. Moreover, they applied a 2D Gaussian-after median filtering to preprocess the image. To detect MAs they used the kNN classification using nine characteristics from extracted candidates. Sinthanayothin *et al* [21] identified MAs by applying region growing and adaptive intensity thresholding, combined with a moat operator. Quellec *et al* [107] used wavelet transform algorithm for MAs detection by matching wavelet sub-bands with the lesion template.

Tavakoli *et al* [19] employed an approach based on top-hat, mathematical morphology and radon transform, applied locally in multi-overlapping windows to detect MAs. Niemeijer *et al* [233] introduced a hybrid red lesion detection method using ideas from Spencer *et al* [249] and Frame *et al* [250]. At the beginning, using pixel classification the red lesion candidates were identified [249]. The MAs and HEs were detected after masking the vessels. Usher *et al* [131] and Gardner *et al* [122] applied NN to locate MAs. Initially, they used local adaptive contrast enhancement to preprocess the image. After that, landmarks (ONH and vessel segmentation), and red lesions (MAs and HEs) were identified using region growth and adaptive intensity thresholding combined with moat operation [21]. Zhang *et al* [251] described a method using multi-scale correlation filtering and threshold values within angiographic retinal image for MA detection. They used moving Gaussian kernels to extract the correlation coefficient for pixels. Larsen *et al* [145] introduced an automatic MAs and HEs detection approach using classification algorithm called RetinaLyze using shape and size of the lesions.

Fleming *et al* [132] applied multiscale and morphological operators to place HEs. For the preprocessing, they corrected the variations in intensity and contrast by applying median filter and histogram equalization. In another study [17], they proposed a region growth algorithm to identify large lesion regions. In their approach, they had difficulty differentiating HEs and MAs. Zhang and Chutatape [252] introduced an approach to detect HEs using SVM. In addition, they applied PCA to detect the features. These features were used as the inputs of SVM to extract HEs. Hipwell *et al* [253] proposed an automated MA detection in red-free retinal images. To remove variation in the illumination, a shading correction was applied to the image.

Another approach in contrast with linear landmarks like vessels, which are directional, is to utilize template matching filters with multiscale Gaussian kernels [107, 235, 251]. According to statistical results, the intensity distribution of MAs is matched to a Gaussian distribution [107] because MAs display Gaussian profiles in all projections. Therefore, a template matching filter highly increased the accuracy of MA detection. Using this idea, Quellec *et al* proposed a wavelet transform method for MA detection. In their algorithm, they detected the MAs using a local template matching filter in the wavelet domain [107]. Zhang *et al* [251]used a multi-scale correlation coefficient based approach. In their method, they used a nonlinear filter with five Gaussian kernels at various standard deviations to detect MA candidates. Ram *et al* [239] described a feature based method that rejects specific classes of clutter while accepting a higher number of true MAs. The potential MAs, obtained after the final step, are labelled a grade that was based on their shape similarity to true MAs. Similar to the morphological approach, this type of algorithm is still limited by ignorance of hidden and unnoticeable structures. Moreover, a lot of parameters are required to be assigned empirically.

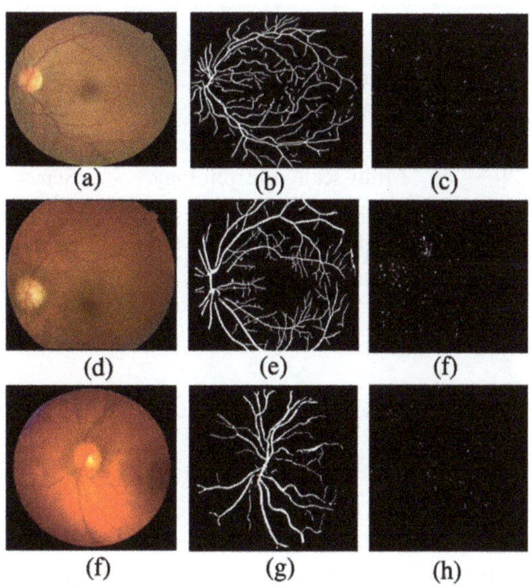

Figure 12.6. MA detection. (a, d, and f) Retinal images from different datasets; (b, e, and g) vessel segmentation and masking results; and (c, f, and h) detected MAs.

Table 12.8. Details of some of the MA and HE detection methods.

Authors	Methods	Performance
MA detection		
Tavakoli *et al* (2013) [19]	Radon transform	Sen. = 92%, Sp. = 90%
Hipwell *et al* (2000) [253]	Matched filter, region growing	Sen. = 81%, Sp. = 93%
Sinthanayothin *et al* (2002) [21]	Moat operator	Sen. = 77.50%, Sp. = 88.70%
Larsen *et al* (2003) [145]	Size and shape	Sen. = 71.4%
Usher *et al* (2004) [131]	Moat operator	Sen. = 95.10%, Sp. = 46.30%
Niemeijer (2005) [233]	Pixel classification using kNN	Sen. = 100%, Sp. = 87%
Fleming *et al* (2006) [17]	Contrast normalization, watershed region growing	Sen. = 85.40%, Sp. = 83.10%
Walter *et al* (2007) [22]	Gaussian filtering, top-hat	Sen. = 88.5%
Quellec *et al* (2008) [107]	Optimal wavelet transform	Sen. = 89.62%, Sp. = 89.50%
Zhang *et al* (2010) [251]	Multi-scale correlation filtering and dynamic thresholding	Sen. = 71.30%
Sanchez (2011) [73]		Sen. = 91%
Antal and Hajdu (2012) [254]	Ensemble-based	AUC= 0.90
Lazar and Hajdu (2013) [235]	Cross section features	—
Pereira (2014) [112]	multi-agent system approach	Sen. = 87%
Seoud (2016) [23]	Dynamic shape features	Sen. = 84.4%
Srivastava (2017) [109]	Filters for feature extraction	—
HE detection		
Gardner (1996) [122]	NN and statistical threshold tuning	Sen. = 73.80%
Zhang and Chutatape (2005) [252]	PCA and SVM	Sen. = 89.10%
Fleming *et al* (2008) [132]	Multi-scale, morphology and SVM	Sen. = 98.60%, Sp. = 95.50%
Hatanaka *et al* (2008) [248]	Brightness correction and thresholding	Sen. = 80%, Sp. = 88%

The brief explanation of the MAs and HEs detection algorithms is introduced in table 12.8.

12.4.2 Exudates segmentation

Here, we briefly review the studies of EXs segmentation. EXs are one of the main signs of DR, because they cause retinal edema and consequently vision loss and blindness. The approaches for identification of EXs [113, 255–257] can be categorized into four classes: pixel clustering, morphology, thresholding/region growing, and classification-based approaches [1]. Here, we concisely introduce these methods.

Several studies [119, 258, 259] have used Fuzzy C-Means cluster-based method for EXs detection. This method suggests pixels with various classes and changing degrees of membership for segmentation purpose [260, 261]. Osareh *et al* [258] first by correcting color and contrast worked on image preprocessing. The preprocessed images were used for a segmentation step using Fuzzy C-Means. The genetic algorithm was applied to choose characteristics from Fuzzy C-Means clustered regions [259]. The optimal characteristics were then used in the NN to detect EXs and non-EXs regions. This approach obtains the differing retinal color. In another study, the EXs were segmented from low quality images using Fuzzy C-Means [119]. The same authors presented EXs detection using Bayesian classifier [119]. Hsu *et al* [69] described an algorithm of EXs segmentation using cluster-based approach. Then, the cluster-based method was utilized to classify the abnormal pixels. The restriction of the above methods is to know considerable clusters which contribute to the EXs detection.

Morphological [120, 262, 263] and thresholding/region growth methods [21, 135] were applied in many studies about EXs detection. Sopharak *et al* [262] presented an EXs segmentation approach for low image quality by applying adjusted morphology. Walter *et al* [263] detected EXs using gray level changes of green channel. After first step detection, their boundaries were determined applying a morphological reconstruction approach. Their method did not differentiate EXs from CWS. Sanchez *et al* [120] used the same approach as Walter *et al* [263] used in preprocessed the green channel and a histogram of the image was modelled using a Gaussian mixture model. Furthermore, a thresholding method was utilized to detect the EXs using Gaussian mixture model. Sinthanayothin *et al* [21] developed a region growing-based method by selecting some threshold values in the gray scale images. To improve the quality of image, adaptive local contrast enhancement was applied. This algorithm located other retinal regions and landmarks such as the ONH, retinal vessels and fovea. Fleming *et al* [183] presented a multi-scale morphology algorithm to localize EXs. They used Gaussian and median filter to preprocess the green channel images for correcting the contrast variations. A watershed region growing algorithm was used to separate the EXs from other landmarks. Finally, They used SVM to classify regions into EXs and non-EXs. Welfer *et al* [264] presented an EXs detection algorithm based on morphology operator. They applied a set of morphological functions to detect EXs. Mookiah *et al* [135] introduced EXs detection by applying color, shape and morphological processing. Their method assumed EXs to have higher contrast than the ONH.

Several studies [122, 233, 265] utilized machine learning approaches to categorize EXs and non-EXs pixels. Gardner *et al* [122] utilized NN to classify EXs applying gray scale of retinal images. Sanchez *et al* [265] segmented EXs according to Fisher's linear discriminant analysis. They used color normalization and contrast enhancement. Niemeijer *et al* [233] introduced an automated method for EXs detection using classification based on a machine learning idea. The pixels with high probability were classified into EX class from non-EXs.

12.5 Artificial intelligence in diabetic retinopathy

In general, understanding and controlling DR or other retinal diseases has become extensively more complicated because of the large number of images and diagnoses. In fact, processing all these patient examinations seems as a big data challenge [266]. Clearly, the new era of diagnosis and clinical data mining immediately needs intelligent systems to control them sufficiently, safely and efficiently.

Recently, through the advancement of supervised learning ideas, the recent application of artificial intelligence (AI) approaches for DR detection and its related lesions have significantly increased in the research community. AI has already revealed proof-of-concept in medical science such as radiology and pathology, which like ophthalmology relies heavily on diagnostic imaging. As its turns out, image processing is the most important application of AI in healthcare [267]. Moreover, AI is particularly suitable in assisting clinical tasks by using efficient algorithms to identify and learn features from large volumes of imaging data, and helping to reduce diagnostic errors. In addition, it can recognize specific patterns or lesions and correlate novel characteristics to obtain innovative scientific insight [268].

The most important branch of AI is machine learning (ML) [268] where the rules are learned directly by algorithms from a series of examples, called training data, instead of being encoded by hand. ML techniques need a set of biomarkers or characteristics to be measured directly from the training data (e.g. labeled lesions in the retinal image). After that, based on a training database of characteristics with known labels, i.e., hand crafted features, a classifier learns to identify the correct label from the newly seen characteristics. Once a few strong classifiers have been established, the effectiveness of such ML models mostly relies on the differentiative power of the chosen characteristics which underpin the classifier performance. There are different reviews that have summarized ML approaches [1, 15, 74, 269–272]. Both Mookiah *et al* [1] and Mansour [74] classified DR diagnosis methods in respect of affiliated methods, such as DR lesion tracking, mathematical morphology, cluster-based, matched filter, and hybrid methods. Faust *et al* [15] surveyed studies that detect lesion characteristics from retinal images, such as the retinal vessels, MAs, HEs, and EXs. Joshi and Karule [269] summarized the first researches on EXs segmentation. Almotiri *et al* [270] came up with a survey of methods to detect the vessels. Almazroa *et al* [271] and Thakur and Juneja [272] studied several approaches for ONH detection.

Recently, the accessibility of large databases and the immense computing power suggested by advanced graphics processing units have research on deep learning (DL) based methods, which have demonstrated magnificent efficiency in all medical sciences. Many DL-based approaches have been established for a variety of works to analyze images to establish automated CAD systems for detection of DR. The relation between AL, ML, and DL has been shown in figure 12.7.

12.5.1 Overview of deep learning

There are multiple DL-based algorithms that have been presented. Some frequently utilized deep architectures for the DR detection purpose include convolutional

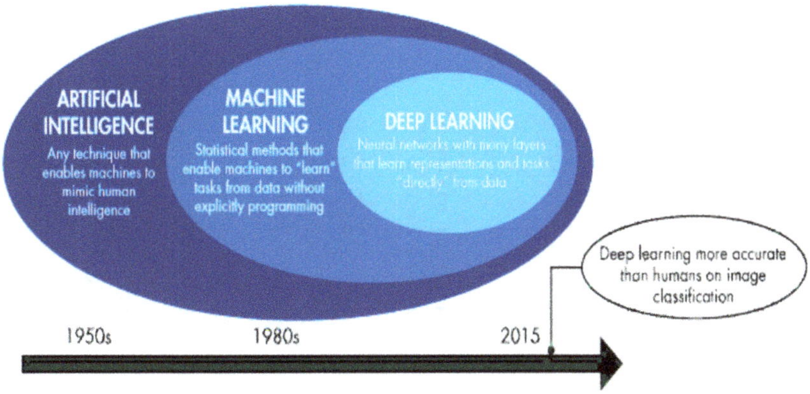

Figure 12.7. Relation between AI, ML and DL.

neural networks (CNNs), autoencoders (AEs), recurrent neural networks (RNNs) and deep belief networks (DBNs). A complete overview of these architectures is found in [273]. There are several DL-based architectures related to DR detection [274–284]. Based on the clinical significance of automated DR detection, we group applications of DL into three broad classes: (i) vessels segmentation, (ii) ONH localization and segmentation, and (iii) retinal lesions detection and classification. In the below sub-sections, we briefly introduce the literature about DL-based algorithms for these tasks. A brief summary of the DL methods in DR detection systems is presented in table 12.9.

12.5.1.1 Retinal blood vessel segmentation

Many of the retinal vessel segmentation approaches are based on CNN architectures [216, 217, 285–289].

Maji *et al* [285] proposed 12 CNN models for this purpose. Each CNN model included three convolutional layers and two fully connected layers. Liskowski and Krawiec [217] introduced a supervised pixel-wise algorithm using deep CNN, which was trained by preprocessed retinal images, and augmented using geometric transformations. Tan *et al* [216] employed a seven-layer CNN model to segment blood vessels, ONH, and fovea.

Some methods use AEs in segmentation of retinal vessels [290, 291]. Roy and Sheet [290] presented an AE-based deep NN model that used the domain adaptation method for training the model. The AE-based deep NN included two hidden layers, which were trained using the source domain databases, using an autoencoding and supervised mechanism. The algorithm introduced by Maji *et al* [285] employed a hybrid DL, which consisted of unsupervised stacked denoising AEs with a total of 500 layers, to segment vessels in retinal images.

12.5.1.2 Optic nerve head detection

As we mentioned at the beginning of this chapter, ONH identification can improve DR detection and classification. In fact, the ONH brightness can cause uncertainty

Table 12.9. Details of some of the MA and HE detection methods.

Authors	Methods	Training	Type	Database	Performance
Maji *et al* [285]	Patch-based ensemble of CNN	End-to-end	Vessel segmentation	DRIVE	AUC = 0.9283 ACC = 94.7
Liskowski and Krawiec [217]	Patch-based CNN	End-to-end	Vessel segmentation	DRIVE, STARE, CHASE	AUC = 0.9790, 0.9928, 0.988 ACC = 95.35, 97.29, 96.96
Maninis *et al* [217]	Fully CNN (FCN)	Transfer learning	Vessel segmentation	DRIVE, STARE	—
Wu *et al* [100]	tracking/patch-based CNN/PCA as classifier	End-to-end	Vessel segmentation	DRIVE	AUC = 0.9701
Dasgupta and Singh [287]	Patch-based FCN	End-to-end	Vessel segmentation	DRIVE	AUC = 0.974 ACC = 95.33
Tan *et al* [216]	Patch-based CNN	End-to-end	Vessel segmentation	DRIVE, STARE	ACC = 94.70, 95.45
Mo and Zhang [289]	Multi-level FCN	End-to-end	Vessel segmentation	DRIVE, STARE, CHASE	AUC = 0.9782, 0.9885, 0.9812 ACC = 95.21, 96.76, 95.99
Maji *et al* [285]	Patch-based AE	Transfer learning	Vessel segmentation	DRIVE	AUC = 0.9195, ACC = 93.27
Li *et al* [291]	Patch-based AE	End-to-end	Vessel segmentation	DRIVE, STARE, CHASE	AUC = 0.9738, –, 0.9716, ACC = 95.27, 96.28, 95.81
Lahiri *et al* [290]	Patch-based AE	Transfer learning	Vessel segmentation	DRIVE	ACC = 95.3
Fu *et al* [288]	Patch-based CNN as RNN	End-to-end	Vessel segmentation	DRIVE, STARE, CHASE	ACC = 95.23, 95.85, 94.89
Lim *et al* [292]	CNN with exaggeration	End-to-end	ONH segmentation	MESSIDOR, SEED-DB	—
Tan *et al* [216]	CNN	End-to-end	ONH segmentation	DRIVE	ACC = 87.90

Sevastopolsky [294]	Modified U-Net CNN	Transfer learning	ONH segmentation	DRION-DB	—
Zilly *et al* [201]	Multi-scale CNN	End-to-end	ONH segmentation	DRISHTI-GS	—
Alghamdi *et al* [295]	Cascade CNN, each model	End-to-end	ONH detection	DRIVE, DIARETDB1, MESSIDOR, STARE,	ACC = 100, 98.88, 99.20, 86.71
Xu *et al* [177]	CNN based on deconvolution	Transfer learning	ONH detection	MESSIDOR, STARE	ACC = 99.43, 89
Haloi [296]	9-layer CNN	End-to-end	MA detection	MESSIDOR, ROC	AUC = 0.982 ACC = 95.4
van Grinsven *et al* [111]	Patches based selective sampling	End-to-end	HE detection	Kaggle, MESSIDOR	AUC = 0.917, 0.979
Gondal *et al* [301]	CNN model	Transfer learning	HE detection	DIARETDB1	—
Quellec *et al* [276]	CNN model	Transfer learning	HE detection	DIARETDB1	AUC = 0.614
Orlando *et al* [114]	HEF + CNN and RF classifier	End-to-end	MA-HE detection	DIARETDB1, e-ophtha, MESSIDOR	AUC = 0.8932
Tavakoli *et al* [2, 6, 16]	Patches CNN	End-to-end	MA detection	DRIVE, DIARETDB1, MESSIDOR	ACC = 95.97
Shan and Li [298]	Patches based AE	Transfer learning	MA detection	DIARETDB	ACC = 91.38
Prentasic and Loncaric [303]	11-layer CNN,	End-to-end	EX detection	DRiDB	—
Gondal *et al* [301]	CNN model	Transfer learning	EX detection	DIARETDB1	
Quellec *et al* [276]	CNN model	Transfer learning	EX detection	DIARETDB1	AUC = 0.974,

for detecting of other bright lesions such as EXs. There are a number of studies that work on both ONH detection and segmentation using the concept of DL [201, 292–295].

Lim *et al* [292] presented a nine-layer CNN model for ONH segmentation. Their algorithm involved different steps: detecting the region around the ONH, enhancement, classification at pixel-level using a CNN model to predict the ONH. Tan *et al* [216] identified the ONH and vessels jointly using pixel patch-based CNN. Zilly *et al* [201] presented an ONH segmentation algorithm based on a multiscale two-layer CNN model. First, the area around the ONH was cropped, and down-sampled. After that, the area was analyzed by entropy filtering to determine the most distinguished points and was passed to the CNN model for final segmentation. Zhang *et al* [293] applied a faster CNN to localize the ONH. After identifying the ONH, the thick vessels inside its area were eliminated by using a Hessian matrix, and its boundaries segmented.

12.5.1.3 Microaneurysms and hemorrhages detection

Many DL-based approaches have been introduced for detection and classification of different types of DR lesions such as MAs, and HEs [2, 6, 16, 111, 114, 274, 275, 296–300].

Haloi [296] introduced a nine-layer CNN model to group each pixel as MA or non-MA and grading the Dr. Each pixel is categorized by taking a window around the candidate and passing it to the CNN model. For training, they applied a data augmentation method to create six windows around each pixel. van Grinsven *et al* [111] worked on detecting HEs. Their main contribution was to overcome the over--represented normal samples created for training a CNN model. To figure out this issue, they employed a dynamic selective sampling idea that selects informative training samples. Their CNN was trained using a dynamic selective sampling strategy as well.

Gondal *et al* [301] and Quellec *et al* [276] also detected HEs and small red dots. Another similar approach was introduced by Orlando *et al* [4]. In their approach, they first extracted red candidates using morphological operations. Next, they extracted CNN and hand-engineered characteristics such as intensity and shape from each candidate to introduce a probability map, which was utilized to make both red lesion and image level decision. Shan and Li [298] applied the stacked sparse AEs to identify MAs. A patch was passed to stacked sparse AEs, which extracted characteristics, and the Softmax classifier labelled it as an MA or non-MA patch. Tavakoli *et al* employed a CNN approach to detect MAs. In their approach, they first extracted candidates using morphological functions. Next, they used CNN classification to classify between MAs and non-MAs. In their study, the authors compared the effect of two preprocessing methods, illumination equalization, and top-hat transformation, on retinal images to detect MAs using a combination of matching based approach and CNN methods [6].

12.5.1.4 Exudates detection

There are different studies on EX detections using DL [276, 301–303]. Prentasic and Loncaric [303] employed a CNN-based algorithm for EXs detection in retinal images. First, they detected the ONH, created a probability map for that. Then they created for both vessels bright-boundary probability maps. Lastly, they applied an 11-layer CNN model to produce an EX probability map and finally segmented the EXs. Gondal *et al* [301] presented an algorithm for EXs detection combined with other DR lesions based on CNN architecture [304]. To identify DR lesions including different types of EXs and HEs, the dense layers were removed from the CNN model. For detecting DR lesions such as EXs at the pixel level, Quellec *et al* [276] proposed a CNN visualization method. The authors did a modification on their CNN based on the sensitivity evaluation done by Simonyan and Zisserman [305], which helped in detecting DR and lesions by optimizing CNN predictions.

12.6 Conclusion

This chapter introduces a detailed study of methods and their results for DR detection and classification using retinal image. DR is a complication of diabetes that damages the retina, and causes vision loss and blindness. A robust DR screening system will remarkably decrease the workload of clinicians. The screening process brings sets of steps, namely identifying and segmenting the retinal landmarks, detecting lesions, feature extraction and classification. All these steps need different methods and techniques. Although considerable accomplishments have been made in digital retinal image analysis, there are a variety of challenges in the selection of suitable approaches which result in high accuracy during DR screening.

References

[1] Mookiah M R K, Acharya U R, Chua C K, Lim C M, Ng E and Laude A 2013 Computer-aided diagnosis of diabetic retinopathy: a review *Comput. Biol. Med.* **43** 2136–55

[2] Tavakoli M, Jazani S and Nazar M 2020 Automated detection of microaneurysms in color fundus images using deep learning with different preprocessing approaches *SPIE* **11318** 113180E

[3] Thomas R L, Dunstan F D, Luzio S D, Chowdhury S R, North R V and Hale S L *et al* 2015 Prevalence of diabetic retinopathy within a national diabetic retinopathy screening service *Br. J. Ophthalmol.* **99** 64–8

[4] Zana F and Klein J C 2001 Segmentation of vessel-like patterns using mathematical morphology and curvature evaluation *IEEE Trans. Image Process.* **10** 1010–9

[5] Nayak J, Bhat P S, Acharya R, Lim C M and Kagathi M 2008 Automated identification of diabetic retinopathy stages using digital fundus images *J. Med. Syst.* **32** 107–15

[6] Tavakoli M and Nazar M 2020 Comparison different vessel segmentation methods in automated microaneurysms detection in retinal images using convolutional neural networks *SPIE* **11317** 113171P

[7] Cade W T 2008 Diabetes-related microvascular and macrovascular diseases in the physical therapy setting *Phys. Ther.* **88** 1322–35

[8] Emdin C A, Rahimi K, Neal B, Callender T, Perkovic V and Patel A 2015 Blood pressure lowering in type 2 diabetes: a systematic review and meta-analysis *JAMA* **313** 603–15

[9] Schaffer S W, Jong C J and Mozaffari M 2012 Role of oxidative stress in diabetes-mediated vascular dysfunction: unifying hypothesis of diabetes revisited *Vascul. Pharmacol.* **57** 139–49

[10] Patz A 1980 Studies on retinal neovascularization. friedenwald lecture *Investig. Ophthalmol. Vis. Sci.* **19** 1133–8

[11] Niemeijer M, van Ginneken B, Russell S R, Suttorp-Schulten M S and Abramoff M D 2007 Automated detection and differentiation of drusen, exudates, and cotton-wool spots in digital color fundus photographs for diabetic retinopathy diagnosis *Investig. Ophthalmol. Vis. Sci.* **48** 2260–7

[12] Tavakoli M, Mehdizadeh A, Pourreza R, Banaee T, Bahreyni Toossi M H and Pourreza H R 2010 Early detection of diabetic retinopathy in fluorescent angiography retinal images using image processing methods *Iran. J. Med. Phys.* **7** 7–14

[13] Antonetti D A, Barber A J, Bronson S K, Freeman W M, Gardner T W and Jefferson L S *et al* 2006 Diabetic retinopathy: seeing beyond glucose-induced microvascular disease *Diabetes* **55** 2401–11

[14] Watkins P J 2003 Retinopathy *Brit. Med. J.* **326** 924–6

[15] Faust O, Acharya R, Ng E Y K, Ng K H and Suri J S 2012 Algorithms for the automated detection of diabetic retinopathy using digital fundus images: a review *J. Med. Syst.* **36** 145–57

[16] Tavakoli M, Nazar M and Mehdizadeh A 2020 The efficacy of microaneurysms detection with and without vessel segmentation in color retinal images *SPIE* **11314** 113143Y

[17] Fleming A D, Philip S, Goatman K A, Olson J A and Sharp P F 2006 Automated assessment of diabetic retinal image quality based on clarity and field definition *Investig. Ophthalmol. Vis. Sci.* **47** 1120–5

[18] Pourreza H R, Bahreyni Toossi M H, Mehdizadeh A, Pourreza R and Tavakoli M 2009 Automatic detection of microaneurysms in color fundus images using a local radon transform method *Iran. J. Med. Phys.* **6** 13–20

[19] Tavakoli M, Shahri R P, Pourreza H, Mehdizadeh A, Banaee T and Toosi M H B 2013 A complementary method for automated detection of microaneurysms in fluorescein angiography fundus images to assess diabetic retinopathy *Pattern Recognit.* **46** 2740–53

[20] Tavakoli M, Naji M, Abdollahi A and Kalantari F 2017 Attenuation correction in spect images using attenuation map estimation with its emission data *Medical Imaging 2017: Physics of Medical Imaging* vol 10132 (Bellingham, WA: International Society for Optics and Photonics) p 101324Z

[21] Sinthanayothin C, Boyce J F, Williamson T H, Cook H L, Mensah E and Lal S *et al* 2002 Automated detection of diabetic retinopathy on digital fundus images *Diabet. Med.* **19** 105–12

[22] Walter T, Massin P, Erginay A, Ordonez R, Jeulin C and Klein J C 2007 Automatic detection of microaneurysms in color fundus images *Med. Image Anal.* **11** 555–66

[23] Seoud L, Hurtut T, Chelbi J, Cheriet F and Langlois J P 2015 Red lesion detection using dynamic shape features for diabetic retinopathy screening *IEEE Trans. Med. Imaging* **35** 1116–26

[24] Group E T D R S R *et al* 1991 Grading diabetic retinopathy from stereoscopic color fundus photographsan extension of the modified airlie house classification: etdrs report number 10 *Ophthalmology* **98** 786–806

[25] Cusick M, Chew E Y, Chan C C, Kruth H S, Murphy R P and Ferris F L III 2003 Histopathology and regression of retinal hard exudates in diabetic retinopathy after reduction of elevated serum lipid levels *Ophthalmology* **110** 2126–33

[26] McLeod D 2005 Why cotton wool spots should not be regarded as retinal nerve fibre layer infarcts *Br. J. Ophthalmol.* **89** 229–37

[27] Chui T Y, Thibos L N, Bradley A and Burns S A 2009 The mechanisms of vision loss associated with a cotton wool spot *Vis. Res.* **49** 2826–34

[28] Vallabha D, Dorairaj R, Namuduri K and Thompson H 2004 Automated detection and classification of vascular abnormalities in diabetic retinopathy Conf. Record of the Thirty-Eighth Asilomar Conf. on Signals, Systems and Computers vol 2 (Piscataway, NJ: IEEE) pp 1625–9

[29] Williams R, Airey M, Baxter H, Forrester J, Kennedy-Martin T and Girach A 2004 Epidemiology of diabetic retinopathy and macular oedema: a systematic review *Eye* **18** 963–83

[30] Giancardo L, Meriaudeau F, Karnowski T P, Li Y, Garg S and Tobin K W Jr *et al* 2012 Exudate-based diabetic macular edema detection in fundus images using publicly available datasets *Med. Image Anal.* **16** 216–26

[31] Taylor S R, Lightman S L, Sugar E A, Jaffe G J, Freeman W R and Altaweel M M *et al* 2012 The impact of macular edema on visual function in intermediate, posterior, and panuveitis *Ocular Immunol. Inflamm.* **20** 171–81

[32] Molina-Casado J M, Carmona E J and Garcia-Feijoo J 2017 Fast detection of the main anatomical structures in digital retinal images based on intra-and inter-structure relational knowledge *Comput. Methods Programs Biomed.* **149** 55–68

[33] Moccia S, De Momi E, El Hadji S and Mattos L S 2018 Blood vessel segmentation algorithmsreview of methods, datasets and evaluation metrics *Comput. Methods Programs Biomed.* **158** 71–91

[34] Wilkinson C, Ferris F L III, Klein R E, Lee P P, Agardh C D and Davis M *et al* 2003 Proposed international clinical diabetic retinopathy and diabetic macular edema disease severity scales *Ophthalmology* **110** 1677–82

[35] Group E T D R S R *et al* 1991 Classification of diabetic retinopathy from fluorescein angiograms: Etdrs report number 11 *Ophthalmology* **98** 807–22

[36] Philip S, Fleming A D, Goatman K A, Fonseca S, Mcnamee P and Scotland G S *et al* 2007 The efficacy of automated disease/no disease grading for diabetic retinopathy in a systematic screening programme *Br. J. Ophthalmol.* **91** 1512–7

[37] Salamat N, Missen M M S and Rashid A 2019 Diabetic retinopathy techniques in retinal images: a review *Artif. Intell. Med.* **97** 168–88

[38] Venkatesan R, Chandakkar P, Li B and Li H K 2012 Classification of diabetic retinopathy images using multi-class multiple-instance learning based on color correlogram features 2012 Annual Int. Conf. of the IEEE Engineering in Medicine and Biology Society (Piscataway, NJ: IEEE) pp 1462–5

[39] Patton N, Aslam T M, MacGillivray T, Deary I J, Dhillon B and Eikelboom R H *et al* 2006 Retinal image analysis: concepts, applications and potential *Prog. Retin. Eye Res.* **25** 99–127

[40] Kanski J J 2009 *Clinical Ophthalmology: A Synopsis* (Amsterdam: Elsevier Health Science)

[41] Zachariah S, Wykes W and Yorston D 2015 Grading diabetic retinopathy (Dr) using the scottish grading protocol *Community Eye Health* **28** 72

[42] Scanlon P H 2017 The english national screening programme for diabetic retinopathy 2003–2016 *Acta Diabetol.* **54** 515–25

[43] Ciulla T A, Amador A G and Zinman B 2003 Diabetic retinopathy and diabetic macular edema: pathophysiology, screening, and novel therapies *Diabet. Care* **26** 2653–64

[44] Abr'amoff M D, Garvin M K and Sonka M 2010 Retinal imaging and image analysis *IEEE Rev. Biomed. Eng.* **3** 169–208

[45] Kauppi T, Kalesnykiene V, Kamarainen J K, Lensu L, Sorri I and Raninen A *et al* 2007 The DIARETDB1 diabetic retinopathy database and evaluation protocol *Proc. of the British Machine Conf.* **1** 1–10
[46] Budai A, Bock R, Maier A, Hornegger J and Michelson G 2013 Robust vessel segmentation in fundus images *Int. J. Biomed. Imaging* **2013** 154860
[47] Hoover A, Kouznetsova V and Goldbaum M 2000 Locating blood vessels in retinal images by piecewise threshold probing of a matched filter response *IEEE Trans. Med. Imaging* **19** 203–10
[48] Niemeijer M, Van Ginneken B, Cree M J, Mizutani A, Quellec G and Sánchez C I *et al* 2009 Retinopathy online challenge: automatic detection of microaneurysms in digital color fundus photographs *IEEE Trans. Med. Imaging* **29** 185–95
[49] Niemeijer M, Staal J, van Ginneken B, Loog M and Abramoff M D 2004 Comparative study of retinal vessel segmentation methods on a new publicly available database *SPIE Medical Imaging* vol 5370 (Bellingham, WA: SPIE) p 648–56
[50] Decenci'ere E, Zhang X, Cazuguel G, Lay B, Cochener B and Trone C *et al* 2014 Feedback on a publicly distributed image database: the messidor database *Image Anal. Stereol.* **33** 231–4
[51] Decenci'ere E, Cazuguel G, Zhang X, Thibault G, Klein J C and Meyer F *et al* 2013 Teleophta: Machine learning and image processing methods for teleophthalmology *Irbm* **34** 196–203
[52] Fraz M M, Remagnino P, Hoppe A, Uyyanonvara B, Rudnicka A R and Owen C G *et al* 2012 Blood vessel segmentation methodologies in retinal images–a survey *Comput. Methods Programs Biomed.* **108** 407–33
[53] Pourreza-Shahri R, Tavakoli M and Kehtarnavaz N 2014 Computationally efficient optic nerve head detection in retinal fundus images *Biomed. Signal Process. Control* **11** 63–73
[54] Kaggle retinal database https://kaggle.com/c/diabetic-retinopathy-detection/data Accessed: 2018-1-08
[55] E-ophtha retinal database http://adcis.net/en/Download-Third-Party/E-Ophtha.html Accessed: 2018-1-08
[56] Fraz M M, Remagnino P, Hoppe A, Uyyanonvara B, Rudnicka A R and Owen C G *et al* 2012 An ensemble classification-based approach applied to retinal blood vessel segmentation *IEEE Trans. Biomed. Eng.* **59** 2538–48
[57] Aria retinal database http://eyecharity.com/aria_online.html Accessed: 2018-2-28
[58] Seed-db retinal database https://seri.com.sg/key-programmes/singapore-epidemiology-of-eye-diseases-seed/ Accessed: 2018-8-08
[59] EyePACS retinal database http://eyepacs.com/eyepacssystem/ Accessed: 2018-3-01
[60] Zhang Z, Yin F S, Liu J, Wong W K, Tan N M and Lee B H *et al* 2010 Origa-light: an online retinal fundus image database for glaucoma analysis and research 2010 Annual Int. Conf. of the IEEE Engineering in Medicine and Biology (Piscataway, NJ: IEEE) pp 3065–8
[61] Esmaeili M, Rabbani H and Dehnavi A M 2012 Automatic optic disk boundary extraction by the use of curvelet transform and deformable variational level set model *Pattern Recognit.* **45** 2832–42
[62] Qureshi R J, Kovacs L, Harangi B, Nagy B, Peto T and Hajdu A 2012 Combining algorithms for automatic detection of optic disc and macula in fundus images *Comput. Vis. Image Understanding* **116** 138–45

[63] Welfer D, Scharcanski J and Marinho D 2013 A morphologic two-stage approach for automated optic disk detection in color eye fundus images *Pattern Recognit. Lett.* **34** 476–85
[64] Harangi B and Hajdu A 2015 Detection of the optic disc in fundus images by combining probability models *Comput. Biol. Med.* **65** 10–24
[65] Akyol K, Sen B and Bayir S 2016 Automatic detection of optic disc in retinal image by using keypoint detection, texture analysis, and visual dictionary techniques *Comput. Math. Method. Med.* **2016** 6814791
[66] Dai B, Wu X and Bu W 2017 Optic disc segmentation based on variational model with multiple energies *Pattern Recognit.* **64** 226–35
[67] Alshayeji M and Al-Roomi S A *et al* 2017 Optic disc detection in retinal fundus images using gravitational law-based edge detection *Med. Biol. Eng. Comput.* **55** 935–48
[68] Basit A and Fraz M M 2015 Optic disc detection and boundary extraction in retinal images *Appl. Opt.* **54** 3440–7
[69] Hsu W, Pallawala P, Lee M L and Eong K G A 2001 *The Role of Domain Knowledge in the Detection of Retinal Hard Exudates* vol 2 (Piscataway, NJ: IEEE) p II
[70] Niemeijer M, Abramoff M D, Van and Ginneken B 2009 Information fusion for diabetic retinopathy cad in digital color fundus photographs *IEEE Trans. Med. Imaging* **28** 775–85
[71] Quellec G, Russell S R and Abra'moff M D 2010 Optimal filter framework for automated, instantaneous detection of lesions in retinal images *IEEE Trans. Med. Imaging* **30** 523–33
[72] Amel F, Mohammed M and Abdelhafid B 2012 Improvement of the hard exudates detection method used for computer-aided diagnosis of diabetic retinopathy *Int. J. Image, Graph. Signal Process.* **4** 19–27
[73] Sánchez C I, Niemeijer M, Dumitrescu A V, Suttorp-Schulten M S, Abramoff M D and van Ginneken B 2011 Evaluation of a computer-aided diagnosis system for diabetic retinopathy screening on public data *Investig. Ophthalmol. Vis. Sci.* **52** 4866–71
[74] Mansour R F 2017 Evolutionary computing enriched computer-aided diagnosis system for diabetic retinopathy: a survey *IEEE Rev. Biomed. Eng.* **10** 334–49
[75] Kumar D, Taylor G W and Wong A 2019 Discovery radiomics with clear-Dr: interpretable computer aided diagnosis of diabetic retinopathy *IEEE Access* **7** 25891–6
[76] Ganesan K, Martis R J, Acharya U R, Chua C K, Min L C and Ng E *et al* 2014 Computer aided diabetic retinopathy detection using trace transforms on digital fundus images *Med. Biol. Eng. Comput.* **52** 663–72
[77] Sim D A, Keane P A, Tufail A, Egan C A, Aiello L P and Silva P S 2015 Automated retinal image analysis for diabetic retinopathy in telemedicine *Curr. Diabet. Rep.* **15** 14
[78] Mandrekar J N 2010 Receiver operating characteristic curve in diagnostic test assessment *J. Thorac. Oncol.* **5** 1315–6
[79] Marín D, Aquino A, Gegúndez-Arias M E and Bravo J M 2010 A new supervised method for blood vessel segmentation in retinal images by using gray-level and moment invariantsbased features *IEEE Trans. Med. Imaging* **30** 146–58
[80] Yan Z, Yang X and Cheng K T 2017 A skeletal similarity metric for quality evaluation of retinal vessel segmentation *IEEE Trans. Med. Imaging* **37** 1045–57
[81] Bharkad S 2017 Automatic segmentation of optic disk in retinal images *Biomed. Signal Process. Control* **31** 483–98
[82] Leontidis G, Al-Diri B and Hunter A 2017 A new unified framework for the early detection of the progression to diabetic retinopathy from fundus images *Comput. Biol. Med.* **90** 98–115

[83] Zou B, Chen C, Zhu C, Duan X and Chen Z 2018 Classified optic disc localization algorithm based on verification model *Comput. Graph.* **70** 281–7
[84] Xiong L and Li H 2016 An approach to locate optic disc in retinal images with pathological changes *Comput. Med. Imaging Graph.* **47** 40–50
[85] Sarathi M P, Dutta M K, Singh A and Travieso C M 2016 Blood vessel inpainting based technique for efficient localization and segmentation of optic disc in digital fundus images *Biomed. Signal Process. Control* **25** 108–17
[86] Lu S 2011 Accurate and efficient optic disc detection and segmentation by a circular transformation *IEEE Trans. Med. Imaging* **30** 2126–33
[87] Lowell J, Hunter A, Steel D, Basu A, Ryder R and Fletcher E *et al* 2004 Optic nerve head segmentation *IEEE Trans. Med. Imaging* **23** 256–64
[88] Hoover A and Goldbaum M 2003 Locating the optic nerve in a retinal image using the fuzzy convergence of the blood vessels *IEEE Trans. Med. Imaging* **22** 951–8
[89] Dash J and Bhoi N 2017 A thresholding based technique to extract retinal blood vessels from fundus images *Future Comput. Inform. J.* **2** 103–9
[90] Zhu C, Zou B, Zhao R, Cui J, Duan X and Chen Z *et al* 2017 Retinal vessel segmentation in colour fundus images using extreme learning machine *Comput. Med. Imaging Graph.* **55** 68–77
[91] Neto L C, Ramalho G L, Neto J F R, Veras R M and Medeiros F N 2017 An unsupervised coarseto-fine algorithm for blood vessel segmentation in fundus images *Expert Syst. Appl.* **78** 182–92
[92] Xu L and Luo S 2010 A novel method for blood vessel detection from retinal images *Biomed. Eng. Online* **9** 14
[93] Nguyen U T, Bhuiyan A, Park L A and Ramamohanarao K 2013 An effective retinal blood vessel segmentation method using multi-scale line detection *Pattern Recognit.* **46** 703–15
[94] Zhang L, Fisher M and Wang W 2015 Retinal vessel segmentation using multi-scale textons derived from keypoints *Comput. Med. Imaging Graph.* **45** 47–56
[95] Fraz M M, Barman S, Remagnino P, Hoppe A, Basit A and Uyyanonvara B *et al* 2012 An approach to localize the retinal blood vessels using bit planes and centerline detection *Comput. Methods Programs Biomed.* **108** 600–16
[96] Marín D, Aquino A, Gegúndez-Arias M E and Bravo J M 2011 A new supervised method for blood vessel segmentation in retinal images by using gray-level and moment invariantsbased features *IEEE Trans. Med. Imaging* **30** 146–58
[97] Akram M U and Khan S A 2013 Multilayered thresholding-based blood vessel segmentation for screening of diabetic retinopathy *Eng. Comput.* **29** 165–73
[98] Vermeer K A, Vos F M, Lemij H G and Vossepoel A M 2004 A model based method for retinal blood vessel detection *Comput. Biol. Med.* **34** 209–19
[99] Chakraborti T, Jha D K, Chowdhury A S and Jiang X 2015 A self-adaptive matched filter for retinal blood vessel detection *Mach. Vis. Appl.* **26** 55–68
[100] Wu D, Zhang M, Liu J C and Bauman W 2006 On the adaptive detection of blood vessels in retinal images *IEEE Trans. Biomed. Eng.* **53** 341–3
[101] Tavakoli M, Mehdizadeh A, Pourreza R, Pourreza H R, Banaee T and Toosi M B 2011 Radon transform technique for linear structures detection: application to vessel detection in fluorescein angiography fundus images *Nuclear Science Symp. Conf. Record* (Piscataway, NJ: IEEE) pp 3051–6

[102] Roychowdhury S, Koozekanani D D and Parhi K K 2015 Blood vessel segmentation of fundus images by major vessel extraction and subimage classification *IEEE J. Biomed. Health Inform.* **19** 1118–28

[103] Tolias Y A and Panas S M 1998 A fuzzy vessel tracking algorithm for retinal images based on fuzzy clustering *IEEE Trans. Med. Imaging* **17** 263–73

[104] Azzopardi G, Strisciuglio N, Vento M and Petkov N 2015 Trainable cosfire filters for vessel delineation with application to retinal images *Med. Image Anal.* **19** 46–57

[105] Zana F and Klein J C 1999 A multimodal registration algorithm of eye fundus images using vessels detection and hough transform *IEEE Trans. Med. Imaging* **18** 419–28

[106] Soares J V, Leandro J J, Cesar R M, Jelinek H F and Cree M J 2006 Retinal vessel segmentation using the 2-d gabor wavelet and supervised classification *IEEE Trans. Med. Imaging* **25** 1214–22

[107] Quellec G, Lamard M, Josselin P M, Cazuguel G, Cochener B and Roux C 2008 Optimal wavelet transform for the detection of microaneurysms in retina photographs *IEEE Trans. Med. Imaging* **27** 1230–41

[108] Wang S, Tang H L, Hu Y, Sanei S, Saleh G M and Peto T *et al* 2016 Localizing microaneurysms in fundus images through singular spectrum analysis *IEEE Trans. Biomed. Eng.* **64** 990–1002

[109] Srivastava R, Duan L, Wong D W, Liu J and Wong T Y 2017 Detecting retinal microaneurysms and hemorrhages with robustness to the presence of blood vessels *Comput. Methods Programs Biomed.* **138** 83–91

[110] Dai L, Fang R, Li H, Hou X, Sheng B and Wu Q *et al* 2018 Clinical report guided retinal microaneurysm detection with multi-sieving deep learning *IEEE Trans. Med. Imaging* **37** 1149–61

[111] van Grinsven M J, van Ginneken B, Hoyng C B, Theelen T and Sánchez C I 2016 Fast convolutional neural network training using selective data sampling: application to hemorrhage detection in color fundus images *IEEE Trans. Med. Imaging* **35** 1273–84

[112] Pereira C, Veiga D, Mahdjoub J, Guessoum Z, Gonçalves L and Ferreira M *et al* 2014 Using a multi-agent system approach for microaneurysm detection in fundus images *Artif. Intell. Med.* **60** 179–88

[113] Liu Q, Zou B, Chen J, Ke W, Yue K and Chen Z *et al* 2017 A location-to-segmentation strategy for automatic exudate segmentation in colour retinal fundus images *Comput. Med. Imaging Graph.* **55** 78–86

[114] Orlando J I, Prokofyeva E, del Fresno M and Blaschko M B 2018 An ensemble deep learning based approach for red lesion detection in fundus images *Comput. Methods Programs Biomed.* **153** 115–27

[115] Zhang X, Thibault G, Decenci'ere E, Marcotegui B, Laÿ B and Danno R *et al* 2014 Exudate detection in color retinal images for mass screening of diabetic retinopathy *Med. Image Anal.* **18** 1026–43

[116] Youssef D and Solouma N H 2012 Accurate detection of blood vessels improves the detection of exudates in color fundus images *Comput. Methods Programs Biomed.* **108** 1052–61

[117] Fraz M M, Jahangir W, Zahid S, Hamayun M M and Barman S A 2017 Multiscale segmentation of exudates in retinal images using contextual cues and ensemble classification *Biomed. Signal Process. Control* **35** 50–62

[118] Amin J, Sharif M, Yasmin M, Ali H and Fernandes S L 2017 A method for the detection and classification of diabetic retinopathy using structural predictors of bright lesions *J. Comput. Sci.* **19** 153–64

[119] Sopharak A, Uyyanonvara B and Barman S 2009 Automatic exudate detection for diabetic retinopathy screening *Sci. Asia* **35** 80–8

[120] Sánchez C I, García M, Mayo A, Lo�ez M I and Hornero R 2009 Retinal image analysis based on mixture models to detect hard exudates *Med. Image Anal.* **13** 650–8

[121] Koh J E, Acharya U R, Hagiwara Y, Raghavendra U, Tan J H and Sree S V *et al* 2017 Diagnosis of retinal health in digital fundus images using continuous wavelet transform (cwt) and entropies *Comput. Biol. Med.* **84** 89–97

[122] Gardner G, Keating D, Williamson T and Elliott A 1996 Automatic detection of diabetic retinopathy using an artificial neural network: a screening tool *Br. J. Ophthalmol.* **80** 940–4

[123] Rubini S S and Kunthavai A 2015 Diabetic retinopathy detection based on eigenvalues of the hessian matrix *Procedia Comput. Sci.* **47** 311–8

[124] Gulshan V, Peng L, Coram M, Stumpe M C, Wu D and Narayanaswamy A *et al* 2016 Development and validation of a deep learning algorithm for detection of diabetic retinopathy in retinal fundus photographs *JAMA* **316** 2402–10

[125] Köse C, Şevik U, IKibaş C and Erdöl H 2012 Simple methods for segmentation and measure-˙ ment of diabetic retinopathy lesions in retinal fundus images *Comput. Methods Programs Biomed.* **107** 274–93

[126] Yang Y, Li T, Li W, Wu H, Fan W and Zhang W 2017 Lesion detection and grading of diabetic retinopathy via two-stages deep convolutional neural networks Int. Conf. on Medical Image Computing and Computer-Assisted Intervention (Berlin: Springer) pp 533–40

[127] Sisodia D S, Nair S and Khobragade P 2017 Diabetic retinal fundus images: preprocessing and feature extraction for early detection of diabetic retinopathy *Biomed. Pharmacol. J.* **10** 615–26

[128] Kumar P S, Deepak R, Sathar A, Sahasranamam V and Kumar R R 2016 Automated detection system for diabetic retinopathy using two field fundus photography *Procedia Comput. Sci.* **93** 486–94

[129] Gupta G, Kulasekaran S, Ram K, Joshi N, Sivaprakasam M and Gandhi R 2017 Local characterization of neovascularization and identification of proliferative diabetic retinopathy in retinal fundus images *Comput. Med. Imaging Graph.* **55** 124–32

[130] Sinthanayothin C, Kongbunkiat V, Phoojaruenchanachai S and Singalavanija A 2003 Automated screening system for diabetic retinopathy 3rd Int. Symp. on Image and Signal Processing and Analysis, 2003. ISPA 2003. Proc. of the vol 2 (Piscataway, NJ: IEEE) pp 915–20

[131] Usher D, Dumskyj M, Himaga M, Williamson T H, Nussey S and Boyce J 2004 Automated detection of diabetic retinopathy in digital retinal images: a tool for diabetic retinopathy screening *Diabet. Med.* **21** 84–90

[132] Fleming A, Goatman K, Williams G, Philip S, Sharp P and Olson J 2008 Automated detection of blot haemorrhages as a sign of referable diabetic retinopathy *Proc. Medical Image Understanding and Analysis*

[133] Perumalsamy N, Prasad N M, Sathya S and Ramasamy K 2007 Software for reading and grading diabetic retinopathy: aravind diabetic retinopathy screening 3.0 *Diabet. Care* **30** 2302–6

[134] Winder R J, Morrow P J, McRitchie I N, Bailie J and Hart P M 2009 Algorithms for digital image processing in diabetic retinopathy *Comput. Med. Imaging Graph.* **33** 608–22

[135] Mookiah M R K, Acharya U R, Martis R J, Chua C K, Lim C M and Ng E *et al* 2013 Evolutionary algorithm based classifier parameter tuning for automatic diabetic retinopathy grading: A hybrid feature extraction approach *Knowl. Based Syst.* **39** 9–22

[136] Abramoff M D and Suttorp-Schulten M S 2005 Web-based screening for diabetic retinopathy in a primary care population: the eyecheck project *Telemed. J. e-Health* **11** 668–74

[137] Abrámoff M D, Niemeijer M, Suttorp-Schulten M S, Viergever M A, Russell S R and Van Ginneken B 2008 Evaluation of a system for automatic detection of diabetic retinopathy from color fundus photographs in a large population of patients with diabetes *Diabet. Care* **31** 193–8

[138] Teng T, Lefley M and Claremont D 2002 Progress towards automated diabetic ocular screening: a review of image analysis and intelligent systems for diabetic retinopathy *Med. Biol. Eng. Comput.* **40** 2–13

[139] Wang H, Hsu W, Goh K G and Lee M L 2000 An effective approach to detect lesions in color retinal images Proc. IEEE Conf. on Computer Vision and Pattern Recognition vol 2 (Piscataway, NJ: IEEE) pp 181–6

[140] Singalavanija A, Supokavej J, Bamroongsuk P, Sinthanayothin C, Phoojaruenchanachai S and Kongbunkiat V 2006 Feasibility study on computer-aided screening for diabetic retinopathy *Jpn. J. Ophthalmol.* **50** 361–6

[141] Gang L, Chutatape O and Krishnan S M 2002 Detection and measurement of retinal vessels in fundus images using amplitude modified second-order gaussian filter *IEEE Trans. Biomed. Eng.* **49** 168–72

[142] Kahai P, Namuduri K R and Thompson H 2006 A decision support framework for automated screening of diabetic retinopathy *Int. J. Biomed. Imaging* **2006** 045806

[143] Yun W L, Acharya U R, Venkatesh Y V, Chee C, Min L C and Ng E Y K 2008 Identification of different stages of diabetic retinopathy using retinal optical images *Inf. Sci.* **178** 106–21

[144] Acharya U R, Lim C M, Ng E Y K, Chee C and Tamura T 2009 Computer-based detection of diabetes retinopathy stages using digital fundus images *Proc. Inst. Mech. Eng. H* 223 545–53

[145] Larsen N, Godt J, Grunkin M, Lund-Andersen H and Larsen M 2003 Automated detection of diabetic retinopathy in a fundus photographic screening population *Investig. Ophthalmol. Vis. Sci.* **44** 767–71

[146] Joshi G D, Sivaswamy J and Krishnadas S 2011 Optic disk and cup segmentation from monocular color retinal images for glaucoma assessment *IEEE Trans. Med. Imaging* **30** 1192–205

[147] Tavakoli M, Kelley P, Nazar M and Kalantari F 2017 Automated fovea detection based on unsupervised retinal vessel segmentation method 2017 IEEE Nuclear Science Symp. and Medical Imaging Conf. (NSS/MIC) (Piscataway, NJ: IEEE) pp 1–7

[148] Sinthanayothin C 1999 Image analysis for automatic diagnosis of diabetic retinopathy *PhD Thesis* (Department of Physics, King's College London)

[149] Gagnon L, Lalonde M, Beaulieu M and Boucher M C 2001 Procedure to detect anatomical structures in optical fundus images *Medical Imaging 2001: Image Processing* vol 4322 (Bellingham, WA: IEEE) pp 1218–25

[150] Hubbard L D, Brothers R J, King W N, Clegg L X, Klein R and Cooper L S *et al* 1999 Methods for evaluation of retinal microvascular abnormalities associated with hypertension/sclerosis in the atherosclerosis risk in communities study *Ophthalmology* **106** 2269–80

[151] Kamble R, Kokare M, Deshmukh G, Hussin F A and Mériaudeau F 2017 Localization of optic disc and fovea in retinal images using intensity based line scanning analysis *Comput. Biol. Med.* **87** 382–96

[152] Elbalaoui A, Fakir M, Boutaounte M and Merbouha A 2017 Automatic localization of the optic disc center in retinal images based on angle detection in curvature scale space *Medical Imaging: Concepts, Methodologies, Tools, and Applications.* (Hershey, PA: IGI Global) pp 679–92

[153] Unver H M, Kökver Y, Duman E and Erdem O A 2019 Statistical edge detection and circular� hough transform for optic disk localization *Appl.Sci.* **9** 350

[154] Meng X, Xi X, Yang L, Zhang G, Yin Y and Chen X 2018 Fast and effective optic disk localization based on convolutional neural network *Neurocomputing* **312** 285–95

[155] Muangnak N, Aimmanee P and Makhanov S 2018 Automatic optic disk detection in retinal images using hybrid vessel phase portrait analysis *Med. Biol. Eng. Comput.* **56** 583–98

[156] Lalonde M, Gagnon L and Boucher M C 2000 Non-recursive paired tracking for vessel extraction from retinal images *Vision Interface.* pp 61–8

[157] Zahoor M N and Fraz M M 2017 Fast optic disc segmentation in retina using polar transform *IEEE Access* **5** 12293–300

[158] Manju K, Sabeenian R and Surendar A 2017 A review on optic disc and cup segmentation *Biomed. Pharmacol. J.* **10** 373–9

[159] Rodrigues L C and Marengoni M 2017 Segmentation of optic disc and blood vessels in retinal images using wavelets, mathematical morphology and hessian-based multi-scale filtering *Biomed. Signal Process. Control* **36** 39–49

[160] Rehman Z U, Naqvi S S, Khan T M, Arsalan M, Khan M A and Khalil M 2019 Multi-parametric optic disc segmentation using superpixel based feature classification *Expert Syst. Appl.* **120** 461–73

[161] Goldbaum M, Moezzi S, Taylor A, Chatterjee S, Boyd J and Hunter E *et al* 1996 Automated diagnosis and image understanding with object extraction, object classification, and inferencing in retinal images *Proc. of 3rd Int. Conf. on Image Processing* vol 3 (Piscataway, NJ: IEEE) pp 695–8

[162] Sinthanayothin C, Boyce J F, Cook H L and Williamson T H 1999 Automated localisation of the optic disc, fovea, and retinal blood vessels from digital colour fundus images *Br. J. Ophthalmol.* **83** 902–10

[163] Walter T and Klein J C 2001 Segmentation of color fundus images of the human retina: Detection of the optic disc and the vascular tree using morphological techniques *Int. Symp. on Medical Data Analysis.* (Berlin: Springer) pp 282–7

[164] Tavakoli M, Nazar M, Golestaneh A and Kalantari F 2017 Automated optic nerve head detection based on different retinal vasculature segmentation methods and mathematical morphology *Nuclear Science Symp. and Medical Imaging Conf. (NSS/MIC)* (Piscataway, NJ: IEEE) pp 1–7

[165] Chrástek R, Wolf M, Donath K, Niemann H, Paulus D and Hothorn T *et al* 2005 Automated segmentation of the optic nerve head for diagnosis of glaucoma *Med. Image Anal.* **9** 297–314

[166] Abdel-Ghafar R and Morris T 2007 Progress towards automated detection and characterization of the optic disc in glaucoma and diabetic retinopathy *Medical Inform. Internet Med.* **32** 19–25

[167] Tavakoli M, Toosi M B, Pourreza R, Banaee T and Pourreza H R 2011 Automated optic nerve head detection in fluorescein angiography fundus images *Nuclear Science Symp. Conf. Record* (Piscataway, NJ: IEEE) pp 3057–60

[168] Sánchez C I, Hornero R, López M I and Poza J 2004 Retinal image analysis to detect and quantify lesions associated with diabetic retinopathy *The 26th Annual Int. Conf. of the Engineering in Medicine and Biology Society* vol 1 (Piscataway, NJ: IEEE) pp 1624–7

[169] Li H and Chutatape O 2004 Automated feature extraction in color retinal images by a model based approach *IEEE Trans. Biomed. Eng.* **51** 246–54

[170] Foracchia M, Grisan E and Ruggeri A 2004 Detection of optic disc in retinal images by means of a geometrical model of vessel structure *IEEE Trans. Med. Imaging* **23** 1189–95

[171] Youssif A A H A R, Ghalwash A Z and Ghoneim A A S A R 2007 Optic disc detection from normalized digital fundus images by means of a vessels' direction matched filter *IEEE Trans. Med. Imaging* **27** 11–8

[172] Lu S and Lim J H 2010 Automatic optic disc detection from retinal images by a line operator *IEEE Trans. Biomed. Eng.* **58** 88–94

[173] Li H and Chutatape O 2001 Automatic location of optic disk in retinal images Proc. 2001 Int. Conf. on Image Processing *(Cat. No. 01CH37205)* vol 2 (Piscataway, NJ: IEEE) pp 837–40

[174] Li H and Chutatape O 2003 A model-based approach for automated feature extraction in fundus images *Innull* (Piscataway, NJ: IEEE) 394

[175] Osareh A, Mirmehdi M, Thomas B and Markham R 2002 Comparison of colour spaces for optic disc localisation in retinal images *Object Recognition Supported by User Interaction for Service Robots* vol 1 (Piscataway, NJ: IEEE) pp 743–6

[176] Osareh A, Mirmehdi M, Thomas B and Markham R 2002 Classification and localisation of diabetic-related eye disease *European Conference on Computer Vision.* (Berlin: Springer) pp 502–16

[177] Xu J, Chutatape O, Sung E, Zheng C and Kuan P C T 2007 Optic disk feature extraction via modified deformable model technique for glaucoma analysis *Pattern Recognit.* **40** 2063–76

[178] Wong D, Liu J, Lim J, Jia X, Yin F and Li H *et al* 2008 Level-set based automatic cup-to-disc ratio determination using retinal fundus images in argali 2008 30th Annual Int. Conf. of the IEEE Engineering in Medicine and Biology Society (Piscataway, NJ: IEEE) pp 2266–9

[179] Lalonde M, Beaulieu M and Gagnon L 2001 Fast and robust optic disc detection using pyramidal decomposition and hausdorff-based template matching *IEEE Trans. Med. Imaging* **20** 1193–200

[180] Ter Haar F 2005 *Automatic localization of the optic disc in digital colour images of the human retina.* (Utrecht: Utrecht University)

[181] Tobin K W, Chaum E, Govindasamy V P and Karnowski T P 2007 Detection of anatomic structures in human retinal imagery *IEEE Trans. Med. Imaging* **26** 1729–39

[182] Niemeijer M, Abràmoff M D and Van Ginneken B 2009 Fast detection of the optic disc and fovea in color fundus photographs *Med. Image Anal.* **13** 859–70

[183] Fleming A D, Goatman K A, Philip S, Olson J A and Sharp P F 2006 Automatic detection of retinal anatomy to assist diabetic retinopathy screening *Phys. Med. Biol.* **52** 331

[184] Niemeijer M, Abràmoff M D and Van Ginneken B 2006 Segmentation of the optic disc, macula and vascular arch in fundus photographs *IEEE Trans. Med. Imaging* **26** 116–27

[185] Abramoff M D and Niemeijer M 2006 The automatic detection of the optic disc location in retinal images using optic disc location regression *2006 International Conference of the IEEE Engineering in Medicine and Biology Society.* (Piscataway, NJ: IEEE) pp 4432–5

[186] Abramoff M D, Alward W L, Greenlee E C, Shuba L, Kim C Y and Fingert J H *et al* 2007 Automated segmentation of the optic disc from stereo color photographs using physiologically plausible features *Investigative Ophthalmol. Vis. Sci.* **48** 1665–73

[187] Hsiao H K, Liu C C, Yu C Y, Kuo S W and Yu S S 2012 A novel optic disc detection scheme on retinal images *Expert Syst. Appl.* **39** 10600–6

[188] Yu H, Barriga E S, Agurto C, Echegaray S, Pattichis M S and Bauman W *et al* 2012 Fast localization and segmentation of optic disk in retinal images using directional matched filtering and level sets *IEEE Trans. Inf. Technol. Biomed.* **16** 644–57

[189] Mahfouz A E and Fahmy A S 2010 Fast localization of the optic disc using projection of image features *IEEE Trans. Image Process.* **19** 3285–9

[190] Estrada R, Allingham M J, Mettu P S, Cousins S W, Tomasi C and Farsiu S 2015 Retinal arteryvein classification via topology estimation *IEEE Trans. Med. Imaging* **34** 2518–34

[191] Dashtbozorg B, Mendonça A M and Campilho A 2014 An automatic graph-based approach for artery/vein classification in retinal images *IEEE Trans. Image Process.* **23** 1073–83

[192] Zhao Y Q, Wang X H, Wang X F and Shih F Y 2014 Retinal vessels segmentation based on level set and region growing *Pattern Recognit.* **47** 2437–46

[193] Kovács G and Hajdu A 2016 A self-calibrating approach for the segmentation of retinal vessels by template matching and contour reconstruction *Med. Image Anal.* **29** 24–46

[194] Lázár I and Hajdu A 2015 Segmentation of retinal vessels by means of directional response vector similarity and region growing *Comput. Biol. Med.* **66** 209–21

[195] Orlando J I, Prokofyeva E and Blaschko M B 2017 A discriminatively trained fully connected conditional random field model for blood vessel segmentation in fundus images *IEEE Trans. Biomed. Eng.* **64** 16–27

[196] Hassan M, Amin M, Murtza I, Khan A and Chaudhry A 2017 Robust hidden markov model based intelligent blood vessel detection of fundus images *Comput. Methods Programs Biomed.* **151** 193–201

[197] Aslani S and Sarnel H 2016 A new supervised retinal vessel segmentation method based on robust hybrid features *Biomed. Signal Process. Control* **30** 1–12

[198] Lupascu C A, Tegolo D and Trucco E 2010 Fabc: retinal vessel segmentation using adaboost *IEEE Trans. Inf. Technol. Biomed.* **14** 1267–74

[199] Vega R, Sanchez-Ante G, Falcon-Morales L E, Sossa H and Guevara E 2015 Retinal vessel extraction using lattice neural networks with dendritic processing *Comput. Biol. Med.* **58** 20–30

[200] Waheed A, Akram M U, Khalid S, Waheed Z, Khan M A and Shaukat A 2015 Hybrid features and mediods classification based robust segmentation of blood vessels *J. Med. Syst.* **39** 128

[201] Zilly J, Buhmann J M and Mahapatra D 2017 Glaucoma detection using entropy sampling and ensemble learning for automatic optic cup and disc segmentation *Comput. Med. Imaging Graph.* **55** 28–41

[202] Annunziata R, Garzelli A, Ballerini L, Mecocci A and Trucco E 2016 Leveraging multiscale hessian-based enhancement with a novel exudate inpainting technique for retinal vessel segmentation *IEEE J. Biomed. Health Inform.* **20** 1129–38
[203] Mendonca A M and Campilho A 2006 Segmentation of retinal blood vessels by combining the detection of centerlines and morphological reconstruction *IEEE Trans. Med. Imaging* **25** 1200–13
[204] Martinez-Perez M E, Hughes A D, Thom S A, Bharath A A and Parker K H 2007 Segmentation of blood vessels from red-free and fluorescein retinal images *Med. Image Anal.* **11** 47–61
[205] Zhang J, Dashtbozorg B, Bekkers E, Pluim J P, Duits R and ter Haar Romeny B M 2016 Robust retinal vessel segmentation via locally adaptive derivative frames in orientation scores *IEEE Trans. Med. Imaging* **35** 2631–44
[206] Lam B S Y and Yan H 2008 A novel vessel segmentation algorithm for pathological retina images based on the divergence of vector fields *IEEE Trans. Med. Imaging* **27** 237–46
[207] Ricci E and Perfetti R 2007 Retinal blood vessel segmentation using line operators and support vector classification *IEEE Trans. Med. Imaging* **26** 1357–65
[208] Wang Y and Lee S C 1997 A fast method for automated detection of blood vessels in retinal images *Conf. Record of the Thirty-First Asilomar Conf. on Signals, Systems and Computers, 1997* vol 2 (Piscataway, NJ: IEEE) pp 1700–4
[209] Jiang X and Mojon D 2003 Adaptive local thresholding by verification-based multi-threshold probing with application to vessel detection in retinal images *IEEE Trans. Pattern Anal. Mach. Intell.* **25** 131–7
[210] Heneghan C, Flynn J, OKeefe M and Cahill M 2002 Characterization of changes in blood vessel width and tortuosity in retinopathy of prematurity using image analysis *Med. Image Anal.* **6** 407–29
[211] Roychowdhury S, Koozekanani D D and Parhi K K 2015 Iterative vessel segmentation of fundus images *IEEE Trans. Biomed. Eng.* **62** 1738–49
[212] Staal J, Abra'moff M D, Niemeijer M, Viergever M A and Van Ginneken B 2004 Ridge-based vessel segmentation in color images of the retina *IEEE Trans. Med. Imaging* **23** 501–9
[213] Li Q, Feng B, Xie L, Liang P, Zhang H and Wang T 2016 A cross-modality learning approach for vessel segmentation in retinal images *IEEE Trans. Med. Imaging* **35** 109–18
[214] Annunziata R and Trucco E 2016 Accelerating convolutional sparse coding for curvilinear structures segmentation by refining scird-ts filter banks *IEEE Trans. Med. Imaging* **35** 2381–92
[215] Wang S, Yin Y, Cao G, Wei B, Zheng Y and Yang G 2015 Hierarchical retinal blood vessel segmentation based on feature and ensemble learning *Neurocomputing* **149** 708–17
[216] Tan J H, Acharya U R, Bhandary S V, Chua K C and Sivaprasad S 2017 Segmentation of optic disc, fovea and retinal vasculature using a single convolutional neural network *J. Comput. Sci.* **20** 70–9
[217] Liskowski P and Krawiec K 2016 Segmenting retinal blood vessels with deep neural networks *IEEE Trans. Med. Imaging* **35** 2369–80
[218] Yue K, Zou B, Chen Z and Liu Q 2019 Retinal vessel segmentation using dense u-net with multiscale inputs *J. Med. Imaging* **6** 034004
[219] Lin Y, Zhang H and Hu G 2018 Automatic retinal vessel segmentation via deeply supervised and smoothly regularized network *IEEE Access* **7** 57717–24

[220] Ren X, Zheng Y, Zhao Y, Luo C, Wang H and Lian J *et al* 2018 Drusen segmentation from retinal images via supervised feature learning *IEEE Access* **6** 2952–61
[221] Oliveira A, Pereira S and Silva C A 2018 Retinal vessel segmentation based on fully convolutional neural networks *Expert Syst. Appl.* **112** 229–42
[222] Singh N, Kaur L and Singh K 2019 Segmentation of retinal blood vessels based on featureoriented dictionary learning and sparse coding using ensemble classification approach *J. Med. Imaging* **6** 044006
[223] Narasimha-Iyer H, Can A, Roysam B, Tanenbaum H L and Majerovics A 2007 Integrated analysis of vascular and nonvascular changes from color retinal fundus image sequences *IEEE Trans. Biomed. Eng.* **54** 1436–45
[224] Goatman K, Charnley A, Webster L and Nussey S 2011 Assessment of automated disease detection in diabetic retinopathy screening using two-field photography *PLoS One* **6** e27524
[225] Köse C, Şevik U and Gençalioğlu O 2008 Automatic segmentation of age-related macular degeneration in retinal fundus images *Comput. Biol. Med.* **38** 611–9
[226] Zhou W, Wu C, Chen D, Yi Y and Du W 2017 Automatic microaneurysm detection using the sparse principal component analysis-based unsupervised classification method *IEEE Access* **5** 2563–72
[227] Al-Bander B, Al-Nuaimy W, Williams B M and Zheng Y 2018 Multiscale sequential convolutional neural networks for simultaneous detection of fovea and optic disc *Biomed. Signal Process. Control* **40** 91–101
[228] GeethaRamani R and Balasubramanian L 2018 Macula segmentation and fovea localization employing image processing and heuristic based clustering for automated retinal screening *Comput. Methods Programs Biomed.* **160** 153–63
[229] Fadzil M A, Izhar L I and Nugroho H A 2010 Determination of foveal avascular zone in diabetic retinopathy digital fundus images *Comput. Biol. Med.* **40** 657–64
[230] Acharya R, Ng Y E and Suri J S 2008 *Image Modeling of the Human Eye.* (Boston, MA: Artech House)
[231] Sekhar S, Al-Nuaimy W and Nandi A K 2008 Automated localisation of retinal optic disk using Hough transform *2008 5th Int. Symp. on Biomedical Imaging: From Nano to Macro* (Piscataway, NJ: IEEE) pp 1577–80
[232] Welfer D, Scharcanski J and Marinho D 2011 Fovea center detection based on the retina anatomy and mathematical morphology *Comput. Methods Programs Biomed.* **104** 397–409
[233] Niemeijer M, Van Ginneken B, Staal J, Suttorp-Schulten M S and Abràmoff M D 2005 Automatic detection of red lesions in digital color fundus photographs *IEEE Trans. Med. Imaging* **24** 584–92
[234] Saleh M D and Eswaran C 2012 An automated decision-support system for non-proliferative diabetic retinopathy disease based on mas and has detection *Comput. Methods Programs Biomed.* **108** 186–96
[235] Lazar I and Hajdu A 2013 Retinal microaneurysm detection through local rotating cross-section profile analysis *IEEE Trans. Med. Imaging* **32** 400–7
[236] Dashtbozorg B, Zhang J, Huang F and ter Haar Romeny B M 2018 Retinal micro-aneurysms detection using local convergence index features *IEEE Trans. Image Process.* **27** 3300–15
[237] Habib M, Welikala R, Hoppe A, Owen C, Rudnicka A and Barman S 2017 Detection of microaneurysms in retinal images using an ensemble classifier *Inform. Med. Unlocked* **9** 44–57

[238] Gegundez-Arias M E, Marin D, Ponte B, Alvarez F, Garrido J and Ortega C *et al* 2017 A tool for automated diabetic retinopathy pre-screening based on retinal image computer analysis *Comput. Biol. Med.* **88** 100–9

[239] Ram K, Joshi G D and Sivaswamy J 2010 A successive clutter-rejection-based approach for early detection of diabetic retinopathy *IEEE Trans. Biomed. Eng.* **58** 664–73

[240] Mizutani A, Muramatsu C, Hatanaka Y, Suemori S, Hara T and Fujita H 2009 Automated microaneurysm detection method based on double ring filter in retinal fundus images *Medical Imaging 2009: Computer-Aided Diagnosis* vol 7260 (Bellingham, WA: International Society for Optics and Photonics) p 72601N

[241] Ravishankar S, Jain A and Mittal A 2009 Automated feature extraction for early detection of diabetic retinopathy in fundus images 2009 IEEE Conf. on Computer Vision and Pattern Recognition (Piscataway, NJ: IEEE) pp 210–7

[242] Shah S A A, Laude A, Faye I and Tang T B 2016 Automated microaneurysm detection in diabetic retinopathy using curvelet transform *J. Biomed. Opt.* **21** 101404

[243] Cao W, Czarnek N, Shan J and Li L 2018 Microaneurysm detection using principal component analysis and machine learning methods *IEEE Trans. Nanobiosci.* **17** 191–8

[244] Manjaramkar A and Kokare M 2018 Statistical geometrical features for microaneurysm detection *J. Digit. Imaging* **31** 224–34

[245] Mumtaz R, Hussain M, Sarwar S, Khan K, Mumtaz S and Mumtaz M 2018 Automatic detection of retinal hemorrhages by exploiting image processing techniques for screening retinal diseases in diabetic patients *Int. J. Diabet. Dev. Countries* **38** 80–7

[246] Biran A, Bidari P S and Raahemifar K 2016 Automatic method for exudates and hemorrhages detection from fundus retinal images *Int. J. Comput. Inform. Eng.* **10** 1599–602

[247] Lahmiri S and Shmuel A 2017 Variational mode decomposition based approach for accurate classification of color fundus images with hemorrhages *Opt. Laser Technol.* **96** 243–8

[248] Hatanaka Y, Nakagawa T, Hayashi Y, Kakogawa M, Sawada A and Kawase K *et al* 2008 Improvement of automatic hemorrhage detection methods using brightness correction on fundus images *Medical Imaging 2008: Computer-Aided Diagnosis* vol 6915 (Bellingham, WA: International Society for Optics and Photonics) p 69153E

[249] Spencer T, Olson J A, McHardy K C, Sharp P F and Forrester J V 1996 An image-processing strategy for the segmentation and quantification of microaneurysms in fluorescein angiograms of the ocular fundus *Comput. Biomed. Res.* **29** 284–302

[250] Frame A J, Undrill P E, Cree M J, Olson J A, McHardy K C and Sharp P F *et al* 1998 A comparison of computer based classification methods applied to the detection of microaneurysms in ophthalmic fluorescein angiograms *Comput. Biol. Med.* **28** 225–38

[251] Zhang B, Wu X, You J, Li Q and Karray F 2010 Detection of microaneurysms using multi-scale correlation coefficients *Pattern Recognit.* **43** 2237–48

[252] Zhang X and Chutatape O 2005 A svm approach for detection of hemorrhages in background diabetic retinopathy Proc. 2005 IEEE Int. Joint Conf. on Neural Networks vol 4 (Piscataway, NJ: IEEE) pp 2435–40

[253] Hipwell J, Strachan F, Olson J, McHardy K, Sharp P and Forrester J 2000 Automated detection of microaneurysms in digital red-free photographs: a diabetic retinopathy screening tool *Diabet. Med.* **17** 588–94

[254] Antal B and Hajdu A 2012 An ensemble-based system for microaneurysm detection and diabetic retinopathy grading *IEEE Trans. Biomed. Eng.* **59** 1720–6

[255] Kusakunniran W, Wu Q, Ritthipravat P and Zhang J 2018 Hard exudates segmentation based on learned initial seeds and iterative graph cut *Comput. Methods Programs Biomed.* **158** 173–83
[256] Zubair M, Ali H and Javed M Y 2016 Automated segmentation of hard exudates using dynamic thresholding to detect diabetic retinopathy in retinal photographs *JMPT* **7** 109–16
[257] Lara Rodríguez L D and Urcid Serrano G 2016 Exudates and blood vessel segmentation in eye fundus images using the fourier and cosine discrete transforms *Comput. Sist.* **20** 697–708
[258] Osareh A, Mirmehdi M, Thomas B and Markham R 2003 Automated identification of diabetic retinal exudates in digital colour images *Br. J. Ophthalmol.* **87** 1220–3
[259] Osareh A, Shadgar B and Markham R 2009 A computational-intelligence-based approach for detection of exudates in diabetic retinopathy images *IEEE Trans. Inf. Technol. Biomed.* **13** 535–45
[260] Bezdek J C 2013 *Pattern Recognition with Fuzzy Objective Function Algorithms.* (Berlin: Springer Science & Business Media)
[261] Sopharak A, Nwe K T, Moe Y A, Dailey M N and Uyyanonvara B *et al* 2008 Automatic exudate detection with a naive Bayes classifier *Int. Conf. on Embedded Systems and Intelligent Technology.* pp 139–42
[262] Sopharak A, Uyyanonvara B, Barman S and Williamson T H 2008 Automatic detection of diabetic retinopathy exudates from non-dilated retinal images using mathematical morphology methods *Comput. Med. Imaging Graph.* **32** 720–7
[263] Walter T, Klein J C, Massin P and Erginay A 2002 A contribution of image processing to the diagnosis of diabetic retinopathy-detection of exudates in color fundus images of the human retina *IEEE Trans. Med. Imaging* **21** 1236–43
[264] Welfer D, Scharcanski J and Marinho D 2010 A coarse-to-fine strategy for automatically detecting exudates in color eye fundus images *Comput. Med. Imaging Graph.* **34** 228–35
[265] Sánchez C I, Hornero R, López M I, Aboy M, Poza J and Abásolo D 2008 A novel automatic image processing algorithm for detection of hard exudates based on retinal image analysis *Med. Eng. Phys.* **30** 350–7
[266] Obermeyer Z and Lee T H 2017 Lost in thought: The limits of the human mind and the future of medicine *New Engl. J. Med.* **377** 1209
[267] Jiang F, Jiang Y, Zhi H, Dong Y, Li H and Ma S *et al* 2017 Artificial intelligence in healthcare: past, present and future *Stroke Vasc. Neurol.* **2** 230–43
[268] Schmidt-Erfurth U, Sadeghipour A, Gerendas B S, Waldstein S M and Bogunovi´c H 2018 Artificial intelligence in retina *Prog. Retin. Eye Res.* **67** 1–29
[269] Joshi S and Karule P 2018 A review on exudates detection methods for diabetic retinopathy *Biomed. Pharmacother.* **97** 1454–60
[270] Almotiri J, Elleithy K and Elleithy A 2018 Retinal vessels segmentation techniques and algorithms: a survey *Appl. Sci.* **8** 155
[271] Almazroa A, Burman R, Raahemifar K and Lakshminarayanan V 2015 Optic disc and optic cup segmentation methodologies for glaucoma image detection: a survey *J. Ophthalmol.* **2015** 180972
[272] Thakur N and Juneja M 2018 Survey on segmentation and classification approaches of optic cup and optic disc for diagnosis of glaucoma *Biomed. Signal Process. Control* **42** 162–89
[273] Litjens G, Kooi T, Bejnordi B E, Setio A A A, Ciompi F and Ghafoorian M *et al* 2017 A survey on deep learning in medical image analysis *Med. Image Anal.* **42** 60–88

[274] Khojasteh P, Aliahmad B and Kumar D K 2018 Fundus images analysis using deep features for detection of exudates, hemorrhages and microaneurysms *BMC Ophthalmol.* **18** 1–13

[275] Parmar R, Lakshmanan R, Purushotham S and Soundrapandiyan R 2019 Detecting diabetic retinopathy from retinal images using CUDA deep neural network *Intelligent Pervasive Computing Systems for Smarter Healthcare* (New York: Wiley) pp 379–96

[276] Quellec G, Charrière K, Boudi Y, Cochener B and Lamard M 2017 Deep image mining for diabetic retinopathy screening *Med. Image Anal.* **39** 178–93

[277] Zhou L, Zhao Y, Yang J, Yu Q and Xu X 2017 Deep multiple instance learning for automatic detection of diabetic retinopathy in retinal images *IET Image Proc.* **12** 563–71

[278] Gargeya R and Leng T 2017 Automated identification of diabetic retinopathy using deep learning *Ophthalmology* **124** 962–9

[279] Son J, Shin J Y, Kim H D, Jung K H, Park K H and Park S J 2020 Development and validation of deep learning models for screening multiple abnormal findings in retinal fundus images *Ophthalmology* **127** 85–94

[280] Kathiresan S, Sait A R W, Gupta D, Lakshmanaprabu S, Khanna A and Pandey H M 2020 Automated detection and classification of fundus diabetic retinopathy images using synergic deep learning model *Pattern Recognit. Lett.* **133** 210–6

[281] Abràmoff M D, Lou Y, Erginay A, Clarida W, Amelon R and Folk J C *et al* 2016 Improved automated detection of diabetic retinopathy on a publicly available dataset through integration of deep learning *Investig. Ophthalmol. Vis. Sci.* **57** 5200–6

[282] Wang Z, Yin Y, Shi J, Fang W, Li H and Wang X 2017 Zoom-in-net: Deep mining lesions for diabetic retinopathy detection *Int. Conf. on Medical Image Computing and Computer-Assisted Intervention* (Berlin: Springer) pp 267–75

[283] Lam C, Yu C, Huang L and Rubin D 2018 Retinal lesion detection with deep learning using image patches *Investig. Ophthalmol. Vis. Sci.* **59** 590–6

[284] Sayres R, Taly A, Rahimy E, Blumer K, Coz D and Hammel N *et al* 2019 Using a deep learning algorithm and integrated gradients explanation to assist grading for diabetic retinopathy *Ophthalmology* **126** 552–64

[285] Maji D, Santara A, Mitra P and Sheet D 2016 Ensemble of deep convolutional neural networks for learning to detect retinal vessels in fundus images. arXiv preprint arXiv:160304833

[286] Maninis K K, Pont-Tuset J, Arbeláez P and Van Gool L 2016 Deep retinal image understanding *Int. Conf. on Medical Image Computing and Computer-Assisted Intervention* (Berlin: Springer) pp 140–8

[287] Dasgupta A and Singh S 2017 A fully convolutional neural network based structured prediction approach towards the retinal vessel segmentation 2017 IEEE 14th Int. Symp. on Biomedical Imaging (ISBI 2017) (Piscataway, NJ: IEEE) pp 248–51

[288] Fu H, Xu Y, Wong D W K and Liu J 2016 Retinal vessel segmentation via deep learning network and fully-connected conditional random fields 2016 IEEE 13th Int. Symp. on Biomedical Imaging (ISBI) (Piscataway, NJ: IEEE) pp 698–701

[289] Mo J and Zhang L 2017 Multi-level deep supervised networks for retinal vessel segmentation *Int. J. Comput. Ass. Rad. Surg.* **12** 2181–93

[290] Lahiri A, Roy A G, Sheet D and Biswas P K 2016 Deep neural ensemble for retinal vessel segmentation in fundus images towards achieving label-free angiography 2016 38th Annual

Int. Conf. of the IEEE Engineering in Medicine and Biology Society (EMBC) (Piscataway, NJ: IEEE) pp 1340–3
[291] Li Q, Feng B, Xie L, Liang P, Zhang H and Wang T 2015 A cross-modality learning approach for vessel segmentation in retinal images *IEEE Trans. Med. Imaging* **35** 109–18
[292] Lim G, Cheng Y, Hsu W and Lee M L 2015 Integrated optic disc and cup segmentation with deep learning 2015 IEEE 27th Int. Conf. on Tools with Artificial Intelligence (ICTAI) (Piscataway, NJ: IEEE) pp 162–9
[293] Zhang D, Zhu W, Zhao H, Shi F and Chen X 2018 Automatic localization and segmentation of optical disk based on faster r-cnn and level set in fundus image *Medical Imaging 2018: Image Processing* vol 10574 (Bellingham, WA: International Society for Optics and Photonics) p 105741U
[294] Sevastopolsky A 2017 Optic disc and cup segmentation methods for glaucoma detection with modification of u-net convolutional neural network *Pattern Recognit. Image Anal.* **27** 618–24
[295] Alghamdi H S, Tang H L, Waheeb S A and Peto T 2016 Automatic optic disc abnormality detection in fundus images: a deep learning approach *Proc. OMIA3 (MICCAI)* pp 17–24
[296] Haloi M 2015 Improved microaneurysm detection using deep neural networks. arXiv preprint arXiv:150504424
[297] Chudzik P, Majumdar S, Calivá F, Al-Diri B and Hunter A 2018 Microaneurysm detection using fully convolutional neural networks *Comput. Methods Programs Biomed.* **158** 185–92
[298] Shan J and Li L 2016 A deep learning method for microaneurysm detection in fundus images 2016 IEEE First Int. Conf. on Connected Health: Applications, Systems and Engineering Technologies (CHASE) (Piscataway, NJ: IEEE) pp 357–8
[299] Lahmiri S 2020 Hybrid deep learning convolutional neural networks and optimal nonlinear support vector machine to detect presence of hemorrhage in retina *Biomed. Signal Process. Control* **60** 101978
[300] Derwin D J, Selvi S T and Singh O J 2020 Secondary observer system for detection of microaneurysms in fundus images using texture descriptors *J. Digit. Imaging* **33** 159–67
[301] Gondal W M, Köhler J M, Grzeszick R, Fink G A and Hirsch M 2017 Weakly-supervised localization of diabetic retinopathy lesions in retinal fundus images 2017 IEEE Int. Conf. on Image Processing (ICIP) (Piscataway, NJ: IEEE) pp 2069–73
[302] Benzamin A and Chakraborty C 2018 Detection of hard exudates in retinal fundus images using deep learning *2018 Joint 7th Int. Conf. on Informatics, Electronics & Vision (ICIEV) and 2018 2nd Int. Conf. on Imaging, Vision & Pattern Recognition (icIVPR).* (Piscataway, NJ: IEEE) pp 465–9
[303] Prentasic P and Loncaric S 2016 Detection of exudates in fundus photographs using deep neural networks and anatomical landmark detection fusion *Comput. Methods Programs Biomed.* **137** 281–92
[304] Asiri N, Hussain M, Al Adel F and Alzaidi N 2019 Deep learning based computer-aided diagnosis systems for diabetic retinopathy: a survey *Artif. Intell. Med.* **99** 101701
[305] Simonyan K and Zisserman A 2014 Very deep convolutional networks for large-scale image recognition. arXiv preprint arXiv:14091556

Lightning Source UK Ltd.
Milton Keynes UK
UKHW051855270122
397741UK00002B/106